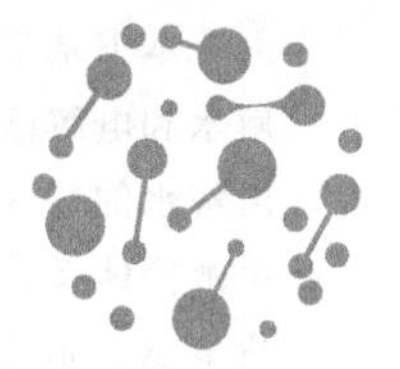

Water Electrolysis and Catalysts

电解水及其催化剂

刘云花　编著

化学工业出版社
·北京·

内容简介

本书基于作者近年来在电解水及电解水催化剂方面的研究以及文献积累，介绍了电解水和电解槽体系的基础知识，并综述了几年来开发的性能优异的催化剂。首先介绍了电解水的基本原理及重要参数，然后介绍了电解槽体系，包括传统的电解槽体系和新型电解槽体系，之后介绍了电催化剂的制备方法，高效电催化剂的构建方法，最后介绍了贵金属、非贵金属以及单原子材料作为电催化剂的研究进展。

本书可供电化学、电催化、材料科学等领域的研究人员及高等院校相关专业学生参考使用。

图书在版编目（CIP）数据

电解水及其催化剂/刘云花编著. —北京：化学工业出版社，2022.8（2023.8重印）

ISBN 978-7-122-41754-1

Ⅰ. ①电… Ⅱ. ①刘… Ⅲ. ①水-电催化剂 Ⅳ. ①TQ426.99

中国版本图书馆CIP数据核字（2022）第110545号

责任编辑：韩霄翠　仇志刚　　文字编辑：公金文　陈小滔
责任校对：宋　夏　　装帧设计：王晓宇

出版发行：化学工业出版社（北京市东城区青年湖南街13号　邮政编码100011）
印　　装：北京科印技术咨询服务有限公司数码印刷分部
710mm×1000mm　1/16　印张15½　彩插7　字数259千字
2023年8月北京第1版第5次印刷

购书咨询：010-64518888　　售后服务：010-64518899
网　　址：http://www.cip.com.cn
凡购买本书，如有缺损质量问题，本社销售中心负责调换。

定　　价：108.00元

氢气（H_2）不仅在石化工业和氨合成工业中发挥重要作用，而且还是一种理想的能源载体和燃料。目前，全球95%以上的H_2来自化石燃料，开发大规模、廉价、清洁、高效的制氢技术是氢能有效利用的关键。其中电解水制氢是生产高纯度H_2的可持续且绿色的方法。在电解水制氢体系中，对催化剂的研究一直是热点，因此本书主要介绍了电解水的基本原理、关键参数以及相关催化剂。旨在帮助电解水行业的初学者理解和熟悉电解水制氢体系的基础知识，并了解催化剂制备的一些方法和研究进展。

本书在笔者近年来对电解水及其催化剂方面的研究以及文献积累基础上编写而成。编写过程中坚持以能给读者有用的借鉴和指导为宗旨，力求内容具有系统性、通俗性和严谨性。目前，为推动电解水制氢体系在工业上的大规模应用，对于电解水的研究主要集中在两个方面：一方面是寻求催化活性良好的非贵金属代替贵金属；另一方面是构建去耦合化的电解槽体系。针对研究的热点，本书首先介绍了电解水的基本原理及电催化的重要参数；然后介绍了电解水槽体系，包括传统的电解槽体系和新型电解槽体系；之后主要介绍了电催化剂的制备方法，紧接着系统地阐述了如何构建高效电催化剂，最后介绍了贵金属、非贵金属以及单原子材料作为电催化剂的研究进展。本书较为系统地阐述了电解水及其催化剂的相关基础知识及研究进展，适用于本科生、研究生以及希望从事电解水制氢研究的学者。

本书得到了贵州省科技计划项目（黔科合基础-ZK［2021］一般050）和贵州理工学院高层次人才科研启动项目（XJGC20190961）的支持。

由于笔者水平有限，书中难免存在不足之处，恳请读者批评指正。

编著者

2022年6月

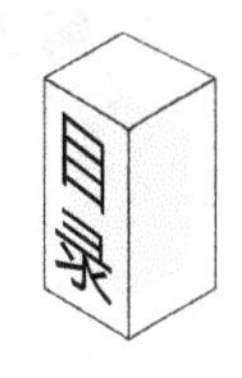
目录

第一章

电解水概述

1.1 引言
1.2 基本原理
1.3 电催化的重要参数
1.4 电解水催化剂

1.1 引言

据国际能源署统计数据，2013 年全球能源的 80%来源于以煤和石油为代表的传统化石燃料[1]。化石燃料的燃烧会排放温室气体或有害气体（诸如 SO_x、NO_x、CO_2 等)，从而产生严重的环境问题，最重要的是其储量有限。为了解决以上问题，需要开发可再生能源代替传统的化石能源。太阳能和风能发电可以替代传统的化石能源，但是太阳能和风能具有季节性、地域性和效率低等问题。此外，由于太阳光具有分散性，如果要广泛应用，必须要有高效的能量捕获和存储单元。除此之外，存储和运输这些能量费用昂贵。经过大量的研究，研究人员已经成功以化学能（比如氢气）的形式存储太阳能。氢气和传统的化石能源一样储存和携带能量。它作为一种高效的能量载体，比能量高，且燃烧不会排放有毒有害气体，燃烧产物只有水。目前生产氢气的方法主要有化石燃料蒸汽重整法制氢、煤气化制氢、水分解制氢(光催化水分解和电解水)[2]。前两种制氢方法以天然气、煤和石油等化石燃料为原料，目前全球 90%的氢气是通过化石燃料的裂解而获得的[3]，耗能大并伴随大量的 CO_2 和氮硫氧化物等废气的产生，加重了全球能源危机，加剧了温室效应和环境污染。尽管水分解制氢的产量仅占全球氢气总量的 4%，但它却是制备氢气最清洁的方法。

目前水分解制氢主要通过光催化水分解和电解水进行，电解水比光化学/光电化学方法占有优势。当有电流通过电解槽时，水分子分解为 H_2 和 O_2，这需要克服一定的能垒。据文献报道，产生 H_2 和 O_2 的理论电位分别为 0V 和 1.23V（相对于 RHE)。但是氧析出反应（OER）是四电子转移过程，具有缓慢的反应动力学，因此水的电解需要很高的活化能垒，需要额外的电势（称为过电位）才能完成。目前，研究者们通过改善电极、催化剂、电解质等方面来降低过电位，其中研究最多的是采用催化剂来加速水分解[4-6]。铂（Pt)、钯（Pd)、钌（Ru）和铱（Ir）等贵金属及其化合物被认为是氢析出反应（HER）和 OER 最佳催化剂，但是这些贵金属成本高、储量有限而限制了其大规模商业化应用。因此需要研发储量丰富、成本低廉的电催化剂来代替贵金属催化剂。过渡金属及其化合物引起了广泛的关注，诸如过渡金属硫族化合物、氧化物、氢氧化物、磷化物、氮化物、碳化物等，以及它们与石墨烯、碳纳米管的复合材料。由于 HER 和 OER 反应是表面反应，因此催化剂的表面形态对催化性能有着重要作用，目前，已经成功开发了具有各种表面形态的纳米材料，比如一维多孔纳米棒、纳米线，二维纳米片/片状结构和三维纳米球等。另外通过杂原子或者离子掺入纳米材料的晶格中也可得到良好的催化活性，但是其催化效率与贵金属催化剂相差甚远。因此，开发具有更好催化效率的新型电活性材料对于电解水产氢技术是十分必要的。一般情况下，粉末状的催化剂需要通过黏结剂，固载在导电表面上形成电极，但是在 OER 和 HER 过程中，气体快速释放导致催化剂膜容易从电极表面剥离，从而使催化性能突然下降。为了克服这一问题，出现了原位生长催化剂的电极，这类电极具有更为优异的稳定性和催化活性。

电解水分解产生氢气和氧气是利用间歇性可再生能源产生清洁且可持续燃料的最有前途的方法之一。近年来，出现了许多电化学水分解的报道，其中大多数报道集中在提高电化学反应本身的效率上。但是，如果不能将电解水分解的产品分开存放，那么即使能有效地产生氢气和氧气，也几乎没有用。因此，在间歇性可再生资源驱动的电解过程中，保持氢气和氧气之间的充分分离面临很大的挑战。解耦电解（阳极产生氧气时阴极还原合适的介体，然后将阴极氢的产生与该介体的再氧化配对）允许在不同时间、以不同速率甚至在完全不同的电化学池中生产 O_2 和 H_2，这为许多挑战提供了解决方案。为了更好地理解电解过程、相关的机理、解耦电解水等，本书将介绍电解水制氢的相关基础知识，概述了解耦电解水领域的最新进展，并介绍了部分优异的电催化剂。

1.2 基本原理

电解是电流通过电解质溶液，在阴极和阳极上引起氧化还原反应的过程。电解水是在水性电解质中，当有电流通过电极时，水分子持续稳定的分解过程。在特定电压下，在阴极发生还原反应产生氢气，在阳极发生氧化反应。电解水的总反应如下：

$$H_2O \rightleftharpoons H_2 + \frac{1}{2}O_2$$

它包括两个半电池反应：①阴极电极表面发生还原反应形成氢气，即氢析出反应（HER）；②阳极电极表面发生氧化反应形成氧气，即氧析出反应（OER）。HER 和 OER 都取决于活性离子（H^+、OH^-）的性质，因此在不同 pH 的电解液中，反应是不同的。表 1.1 列出了在酸性和碱性条件下 OER 和 HER 的反应式。

表 1.1 在酸性和碱性溶液中 HER 和 OER 反应式

反应类型	酸性溶液	碱性溶液
HER	$4H^+ + 4e^- = 2H_2$	$4H_2O + 4e^- = 2H_2 + 4OH^-$
OER	$2H_2O = O_2 + 4H^+ + 4e^-$	$4OH^- = 2H_2O + O_2 + 4e^-$

详细的机理在下一节中讨论。Wang 等人采用循环伏安（CV）法讨论了在电解水过程中发生的反应[7]，使用铂微电极作为工作电极，在纯水和 0.5mol/L H_2SO_4 电解质溶液中分别进行了 CV 扫描。电解过程开始于氧被吸附在电极表面形成的 Pt-O 膜上（*a*），然后氧气释放（*b*）；在阴极扫描过程中电位降低时，形成的 Pt-O 通过氧气还原（*c*）而还原；随着电位的进一步降低，发生氢吸附（*d*），然后释放出氢气（*e*）；最后，氢气在扫描电位增大时被氧化（*f*），从而完成扫描周期。如图 1.1（见文后彩插）所示：

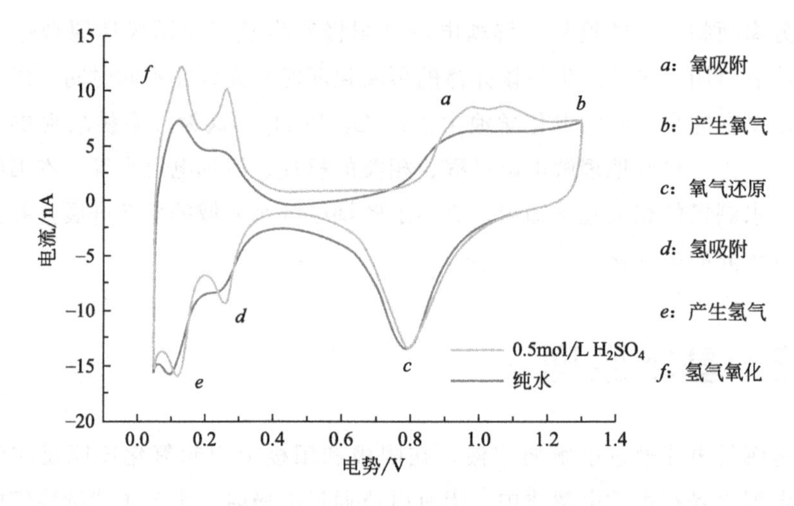

图 1.1 Pt 微型圆盘电极在纯水和 H_2SO_4 中的循环伏安图[7]

扫描速率为 80mV/s，DHE/PEM 系统用作参比电极

当施加的电位达到热力学电位时，水分解产生氢气和氧气。热力学电位通过下面的公式来计算：

$$\Delta G = -nFE^{\ominus} \tag{1.1}$$

式中，ΔG 为吉布斯自由能，J/mol；n 为反应过程中转移的电子数；F 为法拉第常数，数值为 96485C/mol。

通过该方程计算出 HER 的理论电位为 0V，OER 的理论电位为 1.229V，因此电解水的理论电位为 1.229V。在后面的章节中会对反应动力学、催化剂类型以及详细反应机理等进行讨论。

一般情况下，采用三电极电化学装置对催化剂的催化活性进行研究，负载有催化剂的电极作为工作电极，石墨棒或者铂丝为对电极，Hg/HgO（或饱和甘汞、Ag/AgCl）为参比电极。工作电极的准备工作根据催化剂的类型来确定，粉末状催化剂需要用黏结剂与分散溶剂形成悬浊液，然后滴涂到电极上（一般采用玻碳电极，镍箔、泡沫镍、钛片等为催化剂的支撑电极），黏结剂通常使用 Nafion，溶剂为异丙醇和水的混合溶剂。为了提高催化剂的导电性，可以加入一些导电材料，诸如乙炔黑、石墨烯等。一般需要对加入的催化剂、黏结剂和导电剂的质量进行优化。将分散好的催化剂滴涂后，可放置于室温或者低温烘箱进行干燥，催化剂完全干燥后进行电化学性能表征。在导电基底上原位生长的催化剂则不需要这些烦琐的操作，可直接作为工作电极进行电化学性能测试。在进行测试时，即使采用不同的参比电极，也可以通过使用以下的能斯特方程将其全部转换成相对于标准可逆氢电极（RHE）的电势，Ag/AgCl 为参比电极：

$$E_{\mathrm{RHE}}=E_{\mathrm{Ag/AgCl}}+E^{\ominus}_{\mathrm{Ag/AgCl}}+0.059\mathrm{pH} \tag{1.2}$$

式中，$E_{\mathrm{Ag/AgCl}}$为以 Ag/AgCl 为参比电极测试的电势，V；$E^{\ominus}_{\mathrm{Ag/AgCl}}$为 Ag/AgCl 参比电极的标准电极电势，数值为 0.21，V；pH 为测试溶液的酸碱值。

饱和甘汞（SCE）为参比电极：

$$E_{\mathrm{RHE}}=E_{\mathrm{SCE}}+E^{\ominus}_{\mathrm{SCE}}+0.059\mathrm{pH} \tag{1.3}$$

式中，E_{SCE}为以饱和甘汞为参比测试的电势，V；$E^{\ominus}_{\mathrm{SCE}}$为饱和甘汞参比电极的标准电极电势，数值为 0.244，V；pH 为测试溶液的酸碱值。

Hg/HgO 为参比电极：

$$E_{\mathrm{RHE}}=E_{\mathrm{Hg/HgO}}+E^{\ominus}_{\mathrm{Hg/HgO}}+0.059\mathrm{pH} \tag{1.4}$$

式中，$E_{\mathrm{Hg/HgO}}$为 Hg/HgO 为参比电极测试的电势，V；$E^{\ominus}_{\mathrm{Hg/HgO}}$为 Hg/HgO 参比电极的标准电极电势，数值为 0.897，V；pH 为测试溶液的酸碱值。

1.2.1 OER 催化机理

有些催化剂在酸性条件下展现了良好的 OER 性能或 HER 性能，而有些催化剂在碱性条件下具有优异的电催化性能，或者具有良好的双功能催化性能。尽管科研者们提出了不同的反应机理，但不管是在酸性还是碱性条件下，OER都是四电子转移过程。对于典型的催化表面M，其涉及的机理如下：

在碱性条件下：

$$M+OH^- \longrightarrow MOH+e^- \tag{1.5}$$

$$MOH+OH^- \longrightarrow MO+H_2O+e^- \tag{1.6}$$

$$2MO \longrightarrow 2M+O_2 \tag{1.7}$$

$$MO+OH^- \longrightarrow MOOH+e^- \tag{1.8}$$

$$MOOH+OH^- \longrightarrow M+O_2+H_2O+e^- \tag{1.9}$$

总反应：$$4OH^- \longrightarrow 2H_2O+4e^-+O_2$$

在酸性条件下：

$$M+H_2O \longrightarrow MOH+H^++e^- \tag{1.10}$$

$$MOH+OH^- \longrightarrow MO+H_2O+e^- \tag{1.11}$$

$$2MO \longrightarrow 2M+O_2 \tag{1.12}$$

$$MO+H_2O \longrightarrow MOOH+H^++e^- \tag{1.13}$$

$$MOOH+2H_2O \longrightarrow M+2O_2+5H^++5e^- \tag{1.14}$$

总反应：$$2H_2O \longrightarrow 4H^++4e^-+O_2$$

从MO中间体产生O_2一般来说有两种不同的路径。在酸性条件下，一种途径是MO分解产生O_2［式(1.12)］，另一种途径是MOOH与H_2O作用产生O_2［式(1.14)］。在碱性条件下，一种途径同样是MO分解产生O_2［式(1.7)］，但另一种途径是MOOH与OH^-作用产生O_2［式(1.9)］。在这种非均相催化过程中，形成的中间体（MO、MOOH和MOH）中M和O之间的相互作用对于确定总体反应机理至关重要。中间体与催化剂表面的结合强度决定了材料催化OER的效率。但是要在实验上测量和控制被吸附物在催化剂表面上的结合能很困难，因此难以预测新催化剂的催化活性。科研人员已经付出了巨大的努力来确定表面电子结构的性质，这些性质可用于评估催化剂的催化效率并开发新的电催化剂[8-10]。

在其他催化活性描述中，金属d带中心与金属表面化学吸附氧的强度相对于费米能级的关系，在活性评价中起着重要的作用[11]。当吸附质与催化剂表面结合时，其电子与催化剂的s、p、d带电子相互作用。由于在金属-吸附键形成过程中，s带和p带的能量没有明显变化，所以金属的局域

d 带决定键的强度。随着 d 带中心向费米能级偏移，费米能级以上空位的反键轨道数目增加，加强了与吸附质的化学键合[12]。根据 DFT 计算建立了 d 带中心与金属（过渡金属）-氧键强度之间的线性关系，不同 Pt 合金（Pt_3M，M＝Ti，V，Cr，Mn，Fe，Co，Ni）的紫外光电子能谱（UPS）证实了这一观点[13]。

1.2.2 HER 催化机理

在电解过程中，阴极反应即 HER 在酸性介质中通过还原质子或在碱性介质中还原水产生氢气。反应路径有两种，一种是 Volmer-Heyrovsky 反应，另一种是 Volmer-Tafel 反应。

（1）碱性介质

在碱性介质中存在大量的 OH^-，因此反应按照以下步骤进行。

步骤 1　Volmer 反应：水分子与一个电子结合被还原形成氢原子，吸附在电极表面的催化活性位点上。

$$* + H_2O + e^- \rightleftharpoons H_{ad}^* + OH^-$$

步骤 2　Heyrovsky 反应：步骤 1 吸附的氢原子与一个电子和水分子耦合，发生电化学脱附产生氢分子。

$$H_{ad}^* + H_2O + e^- \rightleftharpoons H_2 + OH^- + *$$

步骤 3　Tafel 反应：在这个步骤中，两个吸附的氢相互结合，释放一个氢分子。

$$2H_{ad}^* \rightleftharpoons H_2 + 2*$$

（2）酸性介质

步骤 1　Volmer 反应：H^+ 与一个电子结合，被吸附到催化剂活性位点上。

$$* + H^+ + e^- \rightleftharpoons H_{ad}^*$$

步骤 2　Heyrovsky 反应：步骤 1 吸附的氢与一个电子和另一个 H^+ 耦合，发生电化学脱附。

$$H_{ad}^* + H^+ + e^- \rightleftharpoons H_2 + *$$

步骤 3　Tafel 反应：同碱性介质一样，两种吸附的氢相互结合而释放出氢分子。

$$2H_{ad}^* \rightleftharpoons H_2 + 2*$$

在碱性和酸性介质中，HER 过程均始于将氢吸附到催化活性表面上（Volmer 反应），然后通过脱附（Heyrovsky 反应）或解离解吸（Tafel 反应）过程释放出分子氢。可以从 Tafel 斜率推导出的速率控制步骤（RDS）得到特定催化剂遵循的反应机理。Tafel 斜率代表了特定电催化剂的本质，其值为认识反应机理提供了信息。通过绘制电流密度的对数相对于过电位的曲线（称为塔费尔曲线），可得到相应的 Tafel 斜率。已经发现，对于 Tafel RDS，斜率约为 29mV/dec，而对于 Heyrovsky 和 Volmer 的 RDS，斜率约分别为 39mV/dec 和 118mV/dec。通常在酸性条件下的 HER 反应中，速率控制步骤为 Heyrovsky 步骤，碱性条件下发生的 HER，速率控制步骤为 Volmer 步骤[14,15]。

因此，反应的第一步涉及活性氢物质（H^*）的产生及其在催化表面的吸附（Volmer 步骤），然后通过 Tafel 或 Heyrovsky 步骤产生氢分子解吸（H_2）。活性氢吸附和氢分子解吸的难易程度对于监测反应动力学具有同等重要的意义。因此，采用催化剂表面吸附氢所需要的标准自由能（ΔG_H）来评估活性氢与电极表面的相互作用强度，一般采用如图 1.2 所示的火山图来表示。若 H^* 在电极表面上具有强相互作用（$\Delta G_H>0$）会减慢 Volmer 步骤，使周转频率受到限制。相反，若 H^* 在电极表面上具有弱相互作用（$\Delta G_H<0$）则会加快 Volmer 步骤，而阻碍 Tafel 或 Heyrovsky 步骤。

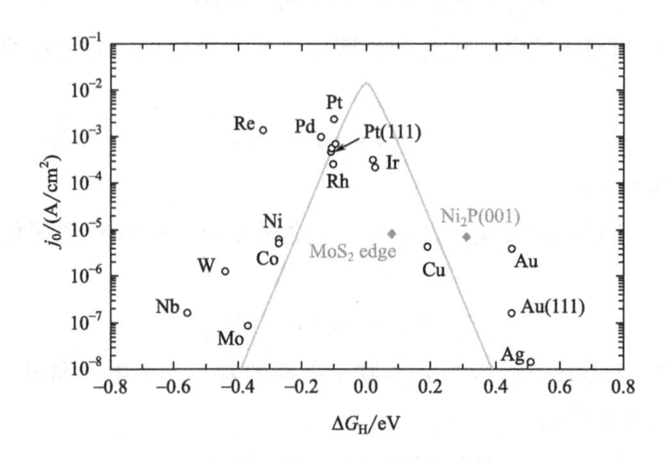

图 1.2　交换电流密度与吉布斯自由能的火山图[16]

具有良好催化活性的催化剂，其反应的中间产物应该具有适中的 ΔG_H 值（几乎为零）[16,17]。从火山图可以看出，Pt 基材料的 ΔG_H 值几乎为零，因此被认为是最优异的 HER 的催化材料。与酸性介质相比，碱性介质中金属—OH 具有强烈的相互作用，需要额外的过电位来提供 HER 所需的能量。

1.3 电催化的重要参数

在催化剂的评价体系中，使用质量活性、塔费尔斜率、过电位（η）、法拉第效率、电化学活性面积（ECSA）、周转频率（TOF）和稳定性等参数来评估不同催化剂的催化性能，且这些参数在探索电催化过程中的相关机制中起着至关重要的作用，下面进行详细讨论。

1.3.1 质量活性

尽管贵金属基的电催化剂具有优异的 HER 和 OER 性能，但其高成本和低储量限制了商业化应用。因此，为减少这些贵金属的使用，科研工作者们做出了许多努力寻找合适的非贵金属催化剂来代替贵金属催化剂。在大多数的研究报告中，这些非贵金属催化剂的性能通过特定区域特定电流所需的过电位进行比较，也就是将相应的阴/阳极电流除以电极的几何表面积。目前所有贵金属基催化剂都需要几乎相同的过电位才能达到基准的电流密度。但催化剂一般都是多孔材料，这种类型的归一化可能无法用于不同催化材料的活性比较。因此，采用质量归一化更合适，即通过将相应的电流（I）除以催化剂质量（m）来表示材料的质量活性：质量活性（mass activity）$=I/m$，单位为 A/g。

1.3.2 塔费尔斜率

Tafel 斜率对于研究电催化剂表面上催化过程（HER 或 OER）的反应机理十分重要。一般情况下，计算 Tafel 斜率是为了研究特定电催化剂的固有性质。从 Tafel 图可以看出 IR 校正的过电位与电流密度之间的关系，其关系式如下：

$$\frac{\mathrm{dlg}(j)}{\mathrm{d}\eta}=\frac{2.303RT}{\alpha nF} \tag{1.15}$$

可以清楚地看到，该斜率与电荷转移系数（α）成反比，而其他参数是常数（R：摩尔气体常数；T：热力学温度；n：该反应涉及的电子数；F：

法拉第常数)。

$$\eta = a + b\lg j \tag{1.16}$$

式中，η、a、b 和 j 分别表示过电位、Tafel 常数、Tafel 斜率和电流密度。通过极化曲线可以绘制电流密度的对数 [$\lg(j)$] 对过电势 (η) 的曲线得到 Tafel 图。之后，通过拟合 Tafel 图的线性部分，可以获得 Tafel 斜率[18]。由于塔费尔斜率与电荷转移系数成反比，因此可作为电催化剂性能评估的参数之一。如前面所述，HER 通过两个步骤进行，即 Volmer 步骤 (吸附活性氢)，然后是 Tafel 或 Heyrovsky 步骤 (氢分子的解吸)。据文献报道，Volmer、Heyrovsky 和 Tafel 反应的理论塔费尔斜率分别为 118mV/dec、39mV/dec 和 29mV/dec[19]。例如，在 0.5mol/L H_2SO_4 溶液中，Pt 催化 HER 反应的 Tafel 斜率为 30mV/dec，说明其反应是通过 Volmer-Tafel 路径，且 Tafel 步骤是速控步骤[20]。

交换电流密度 (j_0) 是在阴极电流和阳极电流相等的反应平衡位置的电流密度。可以通过将线性 Tafel 部分外推到横轴 (Tafel 图的 x 轴) 得到 (图 1.3)。j_0 可以用来描述一个电极反应得失电子的能力，描述影响电极/电解质界面处电荷转移的物理/表面性质、结构以及其他因素，反映一个电极反应进行的难易程度[21]。j_0 的大小取决于电催化剂的催化活性表面。例如，高活性的电催化剂的交换电流密度高于活性较低的其他催化剂。

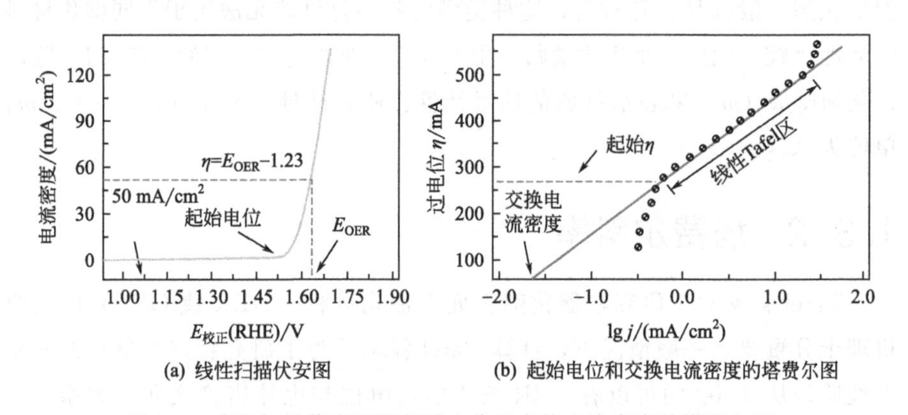

(a) 线性扫描伏安图　(b) 起始电位和交换电流密度的塔费尔图

图 1.3　线性扫描伏安图和起始电位和交换电流密度的塔费尔图

1.3.3 过电位

电解水包含两个关键步骤，即阳极的 OER 和阴极的 HER。实际上，在标准状况下，这些反应在热力学上是不可行的，需要一些额外的能量。根据能斯特方程，OER的标准电势（$E^{\ominus}$）为1.23V（相对于NHE），而HER为0V（相对于NHE）。

$$\text{HER：}2H^{+}(aq)+2e^{-}\longrightarrow H_2(g),E^{\ominus}_{HER}=0.00V$$

$$\text{OER：}2H_2O\longrightarrow O_2(g)+4H^{+}(aq)+4e^{-},E^{\ominus}_{OER}=1.23V$$

在实际电解中，电势的大小取决于OH^-或H^+的浓度即电解质的pH值。由于HER和OER具有较差的能量效率和缓慢的反应动力学，反应所需的电势一般都高于标准电势。因此，需要一些额外的能量来克服反应的能垒，这些额外的所需能量称为HER或OER的过电位（η），即实验过程中得到的电化学反应的电势和热力学标准电势之差。电极的催化活性一般通过线性扫描伏安（LSV）法或者循环伏安（CV）法对电极活性进行初步评估，其中非法拉第电容电流是整个电极电流的核心部分。η是评估材料电化学活性其中一个最重要的电化学指标。在标准状况下，HER和OER的标准电极电位分别是0V和1.23V。过电位是实际催化过程中（HER或OER）的电流的起峰电位与标准电极电位之间的绝对值。在相同电流密度条件下过电位越小说明电催化剂具有更高的催化活性。

一般过电位分为活化过电位、浓度过电位、阻抗过电位等，其中活化过电位可以通过选择合适的电催化剂被大幅度降低。浓度过电位是体溶液和电极表面之间的离子浓度差异造成的，这些离子浓度差异主要来源于缓慢的扩散速率。由于扩散层始终存在，因此浓度过电位只能通过搅拌消除一部分。阻抗过电位，或者叫做接触过电位，发生在测试体系的表界面。通过表界面的阻抗会引起额外的电压降，促使催化剂测试出来的过电位高于其实际值。可以通过IR补偿来排除这种过电位，以此得到电催化剂的实际过电位值。

一般采用标准电化学阻抗谱或电流中断法测量电解液的溶液电阻[22]。在三电极体系中，很多电化学工作站可以直接测得R值或者直接通过能斯特曲线和高频下横坐标最左边的交叉处的值。IR补偿通过方程式$E_{校正}=E_{未校正}-IR$进行表达，其中E是电位，I是该电位下通过系统的电流。从校正方程式可以看出，校正的影响对起峰电位来说比较小，因为该电位下电流比较小。但随着电位的增大，电流也逐渐增大，$E_{校正}$以及极化曲线的偏移会逐渐增大。如图1.4（见文后彩插）所示，材料的极化曲线均通过电化学阻抗谱（EIS）进行了IR补偿。为了得到不变形的曲线，很多研究工作者采用 90%或 95%的 IR 补偿[23]，补偿百分比可根据实际的 LSV 曲线以及 EIS 测试结构进行适当的调整。因此，为了方便比较不同催化剂之间的活性，有必要同时给出没有扣除 IR 降的曲线。图 1.4(d) 给出了 EIS 对应的拟合电路图，阻抗 R_s 对应没有补偿的溶液阻抗，R_1 对应电荷转移阻抗，CPE_1 为双电层电容，R_2 和 CPE_2 可能对应玻碳电极和催化剂之间的接触阻抗。

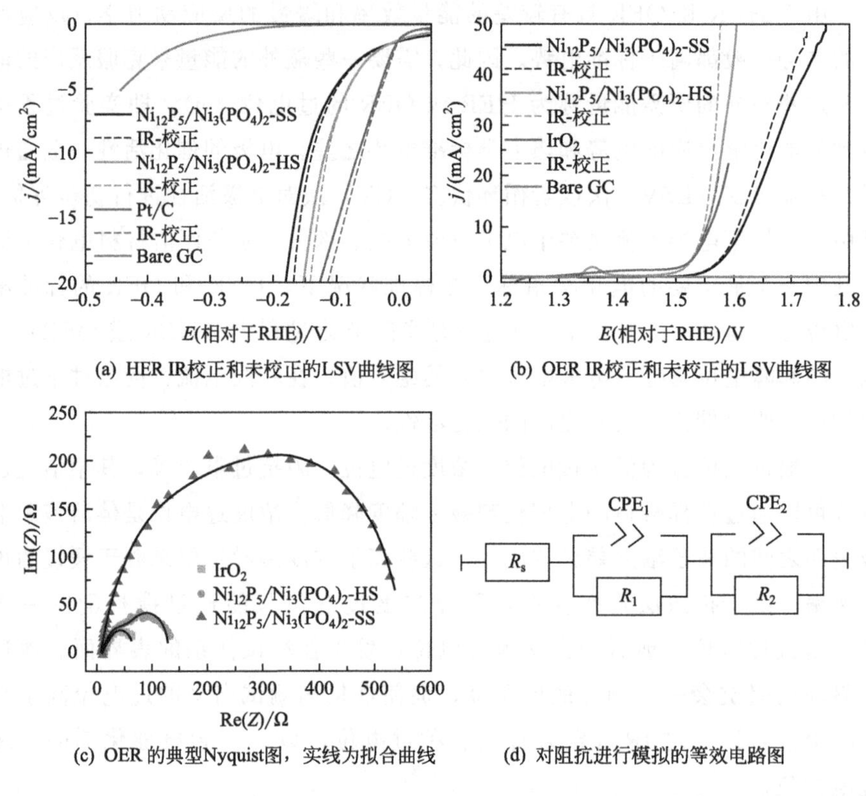

图 1.4　IR 补偿的 LSV、Nyquist 以及等效电路图[33]

目前，已经研发了各种催化剂材料来降低电解水分解的过电位。具体来说，可以通过测量特定的电流密度下所需的过电位来确定催化剂的电催化效率。$10mA/cm^2$ 所需的电势已被设定为太阳能水分解体系的度量标准 [在 1sun 的照射下 (1sun 指在 AM1.5G 的大气因子下，光强度为 $96mW/cm^2$)]，并在效率为 12.3%的太阳能氢装置中可以实现此电流密度[24,25]。因此通常采用电流密度为 $10mA/cm^2$ 时所需的电位来评估和对比催化剂的电催化效率。η_{10} 的值越小，催化材料的效率越高。在某些催化活性较高的电催化剂中，也可以采用较高电流密度下［如 50mA/cm^2（η_{50}）和 100mA/cm^2（η_{100}）］的过电势作为评估与对比的标准[26-28]。另外，为了获得电催化材料的真实电流密度，一般需要消除电极材料的电容背景电流。在 OER 过程中，可以通过设定对称的背景电容电流减去从前向和后向扫描获得的平均电容电流来消除电容电流。

1.3.4 法拉第效率

法拉第效率（FE）也称为库仑效率，它是电化学反应中的重要技术指标之一，是指参与特定电化学反应的电子与所提供电子的数量之比。在 OER 或 HER 中，法拉第效率定义为通过实验释放的分子氧/氢的数量与理论值的分子氧/氢数量之比[2,29]。特定电催化剂的 FE 可以通过两种方法来计算：一种是电化学方法；另一种是使用不同的技术对产生的气体进行定量测试来计算。下面进行详细介绍。

方法 1：使用旋转环形盘电极（RRDE），将催化剂滴涂到玻碳电极上作为圆盘电极，铂环作为环盘电极。优化圆盘电极的电位窗口，保证环盘电极的电势恒定不变，环盘电极的电位取决于电解质的 pH 值，因此需要根据体系的 pH 值对其事先确定。负载有催化剂的圆盘电极催化 OER，环盘电极将其释放的氧气还原为过氧化氢。在测试之前，需要通过 Fe^{2+}/Fe^{3+} 氧化还原探针在不同转速下的响应来确定 RRDE 的收集效率。然后根据以下等式计算 FE：

$$FE=\frac{I_R n_D}{I_D n_R N_{CL}} \tag{1.17}$$

式中，I_R、I_D 分别是环电流和盘电流，mA；n_R 和 n_D 分别是在环和盘上转移的电子数；N_{CL}是电极的预定收集效率。

Jaramillo 研究小组使用该方法评估了 OER 的法拉第效率[30]。他们将 NiO_x 滴涂到圆盘上，并在 1mol/L NaOH 电解质中以 1600r/min 的速度进行测量（图 1.5）。通过测量得到 I_D 为 1.95mA，N_{CL}为 0.19，计算得出 FE≥0.9。一般情况下，需要进行三次实验，然后取其平均的 FE 值。尽管此方法广泛用于许多新型 HER 和 OER 电催化剂的 FE 评估，但仍需要一些灵敏度更高的技术，以便更好地理解和计算 FE。另外可以通过恒电位或恒电流的积分曲线来预判 FE 值，由于在测试过程中催化剂可能发生一些副反应，导致 FE 效率始终低于 100%。

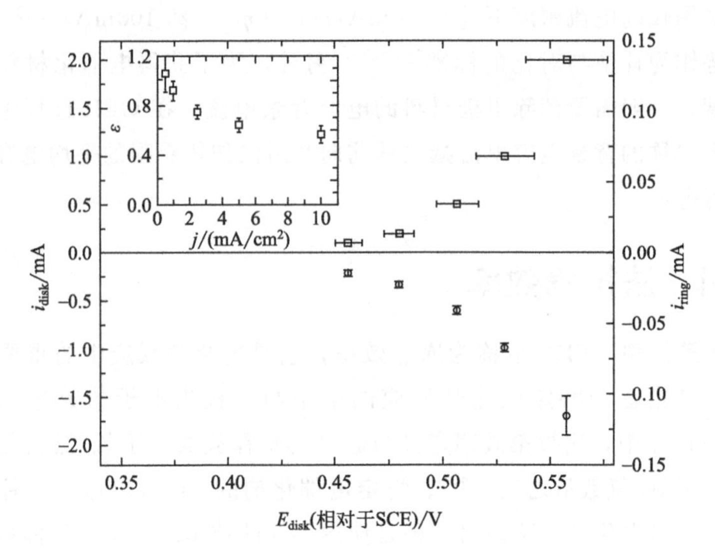

图 1.5 在 1mol/L NaOH 溶液，1600r/min 的转速在 1atm（1atm＝101325Pa）N_2 条件下 NiO_x 催化 OER 的 RRDE

测量了不同电流阶跃下的工作电位（内插图为环盘电流比和电子数量作为盘电流的函数图）

方法 2：该方法是测量 HER 和 OER 的 FE 的最通用方法。首先对反应释放出的气体进行定量测定，计算实验测得值与理论值之比得到相应的 FE。这种方法关键是对反应过程中或反应完成后产生气体的定量测试，通常采用计时电流法或计时电位法进行分析。一般情况下，通过采用气体置换法、色谱法或使用光谱技术进行定量[29]。采用气体置换法是将已标记好的充满水的滴定管倒置连接到电化学反应池，在反应过程中，所产生的气体通过置换水进行收集和量化[31]。色谱法是将气相色谱仪与电化学池集成，直接测试反应过程中放出的气体的量。光谱技术法仅适用于涉及释放氧（从三重态到单重态）的激发并允许通过荧光弛豫的 OER，采用荧光强度来定量氧气。催化剂的类型、电化学池的尺寸、电极的类型以及实验室自身的条件决定了采用其中的哪一种技术来对产生的气体进行定量分析。

1.3.5 电化学活性面积和归一化电流

通常，OER 和 HER 电流通过负载催化剂的质量（m）、催化剂的工作电极的几何表面积（S_{geo}）或催化剂本身的几何表面积（S_{ECSA}）进行归一化[14]。在所有的这些方法中，广泛采用工作电极的表面积归一化，但是它不能表现材料的固有催化性能。另外，不同负载量的催化剂在特定的电流密度下有不同的电势。如果负载为100%，即单层催化剂完全覆盖电极表面，则可以使用这种归一化方法。实际上，这种归一化结果只有助于对表面平坦的电极（如金属箔、玻璃碳板/圆盘等，而不是金属/碳泡沫、碳布等）与已报道的文献进行催化性能比较。但是催化剂的不同负载量会产生不一样的结果，在负载量较低的情况下，较多的催化惰性的电极表面暴露于电解质中，这种情况下S_{geo}将无法提供真实值。由于反应是在催化剂表面进行的，因此在较高负载下，表面的催化剂层参与了反应，但内部的催化剂几乎没有参与反应。因此，在整个工作体系中，应该保持催化剂的负载量一致，最好是形成单层催化剂，以最大程度减小误差。

催化剂的实际表面积可以通过原位或通过非原位表面积分析方法得到。原位分析方法可以通过气体吸附研究（Brunauer-Emmet-Teller，BET）测试得到，也可以通过显微镜（扫描电子显微镜/扫描探针显微镜）测量催化剂的粒径得到。尽管这两种方法都能给出催化剂的平均表面积，但是不是所有的表面积都具有电催化活性，因此对于测试具有活性的比表面积的准确性不高。另外，原位测试方法对催化剂的负载量更为敏感，可以给出催化剂的固有性质。采用测试催化剂的双层电容电化学活性面积（ECSA）来进行原位测试，从而得到更准确的活性面积。ECSA受电催化剂的多孔结构和粗糙度的影响，无法得到ECSA的实际值。但是可以从催化剂的循环伏安图得到双层电容（C_{dl}）：一种是在非法拉第电势区域以不同的扫描速度记录静态循环伏安图，另一种是频率依赖的电化学阻抗图。在测量C_{dl}时，对扫描电势范围的选择非常关键，要求在选择的电势范围内，催化剂不会发生明显的法拉第过程，仅仅是在电极/电解质界面形成双电层电容（图1.6）。以系统的开路电位（OPC）为中心，选择的电位窗口大小一般为0.1V。此处从CV曲线图得到的充电电流（i_c）是扫描速率（v）和双层电容（C_{dl}）的乘积（$i_c=vC_{dl}$）。由此可以看出，充电电流与扫描速率直线的斜率就是双电层电容。例如，Dutta 和 Samantara 研究小组发现在1mol/L KOH电解液中，磷化镍的C_{dl}值为5.6mF[29]；Behera研究小组发现在0.1mol/L H_2SO_4溶液中VS_2纳米材料的C_{dl}值为0.969mF[32]。

同样，C_{dl}值可以从同样的非法拉第电位区域中的电化学阻抗得到，通过施加正弦电位来得到与频率相关的交流阻抗谱。例如，Jaramilo研究小组

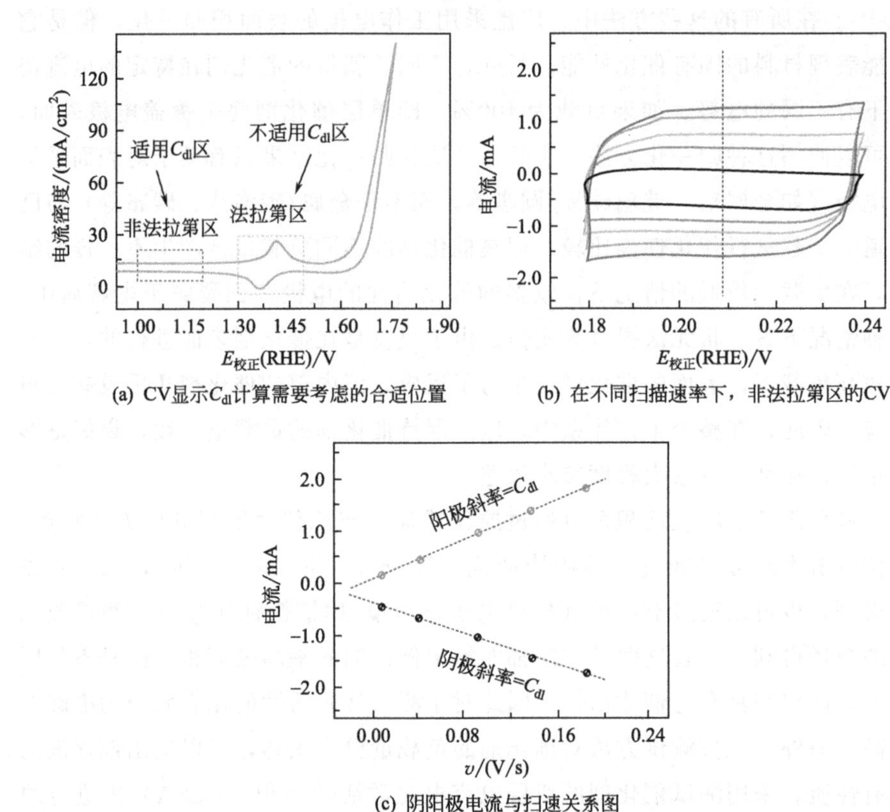

(a) CV显示C_{dl}计算需要考虑的合适位置　(b) 在不同扫描速率下，非法拉第区的CV

(c) 阴阳极电流与扫速关系图

图 1.6　计算活性材料 ECSA 需要的电化学表征和关系图

已在 100Hz 至 100kHz 频率范围内（对于电沉积 NiO_x）的非法拉第区域测试了 Nyquist 阻抗图（虚阻抗与实阻抗图），并对其进行了电路拟合[30]，如图 1.7（见文后彩插）所示。

图中所示的 R_s、R_{ct} 和 CPE 分别对应于系统的溶液电阻、电荷转移电阻和常相位元件（CPE，双层电容）。从下面的公式可以看出，CPE 与施加电势的频率（ω）有关[30,33]：

$$Z_{CPE}=\frac{1}{Q_0(i\omega)^{1-\alpha}} \tag{1.18}$$

式中，i 为 $(-1)^{1/2}$；Q_0 为常数 α（$0\leqslant\alpha\leqslant1$）。在这种情况下，根据下列公式可以看出 Q_0 与 C_{dl} 有关：

$$C_{dl}=\left[Q_0\left(\frac{1}{R_s}+\frac{1}{R_{ct}}\right)^{(\alpha-1)}\right]^{1/\alpha} \tag{1.19}$$

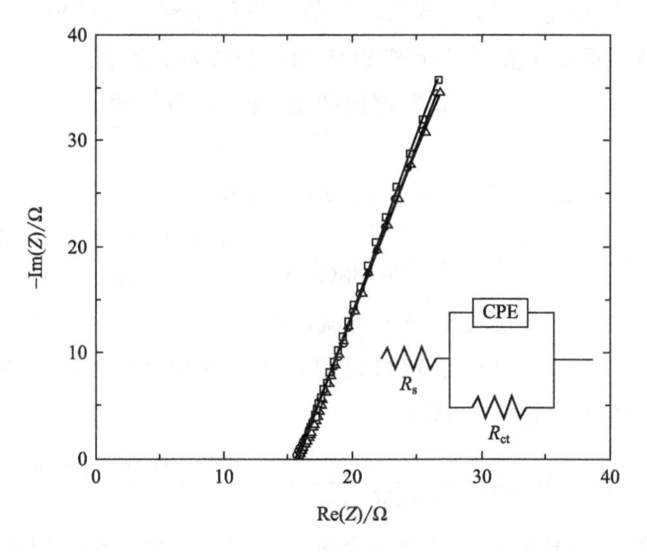

图 1.7 在 1mol/L NaOH 溶液中 NiO_x 在不同电位 −0.1V（黑色）、−0.05V（红色）、0V（蓝色）的阻抗图[30]

从该公式可以看出，当 $\alpha=1$ 时，CPE 相当于一个纯电容器，此时 $C_{dl}=Q_0$；但是当 $\alpha=0$ 时，CPE 的行为就像纯电阻一样，C_{dl}不能定义。ECSA 可以根据双电层电容计算得到，公式如下所示：

$$\mathrm{ECSA}=\frac{C_{dl}}{C_s} \tag{1.20}$$

式中，C_s 是样品的比电容，其值范围大概在 $20\sim60\mu F/cm^2$[34-36]。但从以上两种方法获得的 ECSA 值并不相同（标准误差为±15%）。因此，与文献报道和理论电流密度相比，电极几何表面积的归一化电流更适合。此外，可以根据下列方程式，通过简单地划分电极的几何表面积，可以从 ECSA 中获得粗糙度因子（R_f：每单位表面积的活性位点）。

$$R_f=\frac{\mathrm{ECSA}}{S_{geo}} \tag{1.21}$$

1.3.6 周转频率

电催化剂的每单位活性位点每秒释放出的氢/氧分子的量称为周转频率(TOF)[37,38]，故有必要确定电极材料的反应位点的活性。要计算 TOF，就必须确定电极材料的活性位点数。也就是说，存在于催化剂表面上的活性位点的数量（n）决定了TOF的值，该值可以使用铜的欠电位沉积法或循环伏安图获得。通过n值可以计算TOF值，公式如下所示：

$$\mathrm{TOF}=\frac{\text{每秒释放的 } H_2 \text{ 和 } O_2 \text{ 分子数}}{\text{活性位点数}} \tag{1.22}$$

例如，如果是HER，每秒释放的H_2量为：

$$\text{每秒释放的 } H_2 \text{ 量}=\frac{i}{96485\mathrm{C/mol}}\times\frac{6.022\times10^{23}\mathrm{e^-/mol}}{2\mathrm{e^-}} \tag{1.23}$$

式中，i是在一定过电位下的电流，A。

假设氧化过程是单电子转移过程，从它的循环伏安图中测得的电荷（Q_{cv}）可以确定活性位点的数目[39]：

$$\text{活性位点数}=\frac{Q_{cv}}{96485\mathrm{C/mol}}\times\frac{6.022\times10^{23}\mathrm{e^{-1}/mol}}{1\mathrm{e^-}} \tag{1.24}$$

如果发生OER反应，释放一个氧分子需要4个电子，因此式（1.23）中的分母由4代替2。只需将以上公式表示为：

$$\mathrm{TOF_{HER}}=\frac{i}{2Fn} \tag{1.25}$$

$$\mathrm{TOF_{OER}}=\frac{i}{4Fn} \tag{1.26}$$

式中，i是线性扫描伏安电流，A；F是法拉第常数，96485.3C/cm^2；n是活性位点数，mol。

1.3.7 稳定性

就实际应用和商业化来说，稳定性是电催化剂十分重要的参数。有很多技术可用于评估电催化剂的稳定性，但通常使用的是计时电流法（CA）、计时电位法（CP）、循环伏安法（CV）或者线性循环伏安法（LSV），如图1.8（见文后彩插）。CA法是指在特定的过电势下，一段时间内电流密度变化（i-t曲线图）；CP法则是在恒定电流密度下，在一段时间内过电势变化的测量（v-t曲线图），通常采用10mA/cm^{2}的电流密度来测试其稳定性。测试时间可以从几个小时到几十个小时甚至更长时间。CV法指循环测试几百上千圈后，对材料进行线性扫描伏安测试得到的极化曲线与最初的极化曲线进行比较，如在相同的电流密度下过电位变化越小，说明材料的电催化稳定性越好。近年来，通过恒电流的手段来检测材料的稳定性[40]，在持续的电流变化下，只有结构和稳定性十分优异的催化剂才能保持相应的梯度电位变化，且在短时间内快速达到平衡。

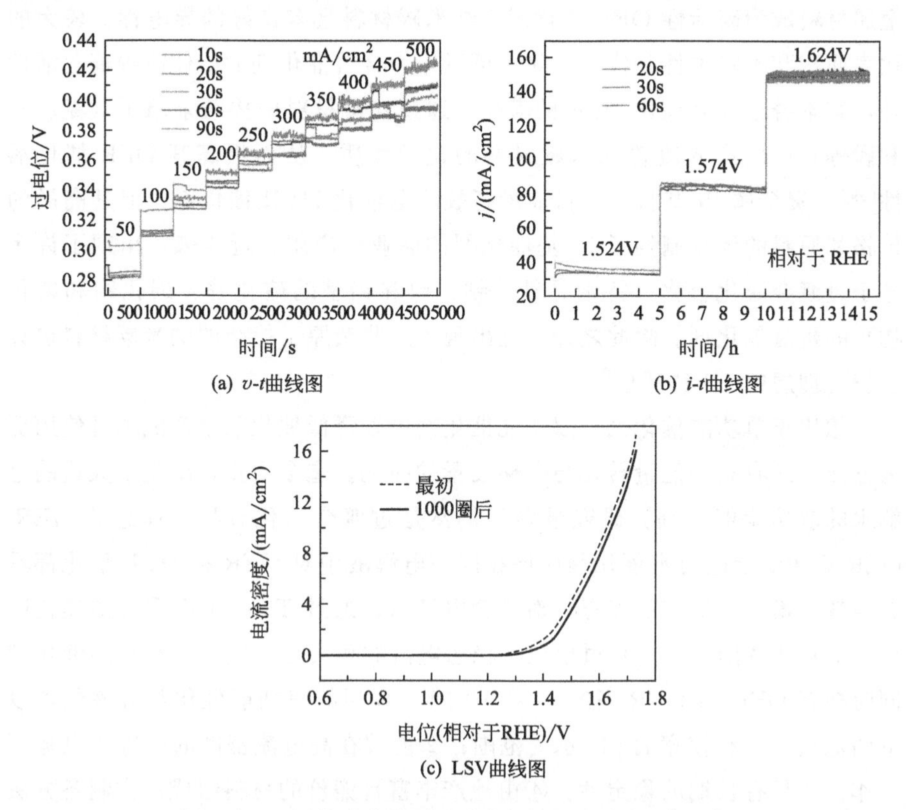

(a) v-t曲线图

(b) i-t曲线图

(c) LSV曲线图

图 1.8　不同的电化学测试评估催化剂的稳定性[41,42]

1.4　电解水催化剂

目前为止，具有电解水活性的催化材料有很多[2,43-53]，包括贵金属铂(Pt)、钌（Ru)、铱（Ir）和铑（Rh）及其化合物等[54,55]；非贵金属包括过渡金属氧化物[56]、磷化物[57]、氮化物[58]、硫化物[59]、氢氧化物[60]、硒化物[61]和碳化物[62]等，如过渡金属镍、钴、铁、钼、钨、锰、铜等；非金属材料，诸如氮、硼、磷、硫掺杂的纳米碳材料[63-67]等。

以 Ir 基和 Ru 基化合物为代表的贵金属催化材料具有优异的 OER 催化活性[68]，但是其价格昂贵且地球储量有限，限制了贵金属催化材料大规模商业化应用。相对于贵金属材料，过渡金属材料在储量、价格和环境友好性等方面都展现出了巨大的优势。因此，过渡金属化合物催化材料有望替代贵金属材料成为高活性 OER 催化剂。纳米碳材料具有良好的导电性、较大的比表面积和耐腐蚀性等优点，除了可以作为金属催化剂的载体改善催化活性外，其本身也具有催化 OER 的潜能。通过在碳材料中掺杂杂原子（氮、硫和磷等）可以有效改善纳米碳材料的化学性质，从而提高其 OER 催化活性[69]。贵金属 Pt 基材料是目前最高效的电催化 HER 催化剂，但其高昂的价格和原料的稀缺制约了 Pt 基催化剂的商业化应用。近年来，出现了许多基于过渡金属化合物（特别是铁、钴、镍和钼基的硫化物、碳化物和磷化物）的析氢催化剂。除此之外，还出现了一些杂原子掺杂的纳米碳材料也具有较为理想的 HER 活性[69]。

相比于单功能催化剂，双功能催化剂可以降低催化剂生产成本且使用更为方便，还有利于能量转化与存储装置的简化。迄今为止，出现了大量的电解水的双功能催化剂，研究最多的仍然是过渡金属化合物。理想的 HER-OER 双功能全电解水催化剂应该在同一电解液中对 OER 和 HER 催化都具有高效、稳定且经济的特点。对于全电解水，通过 Tafel 平面图来描述在阴极（η_c）或者阳极（η_a）过电位上稳态电流的相关性。优异的双功能催化剂同时拥有 OER 和 HER 高的 j_0 和小的 b。理想的双功能催化剂应该包含以下的优点：①有较宽的 pH 承受范围；②能够在高电流密度的中等过电势下工作；③具有长期的稳定性；④由地产丰富且廉价的材料组成；⑤制备方法简单经济。一种单一的催化剂不会同时具有以上的几种条件。因此，为特定的应用选择合适的催化剂是决定未来电解水发展的关键。大多数时候，HER 和 OER 的效率之间的平衡是最重要的。但是在某些情况下，成本可能会超过效率。因此，研究人员非常希望探索具有各种性能的多种双功能电催化剂，以推进下一代能量转换技术的发展。

参 考 文 献

[1] Dargay J, Gately D. Vehicle ownership to 2015: Implications for energy use and emissions. Energy Policy, 1997, 25 (14): 1121-1127.

[2] Zou X, Zhang Y. Noble metal-free hydrogen evolution catalysts for water splitting. Chemical Society Reviews, 2015, 44 (15): 5148-5180.

[3] Reece S Y, Hamel J A, Sung K, et al. Wireless solar water splitting using silicon-based semiconductors and earth-abundant catalysts. Science, 2011, 334 (6056): 645-648.

[4] Bates M K, Jia Q, Ramaswamy N, et al. Composite Ni/NiO-Cr_2O_3 catalyst for alkaline hydro-

gen evolution reaction. The Journal of Physical Chemistry C, 2015, 119 (10): 5467-5477.

[5] De Souza R F, Padilha J C, Gonçalves R S, et al. Electrochemical hydrogen production from water electrolysis using ionic liquid as electrolytes: towards the best device. Journal of Power Sources, 2007, 164 (2): 792-798.

[6] Vilekar S, Fishtik I, Datta R. Kinetics of the hydrogen electrode reaction. Journal of The Electrochemical Society, 2010, 157 (7), B1040-B1050.

[7] Wang Q, Cha CS, Lu J, et al. The electrochemistry of "solid/water" interfaces involved in PEM-H_2O reactors Part Ⅰ. The "Pt/water" interfaces. Physical Chemistry Chemical Physics, 2009, 11 (4): 679-687.

[8] Bockris J O M, Otagawa T. The electrocatalysis of oxygen evolution on perovskites. Journal of The Electrochemical Society, 1984, 131 (2): 290-302.

[9] Suntivich J, May K J, Gasteiger H A, et al. A perovskite oxide optimized for oxygen evolution catalysis from molecular orbital principles. Science, 2011, 334: 1383.

[10] Trasatti S. Physical electrochemistry of ceramic oxides. Electrochimica Acta, 1991, 36 (2): 225-241.

[11] Edmonds T, Mccarroll J J. Impact of surface physics on catalysis. 1978: 261-290.

[12] Koper M T M, Van Santen R A. Interaction of H, O and OH with metal surfaces. Journal of Electroanalytical Chemistry, 1999, 472 (2): 126-136.

[13] Mun B S, Watanabe M, Rossi M, et al. A study of electronic structures of Pt_3M (M=Ti, V, Cr, Fe, Co, Ni) polycrystalline alloys with valence-band photoemission spectroscopy. The Journal of Chemical Physics, 2005, 123 (20): 204717.

[14] Sengeni A, Ede S, Kuppan S, et al. Recent trends and perspectives in electrochemical water splitting with an emphasis on sulfide, selenide, and phosphide catalysts of Fe, Co, and Ni: A review. ACS Catalysis, 2016, 6 (12): 8069-8097.

[15] Kong D, Cha J J, Wang H, et al. First-row transition metal dichalcogenide catalysts for hydrogen evolution reaction. Energy & Environmental Science, 2013, 6 (12): 3553-3558.

[16] Noerskov J K, Bligaard T, Logadottir A, et al. Trends in the exchange current for hydrogen evolution. Journal of the Electrochemical Society, 2005, 152 (3): J23-J26.

[17] Greeley J, Jaramillo T F, Bonde J, et al. Computational high-throughput screening of electrocatalytic materials for hydrogen evolution. Nature Materials, 2006, 5 (11): 909-913.

[18] Smith R D L, Prévot M S, Fagan R D, et al. Water oxidation catalysis: electrocatalytic response to metal stoichiometry in amorphous metal oxide films containing iron, cobalt, and nickel. Journal of the American Chemical Society, 2013, 135 (31): 11580-11586.

[19] Fletcher S. Physical electrochemistry. Fundamentals, techniques, and applications by Eliezer Gileadi. Journal of Solid State Electrochemistry, 2012, 16 (3): 1301-1301.

[20] Li Y, Wang H, Xie L, et al. MoS_2 nanoparticles grown on graphene: An advanced catalyst for the hydrogen evolution reaction. Journal of the American Chemical Society, 2011, 133 (19): 7296-7299.

[21] Anantharaj S, Ede S R, Karthick K, et al. Precision and correctness in the evaluation of electrocatalytic water splitting: revisiting activity parameters with a critical assessment. Energy & Environmental Science, 2018, 11 (4): 744-771.

[22] Cooper K R, Smith M. Electrical test methods for on-line fuel cell ohmic resistance measurement. Journal of Power Sources, 2006, 160 (2): 1088-1095.

[23] Chang J, Lv Q, Li G, et al. Core-shell structured $Ni_{12}P_5/Ni_3(PO_4)_2$ hollow spheres as difunctional and efficient electrocatalysts for overall water electrolysis. Applied Catalysis B: Envi-

ronmental, 2017, 204: 486-496.

[24] Benck J D, Hellstern T R, Kibsgaard J, et al. Catalyzing the hydrogen evolution reaction (HER) with molybdenum sulfide nanomaterials. ACS Catalysis, 2014, 4: 3957-3971.

[25] Gao Q, Zhang W, Shi Z, et al. Structural design and electronic modulation of transition-metal-carbide electrocatalysts toward efficient hydrogen evolution. Advanced Materials, 2019, 31 (2): 1802880.

[26] Chen S, Duan J, Jaroniec M, et al. Three-dimensional N-doped graphene hydrogel/NiCo double hydroxide electrocatalysts for highly efficient oxygen evolution. Angewandte Chemie International Edition, 2013, 52 (51): 13567-13570.

[27] Gong M, Li Y, Wang H, et al. An advanced Ni-Fe layered double hydroxide electrocatalyst for water oxidation. Journal of the American Chemical Society, 2013, 135 (23): 8452-8455.

[28] Lu Z, Xu W, Zhu W, et al. Three-dimensional NiFe layered double hydroxide film for high-efficiency oxygen evolution reaction. Chemical Communications, 2014, 50 (49): 6479-6482.

[29] Shi Y, Zhang B. Recent advances in transition metal phosphide nanomaterials: synthesis and applications in hydrogen evolution reaction. Chemical Society Reviews, 2016, 45 (6): 1529-1541.

[30] Mccrory C, Jung S, Peters J, et al. Benchmarking heterogeneous electrocatalysts for the oxygen evolution reaction. Journal of the American Chemical Society, 2013, 135 (45): 16977-16987.

[31] Dutta A, Mutyala S, Samantara A, et al. Synergistic effect of inactive iron oxide core on active nickel phosphide shell for significant enhancement in the OER activity. ACS Energy Letters, 2018, 3 (1): 141-148.

[32] Das J K, Samantara A K, Nayak A K, et al. VS_2: an efficient catalyst for an electrochemical hydrogen evolution reaction in an acidic medium. Dalton Transactions, 2018, 47 (39): 13792-13799.

[33] Brug G J, Van Den Eeden A L G, Sluyters-Rehbach M, et al. The analysis of electrode impedances complicated by the presence of a constant phase element. Journal of Electroanalytical Chemistry and Interfacial Electrochemistry, 1984, 176 (1): 275-295.

[34] Ma T Y, Dai S, Jaroniec M, et al. Metal-organic framework derived hybrid Co_3O_4-carbon porous nanowire arrays as reversible oxygen evolution electrodes. Journal of the American Chemical Society, 2014, 136 (39): 13925-13931.

[35] Tang C, Wang W, Sun A, et al. Sulfur-decorated molybdenum carbide catalysts for enhanced hydrogen evolution. ACS Catalysis, 2015, 5 (11): 6956-6963.

[36] Zhang C, Huang Y, Yu Y, et al. Sub-1. 1 nm ultrathin porous CoP nanosheets with dominant reactive {200} facets: a high mass activity and efficient electrocatalyst for the hydrogen evolution reaction. Chemical Science, 2017, 8 (4): 2769-2775.

[37] Costentin C, Drouet S, Robert M. Turnover numbers, turnover frequencies, and overpotential in molecular catalysis of electrochemical reactions. Cyclic voltammetry and preparative-scale electrolysis. Journal of the American Chemical Society, 2012, 134 (27): 11235-11242.

[38] Zhao G, Rui K, Dou S X, et al. Heterostructures for electrochemical hydrogen evolution reaction: A review. Advanced Functional Materials, 2018, 28 (43): 1803291.

[39] Merki D, Fierro S, Vrubel H, et al. Amorphous molybdenum sulfide films as catalysts for electrochemical hydrogen production in water. Chemical Science, 2011, 2 (7): 1262-1267.

[40] Lu X, Zhao C. Electrodeposition of hierarchically structured three-dimensional nickel-iron electrodes for efficient oxygen evolution at high current densities. Nature Communications, 2015,

6 (1): 6616.
[41] Tian X, Liu Y, Xiao D, et al. Ultrafast and large scale preparation of superior catalyst for oxygen evolution reaction. Journal of Power Sources, 2017, 365: 320-326.
[42] Li P, Jin Z, Xiao D. A one-step synthesis of Co-P-B/rGO at room temperature with synergistically enhanced electrocatalytic activity in neutral solution. Journal of Materials Chemistry A, 2014, 2 (43): 18420-18427.
[43] Sheng M, Jiang B, Wu B, et al. Approaching the volcano top: Iridium/silicon nanocomposites as efficient electrocatalysts for the hydrogen evolution reaction. ACS Nano, 2019, 13 (3): 2786-2794.
[44] Hou J, Wu Y, Zhang B, et al. Rational design of nanoarray architectures for electrocatalytic water splitting. Advanced Functional Materials, 2019, 29 (20): 1808367.
[45] You B, Tang M T, Tsai C, et al. Enhancing electrocatalytic water splitting by strain engineering. Advanced Materials, 2019, 31 (17): 1807001.
[46] Joo J, Kim T, Lee J, et al. Morphology-controlled metal sulfides and phosphides for electrochemical water splitting. Advanced Materials, 2019, 31 (14): 1806682.
[47] Yan Y, Xia B Y, Zhao B, et al. A review on noble-metal-free bifunctional heterogeneous catalysts for overall electrochemical water splitting. Journal of Materials Chemistry A, 2016, 4 (45): 17587-17603.
[48] Fabbri E, Habereder A, Waltar K, et al. Developments and perspectives of oxide-based catalysts for the oxygen evolution reaction. Catalysis Science & Technology, 2014, 4 (11): 3800-3821.
[49] Gong S, Wang C, Jiang P, et al. O species-decorated graphene shell encapsulating iridium-nickel alloy as an efficient electrocatalyst towards hydrogen evolution reaction. Journal of Materials Chemistry A, 2019, 7 (25): 15079-15088.
[50] Yang W, Chen S. Recent progress in electrode fabrication for electrocatalytic hydrogen evolution reaction: A mini review. Chemical Engineering Journal, 2020, 393: 124726.
[51] Sun H, Yan Z, Liu F, et al. Self-supported transition-metal-based electrocatalysts for hydrogen and oxygen evolution. Advanced Materials, 2020, 32 (3): 1806326.
[52] Zhang T, Zhu Y, Lee J Y. Unconventional noble metal-free catalysts for oxygen evolution in aqueous systems. Journal of Materials Chemistry A, 2018, 6 (18): 8147-8158.
[53] Chi J, Yu H. Water electrolysis based on renewable energy for hydrogen production. Chinese Journal of Catalysis, 2018, 39 (3): 390-394.
[54] Trasatti S. Work function, electronegativity, and electrochemical behaviour of metals: Ⅲ. Electrolytic hydrogen evolution in acid solutions. Journal of Electroanalytical Chemistry and Interfacial Electrochemistry, 1972, 39 (1): 163-184.
[55] Trasatti S. Electrocatalysis by oxides-Attempt at a unifying approach. Journal of Electroanalytical Chemistry and Interfacial Electrochemistry, 1980, 111 (1): 125-131.
[56] Song F, Bai L, Moysiadou A, et al. Transition metal oxides as electrocatalysts for the oxygen evolution reaction in alkaline solutions: An application-inspired renaissance. Journal of the American Chemical Society, 2018, 140 (25): 7748-7759.
[57] Cai J, Song Y, Zang Y, et al. N-induced lattice contraction generally boosts the hydrogen evolution catalysis of P-rich metal phosphides. Science Advances, 2020, 6 (1): eaaw8113.
[58] Jia X, Zhao Y, Chen G, et al. Ni_3FeN nanoparticles derived from ultrathin NiFe-layered double hydroxide nanosheets: An efficient overall water splitting electrocatalyst. Advanced Energy Materials, 2016, 6 (10): 1502585.

[59] Huang G, Xu S, Liu Z, et al. Ultrafine cobalt-doped iron disulfide nanoparticles in ordered mesoporous carbon for efficient hydrogen evolution. Chem Cat Chem, 2020, 12 (3): 788-794.
[60] Yu C, Lu J, Luo L, et al. Bifunctional catalysts for overall water splitting: CoNi oxyhydroxide nanosheets electrodeposited on titanium sheets. Electrochimica Acta, 2019, 301: 449-457.
[61] Tang C, Asiri A M, Sun X. Highly-active oxygen evolution electrocatalyzed by a Fe-doped NiSe nanoflake array electrode. Chemical Communications, 2016, 52 (24): 4529-4532.
[62] Han N, Yang K R, Lu Z, et al. Nitrogen-doped tungsten carbide nanoarray as an efficient bifunctional electrocatalyst for water splitting in acid. Nature Communications, 2018, 9 (1): 924.
[63] Qu K, Zheng Y, Jiao Y, et al. Polydopamine-inspired, dual heteroatom-doped carbon nanotubes for highly efficient overall water splitting. Advanced Energy Materials, 2017, 7 (9): 1602068.
[64] Hu C, Dai L. Multifunctional carbon-based metal-free electrocatalysts for simultaneous oxygen reduction, oxygen evolution, and hydrogen evolution. Advanced Materials, 2017, 29 (9): 1604942.
[65] Pan Y, Sun K, Liu S, et al. Core-shell ZIF-8@ZIF-67-derived CoP nanoparticle-embedded N-doped carbon nanotube hollow polyhedron for efficient overall water splitting. Journal of the American Chemical Society, 2018, 140 (7): 2610-2618.
[66] Wang J, Zhong HX, Wang ZL, et al. Integrated three-dimensional carbon paper/carbon tubes/cobalt-sulfide sheets as an efficient electrode for overall water splitting. ACS Nano, 2016, 10 (2): 2342-2348.
[67] Yu ZY, Duan Y, Gao MR, et al. A one-dimensional porous carbon-supported Ni/Mo_2C dual catalyst for efficient water splitting. Chemical Science, 2017, 8 (2): 968-973.
[68] Li H, Tang Q, He B, et al. Robust electrocatalysts from an alloyed Pt-Ru-M (M=Cr, Fe, Co, Ni, Mo)-decorated Ti mesh for hydrogen evolution by seawater splitting. Journal of Materials Chemistry A, 2016, 4 (17): 6513-6520.
[69] Paul R, Zhu L, Chen H, et al. Recent advances in carbon-based metal-free electrocatalysts. Advanced Materials, 2019, 31 (31): 1806403.

第二章

电解槽体系概述

常规电解水的电解槽通常分为两类：一类是以过渡金属为催化剂，离子交换膜为隔膜的碱性电解槽（AWE）；另一类是使用贵金属或者耐酸的金属为催化剂，质子交换膜为隔膜的酸性电解槽（PEM）[1,2]。近年来，针对常规的电解槽体系存在的一些缺点，出现了解耦电解槽。下面依次进行详细的介绍。

2.1 碱性电解槽体系

自从 1789 年 Troostwijk 和 Diemann 首先发现水的电解现象以来，碱性水电解已成为生产 H_2 比较成熟的技术。因此，碱性电催化是全球商业上使用最广泛的电解技术[3-8]。在 AWE 体系中，电解液由浓度为 20%～30% KOH 溶液组成[9-11]。目前碱性电解槽包括常规碱性电解槽［图 2.1(a)］、“零间隙”碱性电解槽［图 2.1(b)］、GDL 碱性电解槽［图 2.1(c)］和无膜［图 2.4(a)］或解耦碱性电解槽。

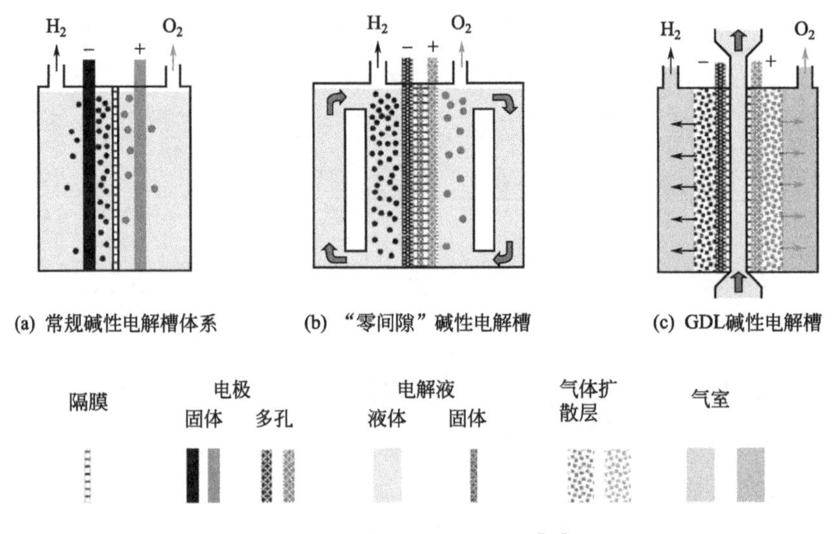

图 2.1　碱性电解槽示意图[12]

在传统的碱性电解槽中，浸入电解质中的阳极和阴极位于集电器的两侧，以便电池槽之间的串联[12]。H_2 和 O_2 在两个电解质腔中产生，电解槽中的离子交换膜可防止气体混合［图 2.1(a)］。这种方法很容易规模化生产 H_2。但是，产生的气泡覆盖在电极表面减小了电极的有效面积，并增加了电解质的电阻，导致电流密度降低。因此目前几乎不考虑使用该装置进行电解水制氢。以镍基材料为电极的 AWE 的问题是 HER 和 OER 催化活性会持续衰减。电解质中的金属阳离子杂质会在催化剂表面产生一层催化活性低的物质，导致 HER 活性降低。因此，Schuhmann 和 Ventosa 等人创造性地提出了一种基于在电解过程中催化剂颗粒原位自组装的方法，以获得具有自愈能力且稳定性好的催化薄膜［图 2.2(a)、(b)］。实验证明，通过用自组装和自修复膜固定催化剂，可以克服电解质中锌杂质对阴极的钝化作用。在电解过程中，以树枝状形式沉积在阴极上的锌杂质增加了 HER 的过电位，但是催化剂的持续自组装和自修复覆盖锌枝晶后，有利于恢复 HER 过电位[13]。

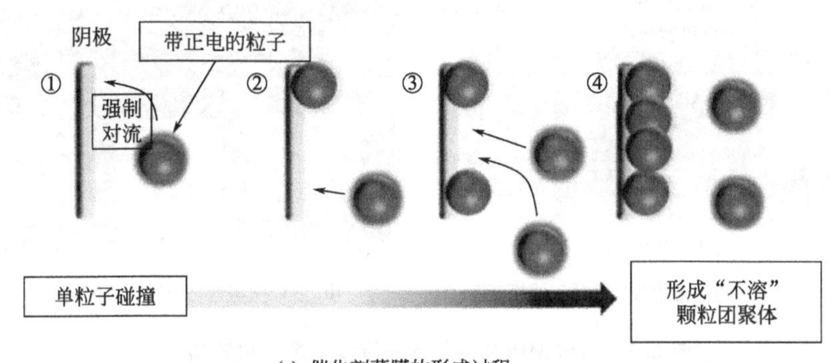

(a) 催化剂薄膜的形成过程

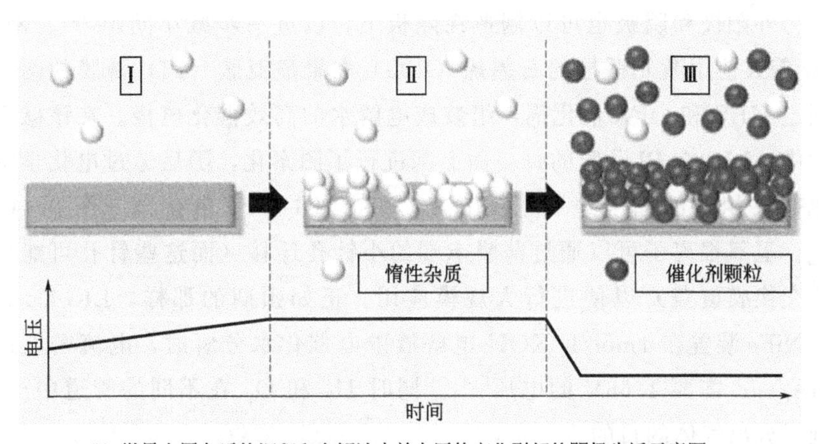

(b) 微量金属杂质的沉积和电解池中总电压的变化引起的阴极失活示意图

图 2.2　催化剂薄膜的形成过程和阴极失活示意图[13]

在“零间隙”碱性电解槽的配置中，电极内的薄纤维素毡可以吸收电解质，从而固定电解质，并夹在两个紧密压在阳极和阴极上的亲水性隔膜之间。阳极和阴极是多孔的材料，以便液体电解质的渗透。多孔结构还可以有效地排除来自电极内部空间的气泡［图 2.1(b)］。例如，据 Dunnill 等人的研究，与传统的碱性电解液中 2mm 的间隙相比，采用零间隙的电池配置可降低 30%的电阻。图 2.3(a) 是传统碱性电解槽的组装器件，图 2.3(b) 是零间隙配置电解槽的组装器件。在所有电流密度下，尤其当电流密度超过 500mA/cm^2 时，零间隙配置电池的性能均优于标准电池。与粗网状电极相比，具有高比表面积的泡沫电极具有较低的欧姆电阻。因此，零间隙配置碱性电解槽具有低成本和高效性[14]。

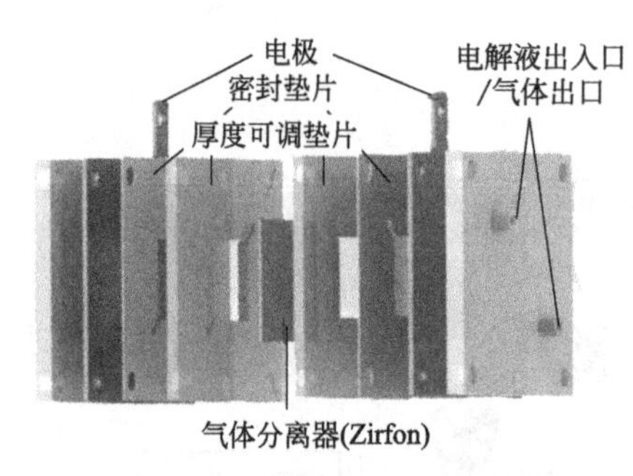

(a) 传统碱性电解槽组装器件的三维模型图

(b) 零间隙碱性电解槽的组装器件

图 2.3　电解槽组装器件的三维模型图和组装器件[14]

另外阳极和阴极也可以制造在隔板上，以进一步减小间隙[15]。例如，Tour 等人使用激光诱导的石墨烯（LIG）在聚酰亚胺（PI）薄膜的两面分别固定 HER 和 OER 催化剂，组装成电解水的高效催化电极。在该碱性电解槽中，LIG 在 PI 薄膜的每一面上都进行了图案化，随后通过电化学沉积催化剂在相应的面上组装了 LIG-Co-P 和 LIG-NiFe 催化剂［图 2.4(a)、(b)］。氢氧根离子可以通过薄膜末端的小针孔迁移（而这些针孔可能会被离子交换膜覆盖）以便进行大规模应用。正如预期的那样，LIG-Co-P 和 LIG-NiFe 装置在 1mol/L KOH 电解液中电催化水分解时，电流密度达到 10mA/cm^2 需要 1.66V 的电压[16]。同时 H_2 和 O_2 在不同的管道中产生，达到了分离气体的目的。

在碱性水电解中，液体电解质的电导率远高于离子交换膜的电导率，因此离子交换膜的使用导致较为严重的欧姆损耗[17,18]。Gillespie 和 Kriek 研究了无膜 DEFT 碱性电解槽[19]，用于提高 H_2 的产量。该电解槽可以克服

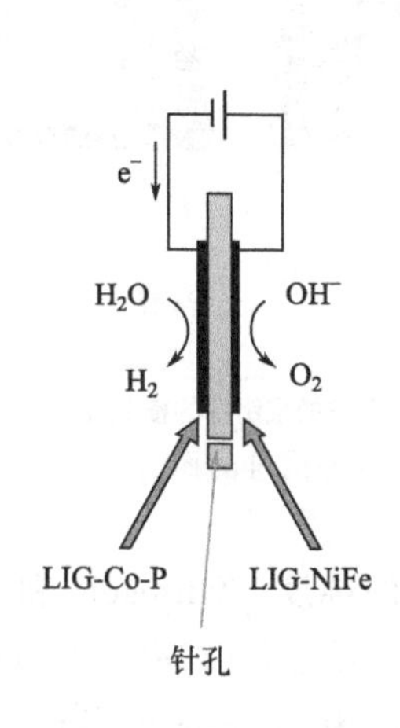

(a) LIG-Co-P和LIG-NiFe电解水制氢的原理示意图

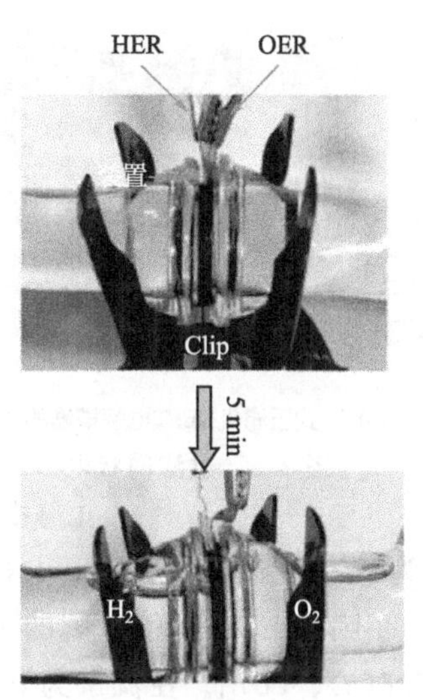

(b) 集成LIG电解槽的光学照片

图 2.4 电解水制氢的原理示意图和电解槽的光学照片

工作电流为 50mA/cm²，5min 后在左侧产生 H_2，右侧产生 O_2

现有技术中电流密度阈值的限制，是产生氢气的理想选择［图 2.5(a)，见文后彩插］。对该技术进行规模型扩大，与原始的测试电堆不一样，该技术是将许多细长的电极封装在压力过滤器组件中［图 2.5(b)、(c)，见文后彩插］。中试装置仅适用低流速，电极的间隙为 2.5mm。中试的性能与之前实验获得的结果一致。用于该性能测试的网状电极的几何面积为 344.32cm²。在流量为 0.04m/s、电解液为 30%KOH、2V 直流电（VDC）、温度为 80℃ 的条件下，NiO 阳极和 Ni 阴极组合的最佳性能可达到 508mA/cm²。遗憾的是，受气液分离系统的影响，与实验室的结果相比，气体质量有所欠缺。

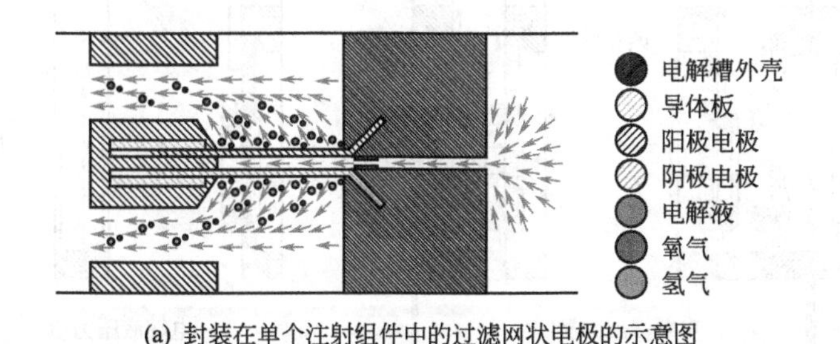

(a) 封装在单个注射组件中的过滤网状电极的示意图

图 2.5

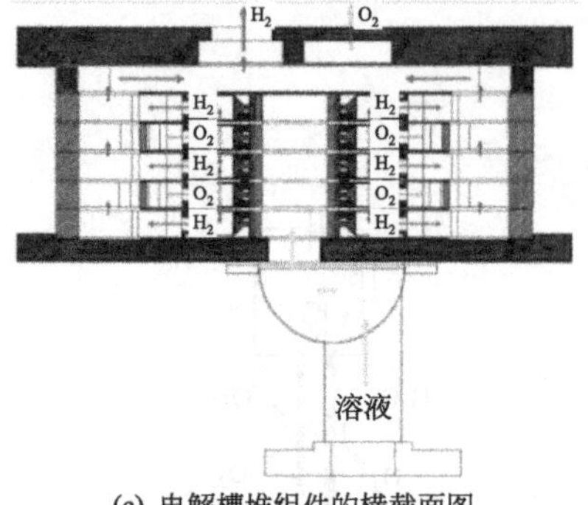

(b) 卧式压滤机DEFT电解槽堆的照片　　(c) 电解槽堆组件的横截面图

图 2.5　过滤网状电极的示意图、卧式 DEFT 电解槽堆的照片和电解槽堆组件的横截面图[19]

为了进一步提高 DEFT 电解槽中产生的气体的纯度，Gillespie 和 Kriek 开发了用于 DEFT 碱性电解的可扩展且简单的单圆压滤机（MCFP）反应器[图 2.6(a)、(b)]。在流量为 0.075m/s、电极间隙为 2.5mm 的条件下，采

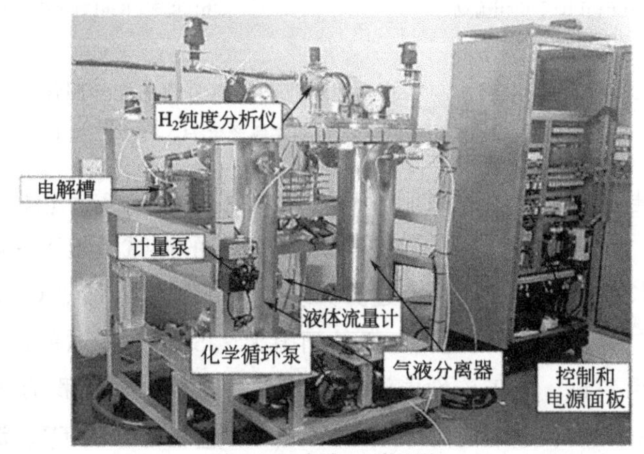

(a) MCFP中试工厂的照片

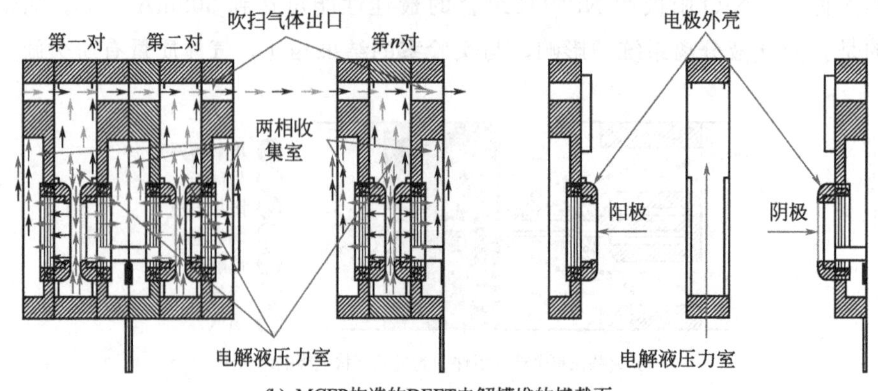

(b) MCFP构造的DEFT电解槽堆的横截面

图 2.6　MCFP 中试工厂的照片和电解槽堆的横截面[20]

用气液分离方法可将 H_2 的纯度提高到 99.81%（体积分数），O_2 的纯度提高到 99.50%（体积分数）。每对 30mm 的圆形网状电极都有独立的加压室和间接注入的电解液。通过采用气体吹扫，可以长时间保持较高的气体纯度。使用 Ni/Ni 催化剂，在流量为 0.075m/s、温度为 60℃和电极间隙为 2.5mm 的条件下，电流密度为 1.14A/cm^2（2.5V，DC）。在相同条件下（使用双层网状电极除外），电流密度还可达到 1.91A/cm^2，证实了多层微孔电极可用于 DEFT[20]。

2.2 酸性电解槽体系

1960 年，通用电气首先提出了一种用于水电解的固态聚合物电解质（SPE）概念，该概念有望克服碱性电解槽的缺点。格鲁布（Grubb）通过使用固态磺化聚苯乙烯膜作为电解质（也称为 PEM 水电解，很少称为 SPE 水电解）实现了这一概念[21,22]。聚合物电解质膜可提供更高的质子传导性、更低的气体交换、更紧凑的系统设计，并可在高压下运行[23-27]。固体聚合物电解质的优点是膜厚度较低（约 20～300μm）。

首先，将催化剂层涂覆在光滑的聚四氟乙烯（PTFE）片上；然后，将 PTFE 片材压在膜上之后将其除去，留下由 Nafion 117 膜表面上的催化剂和 Nafion 离聚物组成的平整的涂层。在膜的表面存在明显的催化剂涂层和膜之间的界限。涂层的厚度可以通过调节催化剂的量来改变[28]。在商用碱性电解槽体系中，分别在 Nafion 117 膜的一面上涂覆一层 Pt/C 和 Nafion 离聚物，在另一面涂上 IrO_2 或 RuO_2 催化剂和 Nafion 离聚物［图 2.7(a)］。

将水加入酸性电解槽的阳极侧，先后通过隔板和集电器。当水到达催化剂层的表面时，水分子分解为质子、电子和双原子氧。随后，生成的质子通过离聚物和膜离开阳极，穿过阴极的一侧，在到达催化层后，与电子耦合形成 H_2。然后，H_2 必须流过阴极集流器和阻挡层，离开电解槽。同时，电子通过集流器，隔板离开阴极催化层，然后离开阴极。O_2 必须通过催化剂层和集流器回流到隔板，然后从电解槽中出来［图 2.7(b)］[9]。

酸性电解槽可以在超过 2A/cm^2 的高电流密度下工作，从而降低运行成本和水电解总成本。其薄膜可以提供良好的质子传导性和高电流密度。聚合物电解质膜的低气体透过率使 PEM 电解槽可以在很宽的功率范围内运

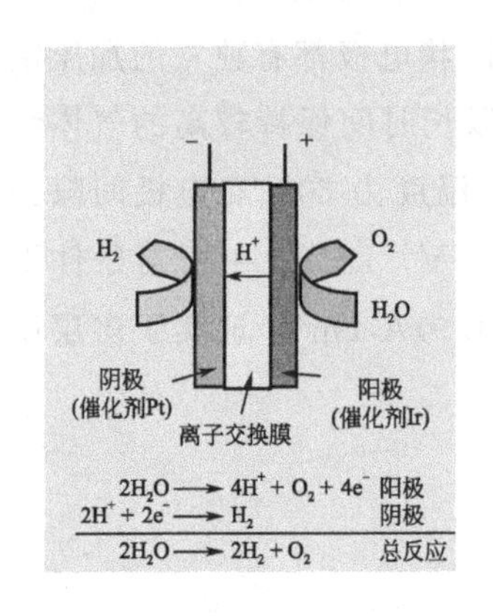

(a) 酸性电解槽的工作原理示意图

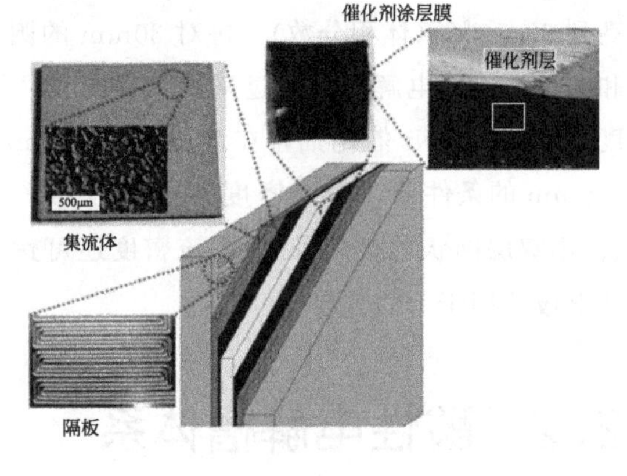

(b) 典型的酸性电解槽的组件

图 2.7　PEM 工作原理示意图和组件[28]

行。但在电解过程中，交叉渗透现象随着操作压力的增大而增大[29]。高压［超过 100bar（$1bar=10^5Pa$）］情况下需要更厚的膜，以减少 H_2 和 O_2 的混合物，确保其临界浓度保持在安全阈值以下［O_2 中 H_2 的含量为 4%（体积分数）］[29]。PEM 电解中的酸性条件需要一些特殊的材料，这些材料需要较高的过电位（大约 2V），具有较强的抗酸（pH 值约为 2）腐蚀性。

2.3　海水电解槽

水电解系统包括两个半反应：阴极上的 HER 和阳极上的 OER。与淡水相比，海水是地球上水电解过程中利用最多的含水电解质。Bennett[30] 研究了海水的电解，由阴极的 HER 和阳极的氯析出反应（ClER）组成[11]。ClER 是双电子过程，产生的氯或次氯酸盐是增值产品[31]。四年后，Trasatti 用不同的阳极进行海水电解，以研究阳极工艺的选择性[32]。2016 年，Dionigi 等人提出了海水电极的化学局限性以及选择性催化海水分解催化剂的设计标准[33]。

在海水电解系统中，膜（比如 Zirfon）应具有较强的韧性，坚固耐用并且不易被阻塞，因为该膜在很大程度上可以阻挡阴离子或阳离子（例如 H^+、Na^+、OH^- 和 Cl^- 等）通过膜迁移[11]。将商业化的氧化钌负载于钛电极（RuO_2/Ti）上作为工作电极，铂电极作为对电极，在阳极发生 ClER 反应，在阴极发生 HER 反应［图 2.8(a)］。次氯酸盐的法拉第效率随阳极上的施加电位线性增加，在 RuO_2/Ti 电极上的施加 1.5V（相对于 RHE）电位时，次氯酸盐的法拉第效率可以达到 99%［图 2.8(b)］[34]。

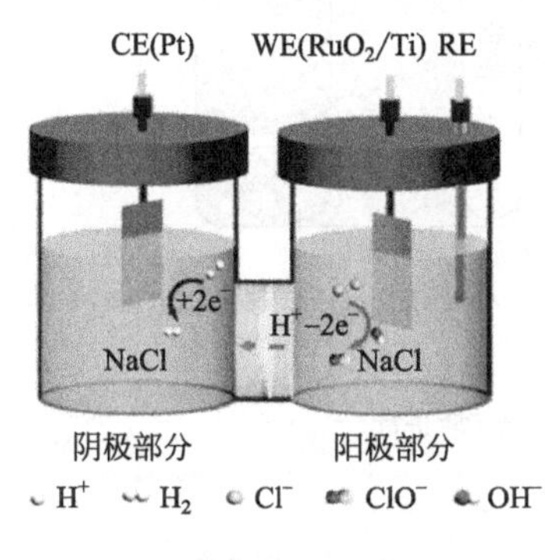

(a) 阴极HER和阳极ClER的电解槽示意图

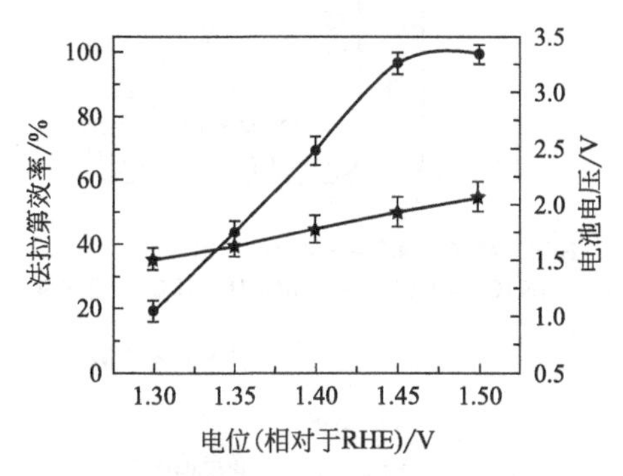

(b) 次氯酸盐的法拉第效率与阴极上的施加电位和电池电压曲线图

图 2.8　海水电解槽及电化学表征[34]

2.4　解耦电解槽

传统的电解槽需要对产生的气体进行分离，增加了成本和系统的复杂性。因此科研学者们利用氧化还原介体将水分解的两个半反应进行解耦，可以完全避免产生的 H_2 和 O_2 发生混合，并有希望进行大规模的实际应用[35-37]。Grader 等人提出了一个分为两步的电化学-热活化化学（E-TAC）循环，用于水的分解。通过 HER 在阴极产生 H_2。如图 2.9(a) 所示传统的 OER 被替换为两个步骤：在第一步中，通过四个单电子氧化反应将 $Ni(OH)_2$ 阳极氧化为 NiOOH；在第二步中氧化的 NiOOH 自发还原为 $Ni(OH)_2$，同时产生 O_2 和阳极的再生。如图 2.9(b) 所示，他们还假设了一个多槽系统，每个槽中都有固定的阳极和阴极，可用于生产纯的 H_2 和 O_2。少量的低温电解液通过电池 A，将生成的 H_2 带到 H_2 分离器。同时，高温电解液通过电池 B 使阳极再生，将生成的 O_2 带到 O_2 分离器。在这种多电池系统中，操作过程中只有冷热电解质发生了移动[38]。

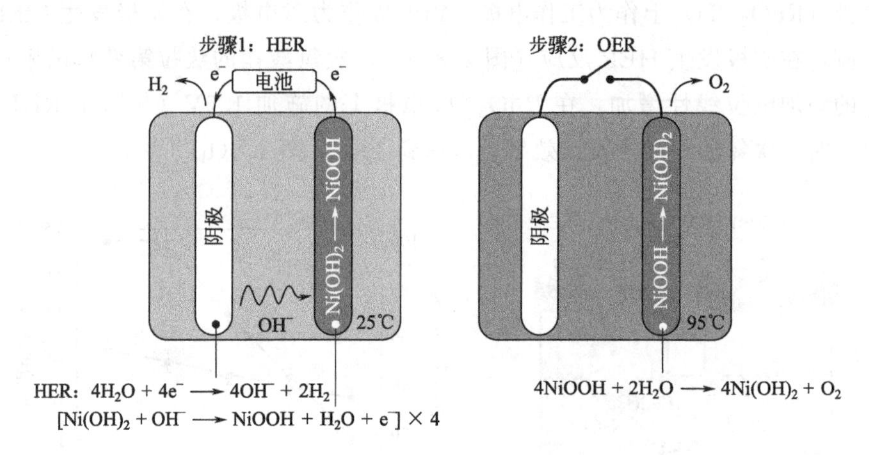

(a) E-TAC的水分解过程分两个连续步骤进行

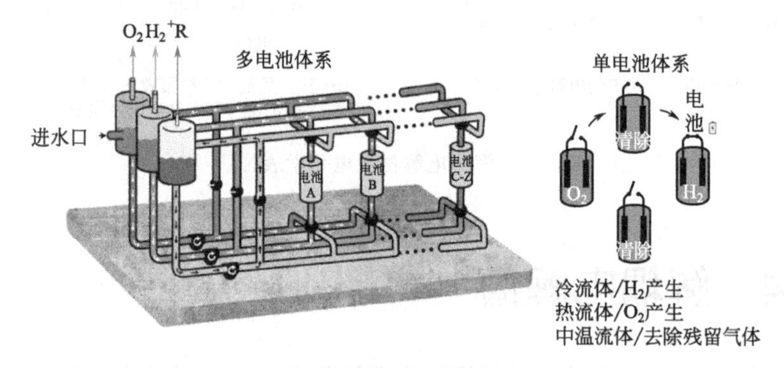

(b) 多电池E-TAC过程的示意图

图 2.9　E-TAC 的水分解的示意图[38]

2016 年 Sun 课题组提出在碱性条件下将氧化生物质反应与解耦水电解相结合的概念[27]，将体系组装成具有阴离子交换膜（AEM）的双室电解槽。使用 5 种具有代表性的生物质乙醇、苯甲醇、糠醛、糠醇和 5-羟甲基糠醛（HMF）作为还原剂，用非贵金属分层多孔双功能电催化剂（Ni_3S_2/NF）催化水电解，得到更具有经济价值的生物质氧化产物和 H_2，整个体系中没有 O_2 的生成，不仅避免了 H_2/O_2 混合物发生爆炸的危险，还避免了活性氧的产生。这是一个具有良好创新性和较高经济价值的方法。图 2.10 是该体系提出的电解槽示意图。

Sun 课题组提出的新型电解槽体系使用了离子交换膜，而构建无膜的碱性电解槽体系具有更重要的意义。Xia 和 Wang 课题组在碱性条件下使用 $Ni(OH)_2$

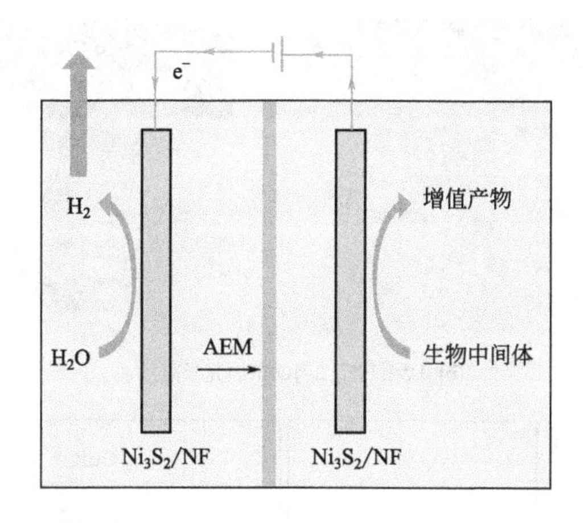

图 2.10　Sun 课题组提出的电解槽示意图[27]

电极作为氧化还原介体，贵金属作为催化剂，成功的对水电解进行了解耦化[39]。与 Cronin 课题组和 Sun 课题组研究的不同之处在于 Xia 课题组使用固体电极作为氧化还原介体。在他们的研究中，电解槽使用三电极体系，利用 $Ni(OH)_2$ 来耦合 OER 和 HER 过程中产生和消耗的电荷，达到了解耦化水电解的效果。如图 2.11（见文后彩插），该体系分为两步：第一步，关闭开关 K1，HER 电极上发生 H_2O 的阴极还原产生 H_2（$2H_2O+2e^- \longrightarrow H_2+2OH^-$），在 $Ni(OH)_2$ 电极发生阳极氧化［$Ni(OH)_2+OH^- -e^- \longrightarrow NiOOH+H_2O$］；第二步，在 NiOOH 电极上发生阴极还原，$Ni^{2+}/Ni^{3+}$ 的氧化还原电位位于 HER 和 OER 起始电位之间。

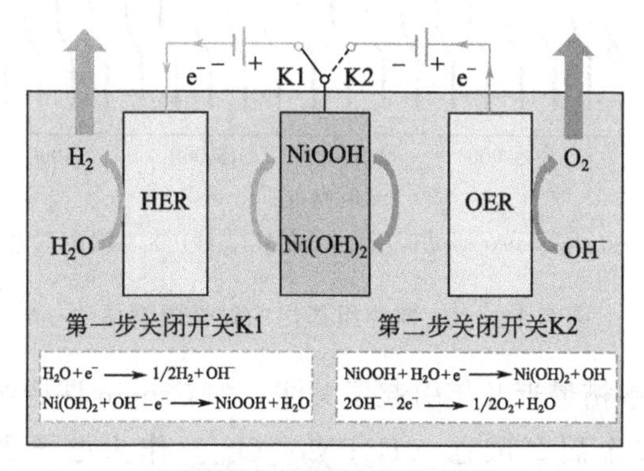

(a) 碱性电解槽的运行机理示意图

图 2.11

(b) 1为HER，2为OER的光学照片

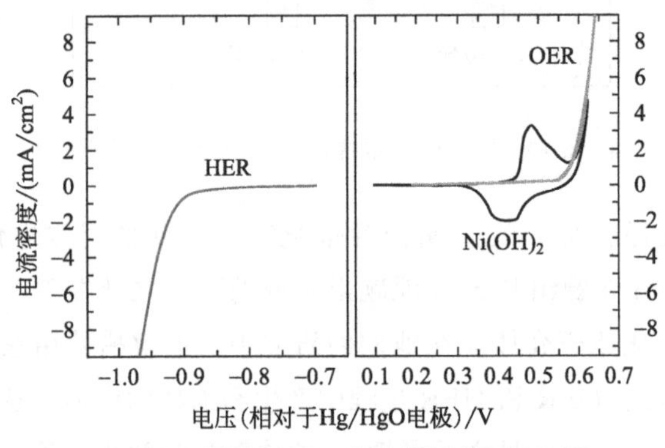

(c) Ni(OH)₂解耦电极的CV图和阴阳两级的LSV图

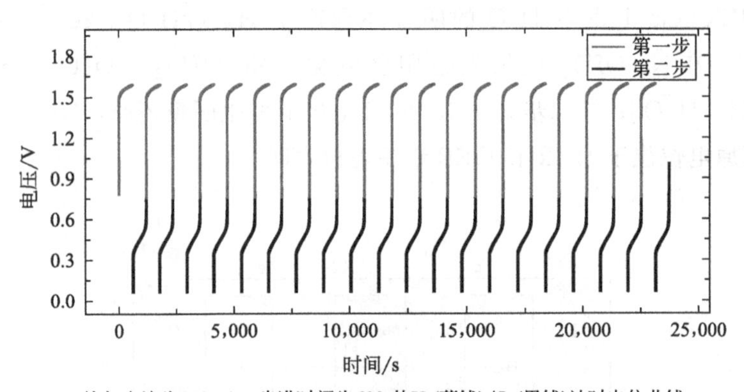

(d) 施加电流为200mA，步进时间为600s的H_2(蓝线)/O_2(黑线)计时电位曲线

图 2.11 Xia 课题组提出的碱性电解槽[39]

解耦酸性电解槽最近几年引起了关注，如 Cronin 课题组[40]在酸性条件下使用大分子多金属氧酸盐（$H_3PMo_{12}O_{40}$）作为电子偶联质子缓冲液(ECPB)［图 2.12(a)］吸收水氧化期间产生的电子，整个体系只发生水的氧化和 ECPB 的还原，没有 H_2 生成。之后在产生 H_2 的同时 ECPB 再被氧

化进而释放这些质子和电子，没有 O_2 生成。达到了在不同的时间产生 H_2 和 O_2 的目的。之后，他们还使用分子量更小的具有可逆氧化还原电位的醌类衍生物（1,4-氢醌衍生物氢醌磺酸钾）代替多氧金属酸盐来实现这一过程［图 2.12(b)］[41]，可以降低 ECPB 的成本，为其在水分解中的进一步广泛应用提供技术支持。作为 ECPB 必须具有以下的条件[40]：

① 由于体系使用的是高浓度溶液，所选择的 ECPB 必须在酸性条件下具有良好的溶解度；

② 必须满足整个电解体系在 ECPB 中能稳定存在；

③ 保证 ECPB 氧化状态和还原状态在酸性体系中稳定存在；

④ ECPB 氧化还原电位要位于 OER 和 HER 的起始电位之间，氧化和还原的可逆性要好，保证良好的转化效率；

⑤ ECPB 具有一定缓冲 pH 变化能力，能适应电解水过程中 pH 的变化。

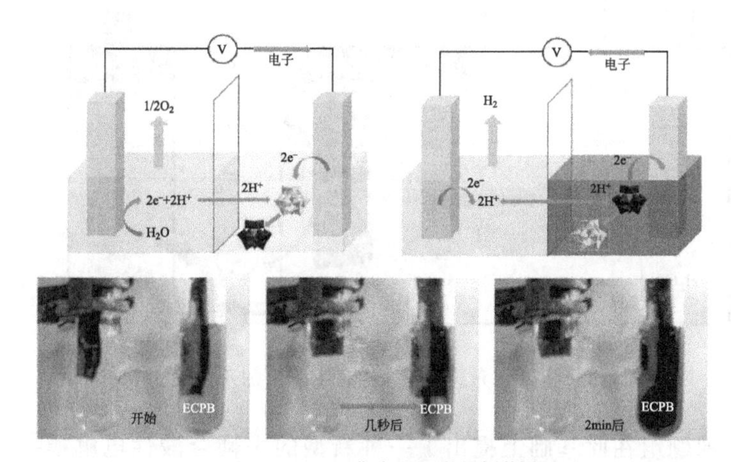

(a) $H_3PMo_{12}O_{40}$作为ECPB的解耦装置图

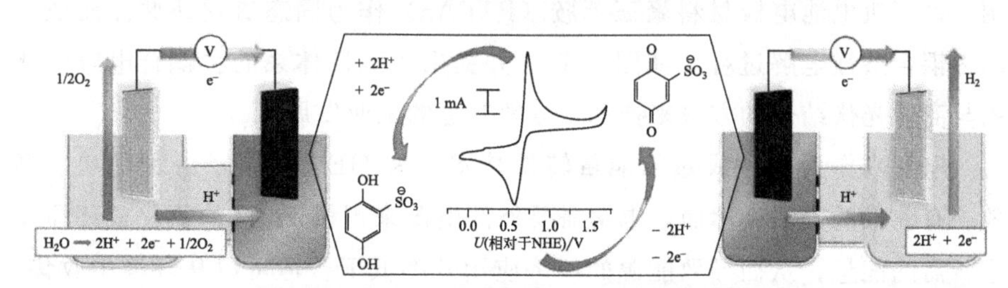

(b) 1,4-氢醌衍生物氢醌磺酸钾作为ECPB的解耦示意图

图 2.12 Cronin 课题组采用 ECPB 的解耦电解槽[41]

使用 ECPBs 去耦合电解水就是将电极、电解液和合适的具有氧化还原性质的缓冲液置于同一个 H 形装置中，用离子交换膜将缓冲液隔开，只允许质子通过。不同于使用 ECPBs 作为去耦合介质，Cronin 课题组研发了另外一种新型的电解装置系统[42]，在常温常压下，通过外加电压在阳极析氧的同时，阴极硅钨酸［$H_4(SiW_{12}O_{40})$］被还原，在低压下放出氧气，再使用氩气将被还原的硅钨酸充入另一个密闭的装置中，采用一定比例的 Pt/C 催化产氢，还原的硅钨酸被氧化，不需要外加的电压。这种方法避免了电解槽内高压气体的产生（导致膜降解的主要原因），并且基本上消除了低电流密度下（可再生能源驱动水分解的特点）气体混合的危险问题。他们证明在相同铂负载量的情况下，铂催化体系产生纯氢的速度可以比现有最先进技术的 PEM 电解槽快 30 倍。其基本的示意图如图 2.13 所示。

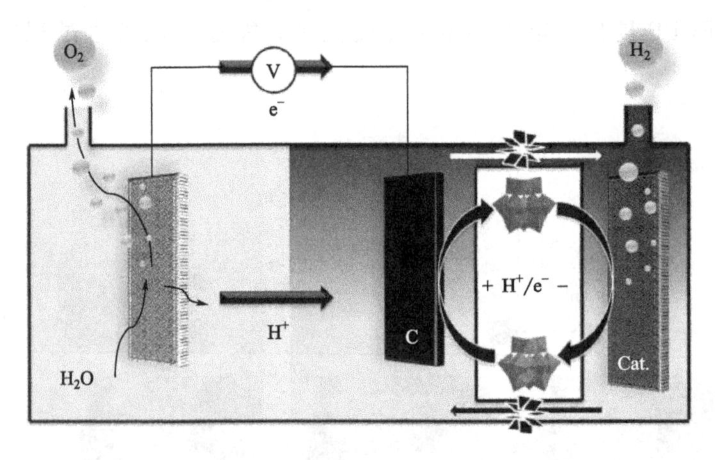

图 2.13　硅钨酸介导的 H_2 从水中析出的示意图[42]

Wang 课题组在此基础上提出了一种新型的去耦合酸性电解槽[43]，使用一种有机电池电极材料聚三苯胺（PTPAn）作为固态氧化还原介质成功的去耦合酸性电解过程，如图 2.14。并提出了酸性体系的去耦合电解水体系与商业光伏结合的方法来推动水分解制氢的工业发展。

电解水作为一种成熟的能量转换技术，为 HER 提供了一种简单、高效、有前景的方法。然而，电解需要外部电源来提供 H_2O 的氧化或还原反应所需的能量，从而导致能源的经济应用效率低下。但可以从环境中收集、储存和转换能量（如风能、热能、太阳能、潮汐能等）直接用于电解。太阳能是一种取之不尽用之不竭的可再生能源，其有效地利用可以降低电解水的

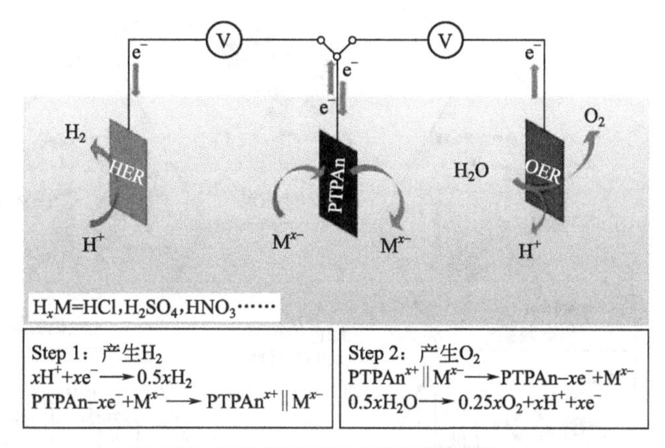

(a) PTPAn作为氧化还原介质的解耦示意图

(b) 电解槽和太阳能电池联用示意图和电解水光学照片图

图 2.14　Wang 课题组提出的去耦合酸性电解槽[43]

能耗。例如，构造一个光电极来吸收太阳能，可以提供一个光电压来有效地减少电解水的外部能量供应。此外，太阳能电池也是一种有效的太阳能转换技术，它可以直接吸收太阳光来转换输出电压，从而有效地降低外部能耗。热电（TE）器件可以利用自然界中来自太阳的热能产生电能，为分解水提供电压。自然界中也存在着大量的风能和潮汐能，可以被摩擦纳米发电机（TENG）捕获来发电，也可以作为电解水的外部能源。因此，建立一个合适的外部能源来驱动水分解体系，减少外部能量消耗，对提高制氢能力具有重要意义。近年来，许多研究者开发了多种高效生产 H_2 的绿色能源系统，如双电极电解水、光电极装置驱动水分解、太阳能电池、TE、TENG 装置等包括热释电和水煤气变换（WGS）反应等（图 2.15），这些绿色能源系统可以有效地驱动水分解产生 H_2[44]。

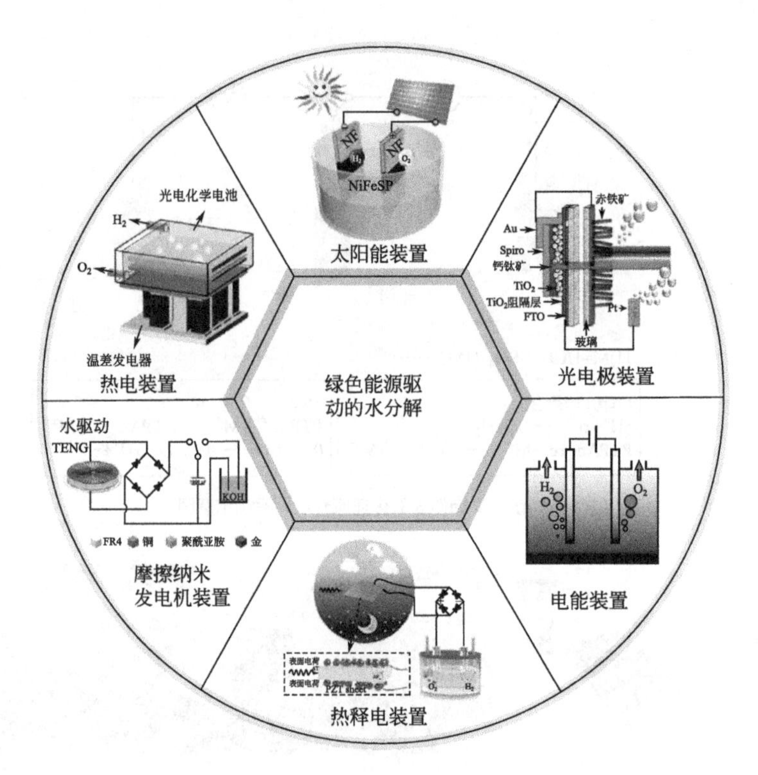

图 2.15　不同绿色能源系统驱动的水分解[44]

参　考　文　献

[1] Vincent I，Bessarabov D. Low cost hydrogen production by anion exchange membrane electrolysis：A review. Renewable and Sustainable Energy Reviews，2018，81：1690-1704.

[2] Ju H，Badwal S，Giddey S. A comprehensive review of carbon and hydrocarbon assisted water electrolysis for hydrogen production. Applied Energy，2018，231：502-533.

[3] Panda C，Menezes P W，Zheng M，et al. In situ formation of nanostructured core-shell Cu_3N-CuO to promote alkaline water electrolysis. ACS Energy Letters，2019，4 (3)：747-754.

[4] Fortin P，Khoza T，Cao X，et al. High-performance alkaline water electrolysis using Aemion™ anion exchange membranes. Journal of Power Sources，2020，451：227814.

[5] Wei J，Zhou M，Long A，et al. Heterostructured electrocatalysts for hydrogen evolution reaction under alkaline conditions. Nano-Micro Letters，2018，10：75.

[6] Aili D，Kraglund M R，Tavacoli J，et al. Polysulfone-polyvinylpyrrolidone blend membranes as electrolytes in alkaline water electrolysis. Journal of Membrane Science，2020，598：117674.

[7] Zhang T，Yang K，Wang C，et al. Nanometric Ni_5P_4 clusters nested on $NiCo_2O_4$ for efficient hydrogen production via alkaline water electrolysis. Advanced Energy Materials，2018，8 (29)：1801690.

[8] Zhang K，Mcdonald M B，Genina I E A，et al. A highly conductive and mechanically robust OH^- conducting membrane for alkaline water electrolysis. Chemistry of Materials，2018，30

(18)：6420-6430.

[9] Carmo M，Fritz D L，Mergel J，et al. A comprehensive review on PEM water electrolysis. International Journal of Hydrogen Energy，2013，38 (12)：4901-4934.

[10] Kovač A，Marciuš D，Budin L. Solar hydrogen production via alkaline water electrolysis. International Journal of Hydrogen Energy，2019，44 (20)：9841-9848.

[11] Tong W，Forster M，Dionigi F，et al. Electrolysis of low-grade and saline surface water. Nature Energy，2020，5 (5)：367-377.

[12] Marini S，Salvi P，Nelli P，et al. Advanced alkaline water electrolysis. Electrochimica Acta，2012，82：384-391.

[13] Barwe S，Mei B，Masa J，et al. Overcoming cathode poisoning from electrolyte impurities in alkaline electrolysis by means of self-healing electrocatalyst films. Nano Energy，2018，53：763-768.

[14] Phillips R，Edwards A，Rome B，et al. Minimising the ohmic resistance of an alkaline electrolysis cell through effective cell design. International Journal of Hydrogen Energy，2017，42 (38)：23986-23994.

[15] Gabler A，Müller C I，Rauscher T，et al. Ultrashort-pulse laser structured titanium surfaces with sputter-coated platinum catalyst as hydrogen evolution electrodes for alkaline water electrolysis. International Journal of Hydrogen Energy，2018，43 (15)：7216-7226.

[16] Zhang J，Zhang C，Sha J，et al. Efficient Water-splitting electrodes based on laser-induced graphene. ACS Applied Materials & Interfaces，2017，9 (32)：26840-26847.

[17] H. Hashemi S M，Karnakov P，Hadikhani P，et al. A versatile and membrane-less electrochemical reactor for the electrolysis of water and brine. Energy & Environmental Science，2019，12 (5)：1592-1604.

[18] Bui J C，Davis J T，Esposito D V. 3D-Printed electrodes for membraneless water electrolysis. Sustainable Energy & Fuels，2020，4 (1)：213-225.

[19] Gillespie M I，Kriek R J. Hydrogen production from a rectangular horizontal filter press Divergent Electrode-Flow-Through (DEFT^{TM}) alkaline electrolysis stack. Journal of Power Sources，2017，372：252-259.

[20] Gillespie M I，Kriek R J. Scalable hydrogen production from a mono-circular filter press Divergent Electrode-Flow-Through alkaline electrolysis stack. Journal of Power Sources，2018，397：204-213.

[21] Grubb W T. Ionic migration in ion-exchange membranes. The Journal of Physical Chemistry，1959，63 (1)：55-58.

[22] Park E J，Capuano C B，Ayers K E，et al. Chemically durable polymer electrolytes for solid-state alkaline water electrolysis. Journal of Power Sources，2018，375：367-372.

[23] Zhu M，Shao Q，Qian Y，et al. Superior overall water splitting electrocatalysis in acidic conditions enabled by bimetallic Ir-Ag nanotubes. Nano Energy，2019，56：330-337.

[24] Fu L，Yang F，Cheng G，et al. Ultrathin Ir nanowires as high-performance electrocatalysts for efficient water splitting in acidic media. Nanoscale，2018，10 (4)：1892-1897.

[25] Liu C，Carmo M，Bender G，et al. Performance enhancement of PEM electrolyzers through iridium-coated titanium porous transport layers. Electrochemistry Communications，2018，97：96-99.

[26] Yu J W，Jung G B，Su Y J，et al. Proton exchange membrane water electrolysis system-membrane electrode assembly with additive. International Journal of Hydrogen Energy，2019，44 (30)：15721-15726.

[27] Zinser A, Papakonstantinou G, Sundmacher K. Analysis of mass transport processes in the anodic porous transport layer in PEM water electrolysers. International Journal of Hydrogen Energy, 2019, 44 (52): 28077-28087.
[28] Xu W, Scott K. The effects of ionomer content on PEM water electrolyser membrane electrode assembly performance. International Journal of Hydrogen Energy, 2010, 35 (21): 12029-12037.
[29] Millet P, Ngameni R, Grigoriev S A, et al. Scientific and engineering issues related to PEM technology: Water electrolysers, fuel cells and unitized regenerative systems. International Journal of Hydrogen Energy, 2011, 36 (6): 4156-4163.
[30] Bennett J E. Electrodes for generation of hydrogen and oxygen from seawater. International Journal of Hydrogen Energy, 1980, 5 (4): 401-408.
[31] Kuang Y, Kenney M J, Meng Y, et al. Solar-driven, highly sustained splitting of seawater into hydrogen and oxygen fuels. Proceedings of the National Academy of Sciences, 2019, 116 (14): 6624-6629.
[32] Trasatti S. Electrocatalysis in the anodic evolution of oxygen and chlorine. Electrochimica Acta, 1984, 29 (11): 1503-1512.
[33] Dionigi F, Reier T, Pawolek Z, et al. Design criteria, operating conditions, and nickel-iron hydroxide catalyst materials for selective seawater electrolysis. Chem Sus Chem, 2016, 9 (9): 962-972.
[34] Quan F, Zhan G, Shang H, et al. Highly efficient electrochemical conversion of CO_2 and NaCl to CO and NaClO. Green Chemistry, 2019, 21 (12): 3256-3262.
[35] Wallace A G, Symes M D. Decoupling strategies in electrochemical water splitting and beyond. Joule, 2018, 2 (8): 1390-1395.
[36] Lei C, Wang Y, Hou Y, et al. Efficient alkaline hydrogen evolution on atomically dispersed $Ni\text{-}N_x$ Species anchored porous carbon with embedded Ni nanoparticles by accelerating water dissociation kinetics. Energy & Environmental Science, 2019, 12 (1): 149-156.
[37] Liu X, Chi J, Dong B, et al. Recent progress in decoupled H_2 and O_2 production from electrolytic water splitting. Chem Electro Chem, 2019, 6 (8): 2157-2166.
[38] Dotan H, Landman A, Sheehan S W, et al. Decoupled hydrogen and oxygen evolution by a two-step electrochemical-chemical cycle for efficient overall water splitting. Nature Energy, 2019, 4 (9): 786-795.
[39] Chen L, Dong X, Wang Y, et al. Separating hydrogen and oxygen evolution in alkaline water electrolysis using nickel hydroxide. Nature Communications, 2016, 7 (1): 11741.
[40] Symes M D, Cronin L. Decoupling hydrogen and oxygen evolution during electrolytic water splitting using an electron-coupled-proton buffer. Nature Chemistry, 2013, 5 (5): 403-409.
[41] Rausch B, Symes M D, Cronin L. A bio-inspired, small molecule electron-coupled-proton buffer for decoupling the half-reactions of electrolytic water splitting. Journal of the American Chemical Society, 2013, 135 (37): 13656-13659.
[42] Rausch B, Symes M D, Chisholm G, et al. Decoupled catalytic hydrogen evolution from a molecular metal oxide redox mediator in water splitting. Science, 2014, 345 (6202): 1326-1330.
[43] Ma Y, Dong X, Wang Y, et al. Decoupling hydrogen and oxygen production in acidic water electrolysis using a polytriphenylamine-based battery electrode. Angewandte Chemie International Edition, 2018, 57 (11): 2904-2908.
[44] Li X, Zhao L, Yu J, et al. Water splitting: from electrode to green energy system. Nano-Micro Letters, 2020, 12 (1): 131.

第三章

电解水催化剂的制备方法

开发地球储量丰富、高效、稳定性好的电催化剂在电解水商业化中起着至关重要的作用。在过去的几十年中，HER 和 OER 的非贵金属电催化剂的开发和利用得到了广泛的研究。磷化物、硫化物、碳化物、氮化物、硒化物及合金等催化剂主要用于 HER，对于 OER 电催化剂，研究最多的是氧化物、氢氧化物或羟基氧化物。可推广到工业领域使用的催化剂应该满足这些要求：①具有与贵金属基电催化剂性能相当甚至更高的催化活性；②具有优异的稳定性，电解数千小时仍具有良好的催化活性；③制备方法具有普适性且成本低。

一维（1D）和二维（2D）纳米材料具有独特的结构、电子和物理化学特征，包括高比表面积、高长径比、高表面不饱和原子密度和高电子迁移率，在电化学水分解领域比其他同类产品比如 0 维（0D）和块体材料具有更为显著的优势[1-3]。与 0D 纳米颗粒相比，1D 和 2D 纳米材料具有高度的各向异性，具有更好的导电性和传质性能，同时具有更好的抗团聚和溶解性能。通过控制合成条件，可以选择性地暴露高活性的晶面以提高电催化性能。与 3D 紧密堆积的纳米粒子相比，1D 和 2D 纳米结构的表面原子高度不饱和，提供了更多的高活性的催化位点。然而，当纳米结构以 1D 或 2D 形式存在时，会产生明显不同的结构和电子性质。例如，对于具有丰富边缘位点的 RuCu 纳米片，通过 DFT 计算了其表面电子活化能[2]。由于二维 RuCu 纳米片的超薄性质，使得纳米片边缘附近的晶格发生了明显的畸变，受益于二维纳米片的扭曲晶格，降低了键解离的中间能量，克服了库仑势垒，产生了活跃的电子转移，促进了水分解过程，这是 0D 金属纳米粒子所不具备的。随着纳米技术的发展和对低维纳米结构的深入理解，1D 和 2D 纳米结构材料的合成方法已朝多样化发展。因此这里重点讨论几种用于合成 1D 和 2D 纳米结构材料的典型方法，包括有机配体辅助合成法、水热和溶剂热合成法、电沉积法、共沉淀法以及静电纺丝法等。

3.1 有机配体辅助合成法

有机配体辅助合成是指在单一或混合的有机溶剂中发生反应，该溶剂包含分散良好的金属前驱体和还原剂。有机溶剂由可以与金属离子/原子配位并调节金属纳米晶体生长的长碳链分子组成。作为一种典型的基于溶液的合成方法，有机配体辅助合成可以较为容易地控制溶剂、前驱体、还原剂和温度，为控制金属纳米晶体的生长提供了很多可能性。金属前驱体（至少一种有机金属前驱体）很好地分散在有机溶剂中，有助于合成高质量的金属纳米结构。油酸胺（OAm）是有机配体辅助合成中最常用的有机溶剂之一，它也可以用作表面活性剂和还原剂。实际上，OAm 通常与另一类有机溶剂（油酸、1-十八烯、二苯醚等）一起使用，以控制金属纳米结构的生长方向和形态。用 OAm 作为有机溶剂（也用作表面活性剂和还原剂），已经制备了各种金属纳米结构，包括 1D 纳米线［例如 Pt 基的纳米线（NWs）[4-8]］和 2D 纳米片（NSs/NPs）（例如 Ru-Ni NSs[9]、Rh NSs[10] 和 Pt-Pb NPs[11]）。Jang 等人以［Rh（CO)$_2$Cl]$_2$ 为前驱体，OAm 为溶剂，在无搅拌的条件下，50℃反应 10 天，制备了厚度为 1.3nm 的超薄 Rh NSs［图 3.1(a)］[10]。他们提出由金属-金属相互作用形成的 Rh^+ 链结构首先与 OAm 配位，然后通过范德华力由配位链形成层状结构，Rh^+ 还原为 Rh 从而形成 Rh NSs［图 3.1(b)］。Bu 课题组在 OAm 中，以乙酰丙酮铂和乙酰丙酮钴为金属前驱体，合成了具有富 Pt 晶面和有序金属间化合物结构的分级高指数 Pt-Co NWs[5]。

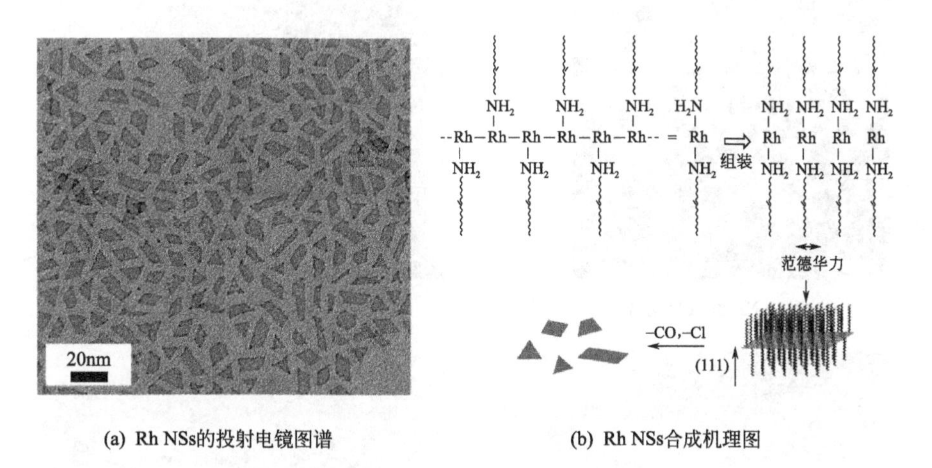

(a) Rh NSs的投射电镜图谱　　(b) Rh NSs合成机理图

图 3.1　Rh NSs 的投射电镜图谱和合成机理图[10]

另外，添加第二种有机溶剂作为封端和结构导向剂，是有效调节金属纳米结构生长的重要策略。Bu 等人以 OAm 和 1-十八烯（ODE）为溶剂，制备了 Pt-Pb 六方纳米片[11]［图 3.2(a)］。实验表明，OAm 和 ODE 对于 Pt-Pb 纳米片的合成是不可或缺的。在只有 OAm 或 ODE 存在下形成的是立方纳米晶或聚集纳米颗粒。另外通过改变 OAm 与 ODE 的体积比，可以调控 Rh 的形貌从四面体（只有 OAm）到凹四面体（OAm/ODE=1）和纳米片（OAm/ODE=1.5）[12]。除上述有机结构导向剂外，前驱体、还原剂、小分子和温度等因素都会对合成的金属纳米晶的生长取向和形貌产生显著影响[7]。这虽然使金属纳米晶生长的控制复杂化，但为金属纳米结构和形貌的调控提供了平台。Huang 等研究表明，在 OAm 中 Cu（acac）$_2$ 的前驱体中加入小分子 NH_4Cl 可得到 Cu 纳米立方体［图 3.2(b)］，而在 170℃下用 $RuCl_3 \cdot H_2O$ 取代 NH_4Cl 反应 3h 可得到 Cu NWs［图 3.2(c)］。如果在 Cu NWs 合成后进一步升温至 210℃，并维持 12h，Ru 被还原，最终形成 RuCu 纳米管（NTs）［图 3.2(d)］[13]。这都说明了小分子和温度对金属纳米晶结构的影响。Pt-Pb 纳米片的合成中如果将还原剂抗坏血酸（AA）替换为葡萄糖或柠檬酸，则无法获得六边形 Pt-Pb 纳米片[11]。通过以上的讨论可以看出，在优化的合成条件下，有机配体辅助合成为获得多样化的 1D 和 2D 金属纳米结构提供了前所未有的机遇。

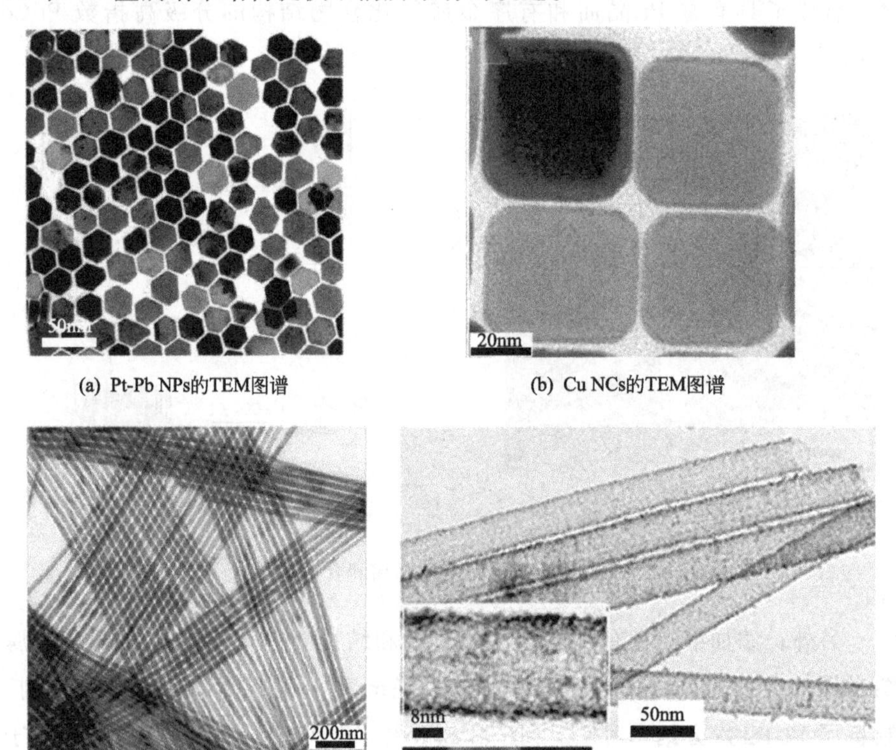

(a) Pt-Pb NPs的TEM图谱　(b) Cu NCs的TEM图谱

(c) Cu NWs的TEM图谱　(d) RuCu NTs的TEM图谱

图 3.2　Pt-Pb、CuNCs、Cu NWs 和 RuCu NTs 金属纳米材料的 TEM 图[13]

3.2 水热和溶剂热合成法

水热和溶剂热合成是无机合成的重要分支。水热合成是指在温度高于水沸点的水溶液中通过化学反应合成，溶剂热合成是在非水溶液中较高温度下进行。这些技术已广泛用于材料的合成、废物处理以及模拟地热和生物水热过程。水热和溶剂热合成的研究主要集中在反应物的反应性、合成反应的规律和条件，以及它们的产物结构和性质的关系。水热反应和固态反应的主要区别在于“反应性”，反映在它们不同的反应机制上。固态反应取决于原料在界面处的扩散，而在水热和溶剂热反应中，反应物离子和/或分子在溶液中反应。显然，即使使用相同的反应物，反应机理不同也可能导致产物的最终结构不同。更重要的是，许多固态合成无法制备的具有特殊结构和性能的化合物或材料，可以通过水热和溶剂热反应获得。在某些情况下，水热或溶剂热反应通过降低反应温度为固态反应提供了另一种温和的合成方法。非理想和非平衡状态是水热和溶剂热反应系统的关键特征。水或其他溶剂可在高温高压条件下活化，水热反应和溶剂热反应已成为制备大多数无机功能材料、特殊组分、结构、凝聚态和特殊形貌的重要途径，如纳米材料和超细粉体、水凝胶、非晶态、无机膜和单晶等。水热和溶剂热的另一个特点是可操作性和可调谐性，因此便于开发一些新的水热反应。与其他合成技术相比，水热法和溶剂热法各有优势。基于水热和溶剂热化学，我们可以进行由于反应物在高温下蒸发而无法在固态反应中发生的独特合成反应，制备特殊价态、亚稳态结构、凝聚态和聚集态的新材料，合成低熔点、高蒸气压、低热稳定性的亚稳相或材料，生长出热力学平衡缺陷、形貌可控、粒径可控的完美单晶，并可以在合成反应中直接进行离子掺杂。

水热和溶剂热合成实验的基本设备是高压釜。水热和溶剂热合成的发展很大程度上取决于可用的设备。水热和溶剂热条件下的晶体生长或材料加工要求高压容器具有优异的耐腐蚀性、耐高温、高压的能力。针对水热合成的目标和容差，有多种不同的高压容器。这里简单讨论水热合成中常用的高压釜和水热反应器。一般来说，理想的水热反应釜应具备这些主要特征：①应具有较高的机械强度，以承受长时间的高压和高温实验；②应具有优良的耐酸、耐碱、耐氧化性；③应具有简单的机械结构，便于操作和维护；④应具有良好的密封性能，以获得所需的温度和压力；⑤应具有合适的尺寸和形状，以获得期望的温度梯度。选择合适的高压釜，首要考虑的参数是实验压力和温度条件以及在水热溶液中的耐蚀性。一些水热反应，其中试剂或溶剂无腐蚀性，可直接在水热釜中进行。相比之下，在大多数水热反应中，所用的试剂或溶剂具有很强的腐蚀性，会在高温高压范围内腐蚀高压釜，因此它要求反应釜含有惰性内衬或罐子。一般用作内衬的材料包括聚四氟乙烯、石英、石墨、钛、银等。图 3.3 是温和的水热和溶剂热反应常用的聚四氟乙烯内衬不锈钢高压釜，这类反应釜可承受的最高温度为 270℃，压力约 150MPa。

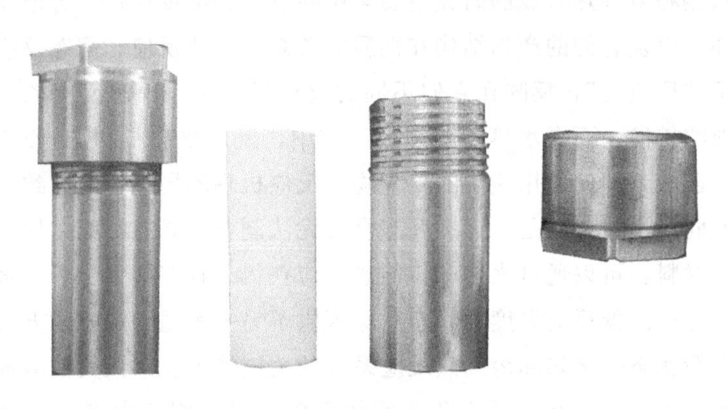

图 3.3 用于水热和溶剂热合成的通用高压釜

水热或溶剂热是合成纳米材料的一种低成本且便捷的方法。将含有前驱体、还原剂和表面活性剂等的溶液转移到高压釜中，在高温高压下进行反应。在常温常压条件下无法进行的反应，在高温高压下很容易发生并且可以获得独特的材料结构。该方法通常采用水、有机溶剂或两者的混合物作为反应介质。通过控制溶剂、金属前驱体、表面活性剂、还原剂、温度、反应时间、pH 值等因素，得到形貌、生长取向、结晶度、尺寸等可控的纳米材料。水热法或溶剂热法已被广泛用于合成 1D 和 2D 纳米材料。例如，Duan 等人[14]以苯甲醇和甲醛的混合物为溶剂，聚乙烯吡咯烷酮（PVP）为表面活性剂，通过溶剂热法合成了单层 Rh NSs。所合成的 Rh NSs 的厚度小于 4Å（$1Å=1\times10^{-10}m$），结构分析表明 Rh NSs 由 Rh 单原子层状片层组成。Dai 等人采用溶剂热法，在甲酰胺中加热 Pt $(acac)_2$、Cu $(acac)_2$、PVP 和 KI 的混合物，合成了 PtCu 合金 NSs[15]。研究发现，I^- 对 PtCu NSs 的合成至关重要，I^- 可以调节 Pt 的还原动力学，使 Cu^{2+} 还原为 Cu^+ 形成 CuI，原位形成的 CuI 在材料合成过程中起着重要作用。该材料在酸性条件下具有良好的 HER 催化活性。另外以二十八烷基二甲基氯化铵（DO-

DAC）为模板，AA 为还原剂，在水中水热合成 PdPt NWs[16]，其合成路线如图 3.4（见文后彩插）所示。

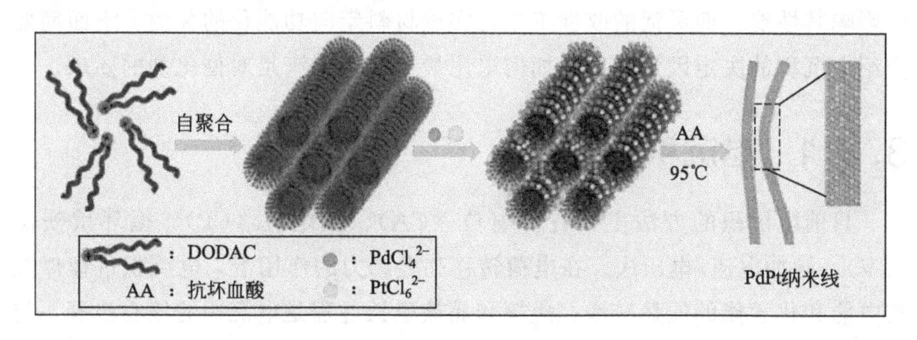

图 3.4　PdPt NWs 合成路线[16]

3.3　电沉积法

电化学技术起源于 19 世纪，是将电荷的运动与化学能的转化结合起来的一项突破性技术。其中电化学反应由阴极的还原反应和阳极的氧化反应组成，通过在外加电能驱动下发生电荷转移来实现[17]。电化学技术不仅是一种有效评估电极材料性能的分析方法，而且还逐渐完善为一种材料合成技术，称为电沉积。在负载电力的驱动下，电沉积具有实时响应快速、反应温和等优点，是一种低成本、环保的合成方法[18,19]。电沉积技术具有优异的技术灵活性，对制备功能性材料尤为关键。特别是在制备自支撑催化剂时，电沉积可以在短时间内实现催化剂均匀而有力的生长，避免了引入黏结剂，降低了电子传递电阻，提高了工作稳定性，对激发催化剂的催化性能尤为重要[20,21]。通常通过改变电解液的组成和简单的工艺调整，采用电沉积技术能够得到具有理想纳米结构和组成的最佳催化剂，顺应了采用简单合成法制备新型纳米功能材料的发展趋势。无定形化合物具有丰富的缺陷，这些缺陷存在大量的活性位点，同时无定形化合物具有独特的电子结构，从而使材料的催化性能得到质的提升[22,23]，这类无定形化合物可以通过电沉积合成。除了通过一步电沉积合成催化剂外，许多研究还通过电沉积合成或修饰前驱体。通过与其他合成技术结合，大大丰富了催化剂的制备方法，赋予了催化剂更优异的性能[24-26]。

作为一种历史性的合成技术，电沉积系统一直在不断发展以满足材料的多维需求。毫无疑问的是材料优异的电催化性能其微观结构和成分密切相关，这些可以通过优化电沉积工艺来实现。对于催化剂的制备，各种因素都会影响其性能，而灵活的改性工艺是实现材料定向功能化的关键。下面简要介绍电沉积的决定因素，以及如何采用最佳工艺来满足对催化剂的要求。

3.3.1 电沉积的方式

目前电沉积的方法主要有恒电位（CA）、恒电流（CP）、循环伏安法（CV）、脉冲电流/电压法。在电荷转移和电场力的作用下，电沉积过程伴随着电能和化学能的能量转换。成核和晶核生长过程是电沉积的核心步骤，与电输入直接相关，相互作用可用 Kelvin 方程来描述：

$$\tau = 2\delta V / z e_0 |\eta| \tag{3.1}$$

其中，τ、δ、η、V、z 分别表示核半径、表面能、过电位、原子体积和转移电子数。通常采用标准的三电极体系进行电沉积反应。目标反应在工作电极（WE）/电解质界面进行，图 3.5 是典型的电沉积的输入模式。

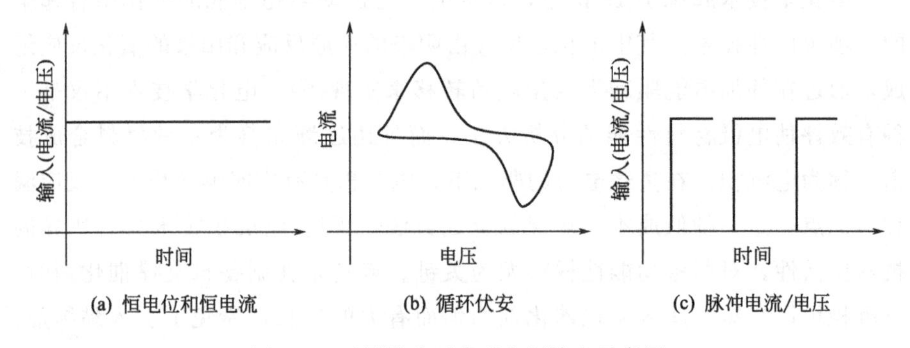

图 3.5　不同模式下典型电源输入示意图

3.3.2 电解液

除了一些特殊的熔盐体系，传统的溶液中都含有溶剂和溶质。水是应用最广泛的溶剂，大多数经典电解液是将盐类溶解于水介质中得到的，通过调节溶质可以设计金属/非金属化合物的组成。但其缺点是体系的电位窗口和使用的温度范围较窄，限制了在一些高标准材料合成领域中的应用。为了沉积目标物质，需要寻找具有更宽电化学窗口的替代溶剂[27]。高温熔盐一般依赖于高温和高压，应用起来较为困难[28]。离子液体（ILs）具有环境友好、化学和热稳定性好等优点（图 3.6），近年来受到越来越多的关注，已经成功用于合成具有高活性的 HER 和 OER 的二元/三元合金或化合物[29-31]。

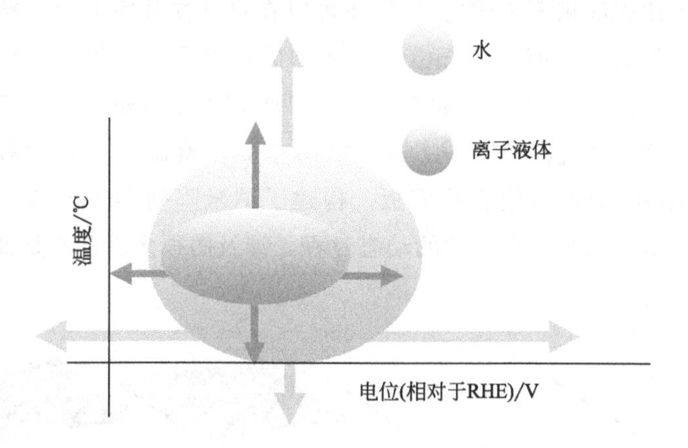

图 3.6　离子液体和水的性质比较（温度和电化学窗口）

所有的电化学反应都是以溶解在溶剂中的阴离子和阳离子作为反应物。离子的扩散行为会引起浓度梯度，因此需要根据实验情况调整离子浓度。为了获得纳米材料，增加催化活性面积，通常将溶质浓度调节到较低水平，以增强沉积层的扩散控制，从而促进催化活性[32]。另外不同物种之间的反应势垒将导致共沉积体系不一致性，络合剂可以减小或消除这一现象[33]。比如，Mo 基合金和 W 基催化剂都具有独特的电子环境和氧空位，因而表现出良好的 OER 和 HER 活性，其合金或化合物需要柠檬酸盐作为络合剂才能实现共沉积[34]。温度影响反应的热力学特性，进而促进较高的反应速率和能量转换效率[35]。此外，电解液的 pH 直接决定了反应的类型，通过控制 pH 值对于获得目标产物至关重要（在酸性条件下容易得到金属相固溶体，而在碱性条件下有利于形成络合物）。电沉积时间的长短决定镀层的厚度或纳米结构的尺寸，只有适量才能暴露最活跃的位点，以获得更好的催化活性[36]。

3.3.3　基底和模板

从早期的贵金属催化剂制备研究到目前过渡金属基催化剂，原位生长自支撑结构一直是研究的热点，具有纳米多孔结构的催化剂可以通过采用多维基材或模板为基底得到。不仅有传统的二维基底，如金属箔和氟掺杂 SnO_2

(FTO) 玻璃，还有许多类型的碳基基底包括碳布（CC)、碳纸（CP）和非贵金属泡沫（Ni/Cu/Fe）(图 3.7)，都具有高的电导率和丰富的微孔，从而产生较大的比表面积[22,25,37]。另外还可以在自支撑基底上生长预设模板进一步增大材料的比表面积。ZnO 纳米棒和硅铝多孔氧化物是生长纳米阵列的典型自牺牲模板[38,39]。比如 Dong 等人[40]首先用水热法在高密度泡沫镍表面先生长 ZnO 模板，然后电沉积 NiMo 前驱体覆盖于 ZnO 模板上，最后用高浓度 KOH 溶液浸泡去除模板，构建了纳米阵列 $Ni(OH)_2/NiMoO_x$。总的来说，电沉积是一个复杂的动态过程，涉及的很多因素不是孤立的，而是相互作用的。

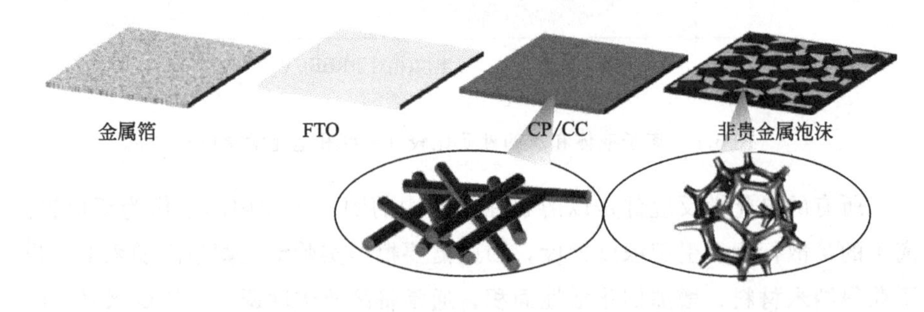

图 3.7 常用的电沉积基底

目前为止，通过电沉积方法已经合成了许多性能优异的 HER 和 OER 催化剂，比如金属合金、过渡金属氢氧化物、硫化物、磷化物、磷酸盐、硼酸盐等。与传统熔炼前驱体或液相还原合金的策略相比，电沉积可以降低能耗，扩大应用范围（常温、常压、金属盐反应物丰富)，具有更灵活的相变和熵变，有利于提高催化剂活性[41]。对于纯金属基镀层的制备，电沉积技术可以通过引入添加剂和改变沉积方式改变晶态和纳米结构完成。然而，非贵金属的纯金属的电子环境不理想，导致其 H 吸附能过强或过弱，从而限制了其催化性能[42]。但是，多金属原子之间的强协同作用可以弥补该缺陷，电沉积制备的合金能够调控催化剂电子环境而受到广泛的关注。

3.4 共沉淀法

一般多组分催化剂可以用共沉淀法来制备。共沉淀是指在同一溶液中，通过吸附或机械包裹等作用形成混合晶体，从而产生沉淀的过程。在许多情况下，尽管两种待沉淀组分的溶解度可能存在很大差异，但在沉淀条件下基本上是不溶的。共沉淀非常适合产生均匀分布的催化剂组分，或产生具有明确化学计量的前驱体，这些前驱体可以很容易地转化为活性催化剂。如果前驱体是目标催化剂的化学计量的化合物，那么通过煅烧、还原等步骤生成的目标催化剂，通常会产生非常小且紧密混合的微晶。这种具有良好分散性的催化剂成分是其他制备手段难以实现的。但共沉淀法也存在一定的弊端，比如对技术要求较高，在沉淀过程中难以跟踪沉淀产物的质量，如果沉淀不连续进行，在整个沉淀过程中难以保持良好的催化剂质量。尽管共沉淀法工艺存在诸多弊端，但仍是制备固体催化剂的重要技术。

基本上，所有的工艺参数（无论是固定的还是可变的）都会影响最终催化剂的质量。通常，要制备具有特定性质的催化剂（包括特定的化学成分、纯度、粒度、表面积、孔径、孔体积、与母液的分离性等），对下游工艺诸如干燥、造粒或煅烧步骤具有一定的要求。因此，必须对参数进行优化，以制备所需要的催化剂。图 3.8 总结了沉淀过程中可以优化的参数，以及受这些参数影响的主要性质。下面将讨论一些工艺参数对所得催化剂性质的影响。这里所述的情况在特殊情况下可能会有所不同，因为沉淀参数的准确选择通常是由长期的实验优化产生的。

3.4.1 原材料的影响

通常情况下，原材料的阴离子容易分解为挥发性物质。一般采用硝酸盐、氨或碳酸钠作为沉淀剂，有时候也会采用草酸盐。在沉淀过程中如果存在可能被吸留的离子，且对最终合成的催化剂的催化性能会产生不利影响，需要重复洗涤除去被吸留的离子。比如在许多催化反应中，氯离子或硫酸盐等离子是不利于催化反应的，因此在沉淀中应尽量避免。Matijevic 研究了许多金属的沉淀，主要是氢氧化物，证明了沉淀溶液中离子的性质会强烈影响最终材料的性质[43]，阴离子不仅会影响材料的形态和尺寸，甚至会产生不同的相成分。

3.4.2 浓度和组成的影响

沉淀反应一般在高浓度的金属离子条件下进行。过饱和程度越高，沉淀越快，如果发生均相成核，在较高过饱和度下成核速率增加，通常在较高浓

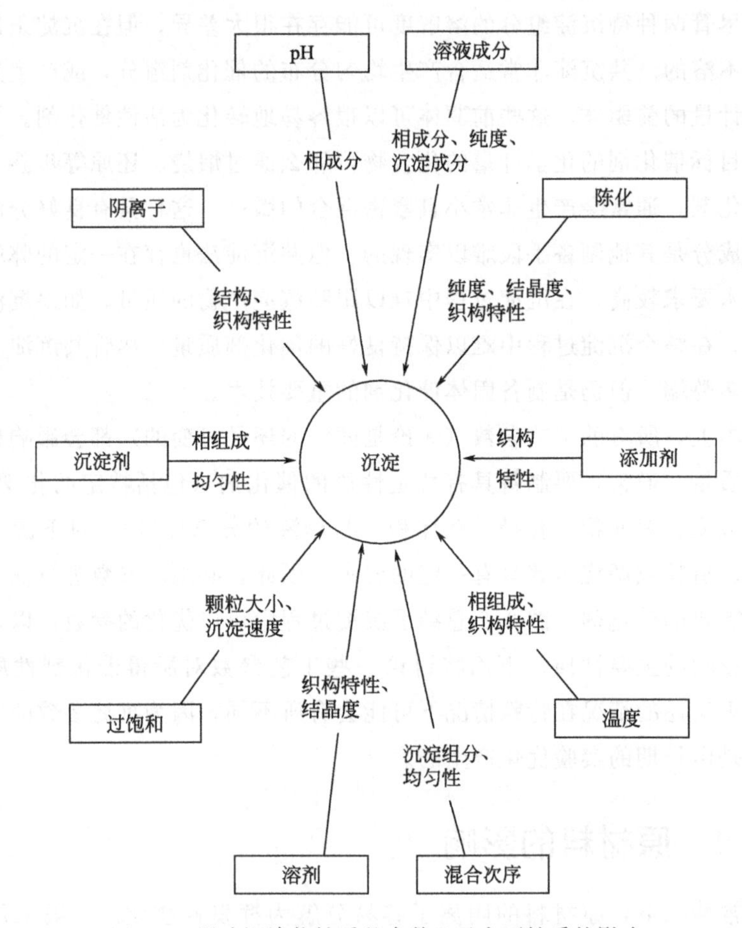

图 3.8　影响沉淀物性质的参数以及主要性质的影响

度下形成较小粒径和较高比表面积的产物。但如果沉淀反应必须在低浓度下进行（比如需要得到较大的初级粒子），那么沉淀反应要么在含有晶种的连续系统中进行，要么溶液中含有晶种。采用共沉淀法制备催化剂，溶液的组成有时候决定最终产物的组成。在共沉淀过程中，如果不同化合物的溶解度差异很大并且沉淀不完全，或者除了化学计量化合物之外，只有一种组分形成不溶性沉淀，则共沉淀过程中通常会发生溶液组成的偏差。

3.4.3　溶剂效应

有机溶剂比水贵很多，因此在沉淀过程中大多使用水作为溶剂。另外大多数金属盐在有机溶剂中的溶解度比在水中要低很多，如果要得到相同的产率，需要大量的有机溶剂。更重要的是，有机溶剂的使用带来了更大的环境污染。但也有报道称有机溶剂对某些材料的沉淀是有利的，低溶解度会导致极低的过饱和度，结晶缓慢，这样可以改变颗粒的粒度分布。另外还有反应是在反胶束体系中（微乳液）进行的，在该体系中需要加入表面活性剂，形成油包水的微乳液，沉淀在油相中的水滴环境中进行。沉淀可以通过第二微乳液来触发（该第二微乳液含有溶解在水相中的沉淀剂），也可以通过微乳液传递反应气体引发。在微乳液中，油不是严格意义上的溶剂，而是作为分散水滴的介质，尽管这个反应过程看起来很有研究意义，但是目前还没有报道过由微乳液合成的商品化催化剂。

3.4.4 温度和 pH 的影响

由于成核速率对温度变化极为敏感，因此沉淀温度是控制沉淀物性质（如初级微晶尺寸、表面积甚至是相的形成）的决定性因素。然而，对于一个特定的反应，要说明如何调整反应温度才能获得具有特定性能的催化剂很难，必须通过实验来确定最佳的温度。一般来说，大多数沉淀反应的温度都在室温以上。由于 pH 值直接控制过饱和度（至少在氢氧化物沉淀的情况下），这应该是沉淀过程中的关键因素之一。相对于其他参数，pH 值的影响并不简单，必须进行实验来确认。

3.5 静电纺丝法

静电纺丝技术是利用静电力从聚合物溶液或熔体中产生细小纤维的独特方法，由此产生的纤维比常规纺丝工艺得到的纤维直径更小（从纳米到微米），比表面积更大。静电纺丝需要在几十千伏范围内的直流电压下才能产生。其原理是基于强相互排斥力克服带电聚合物液体中较弱的表面张力。目前，有垂直和水平两种标准的静电纺丝装置。随着这项技术的发展，出现了更精密的系统以更可控、更高效的方式制备更复杂的纳米纤维结构[44]。电纺是在常温常压下进行的，其典型的设备如图 3.9 所示。基本上，静电纺丝系统由三大部分组成：高压电源、喷丝头（例如一个吸管尖）和接地接收器（通常是金属屏、金属板或旋转芯棒）。利用高压电源向聚合物溶液或熔体中注入一定极性的电荷，然后加速喷向相反极性的接收器[45]。在进行纺丝前，大部分聚合物都溶解在一些溶剂中形成聚合物溶液，然后将聚合物流体引入毛细管中进行静电纺丝。但有些聚合物可能会散发出不愉快甚至有害的气味，因此需要在有通风系统的腔室内进行。在静电纺丝过程中，毛细管末端的聚合物溶液受到电场的作用，当施加的电场达到临界值时，排斥电力克服表面张力，最终带电的溶液射流从泰勒锥的尖端喷出，在毛细管尖端和接收器之间的空间中发生不稳定且快速的射流，溶剂蒸发，留下聚合物。

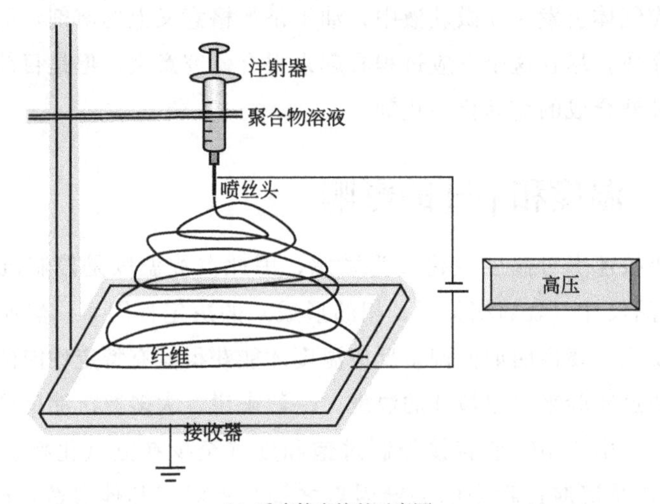

(a) 垂直静电纺丝示意图

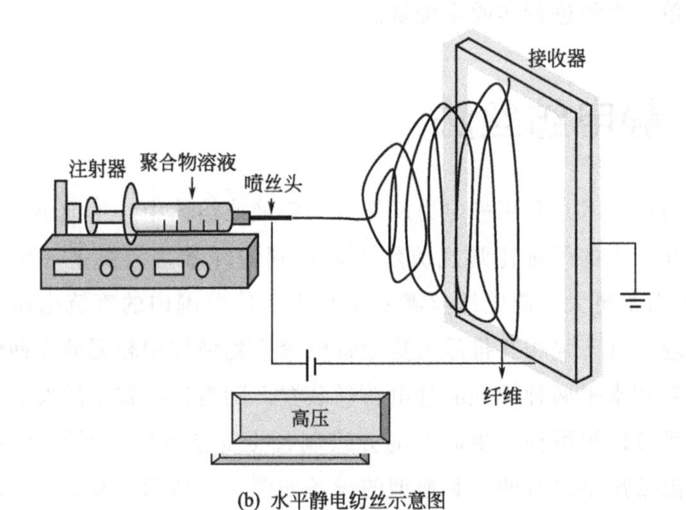

(b) 水平静电纺丝示意图

图 3.9　静电纺丝设备的示意图

可用于静电纺丝的聚合物有很多种，能够形成亚微米范围内的细纳米纤维，在各个领域都有应用。据报道，电纺纳米纤维来自各种合成聚合物、天然聚合物或者两者的混合物。静电纺丝过程由许多参数控制，诸如溶液参数、工艺参数和环境参数。溶液参数包括浓度、黏度、电导率、分子量和表面张力；工艺参数包括施加的电场、喷丝头尖端到接收器的距离以及进料或流速，这些参数都会影响静电纺丝的纤维状态。通过控制这些参数，可以获得想要的形态和直径的纳米纤维。除了这些变量参数外，环境参数（即周围环境的湿度和温度）严重影响电纺丝纳米纤维的形态和直径。下面对这些影响参数进行简单的介绍。

3.5.1 溶液参数

（1）浓度

研究发现，在较低的溶液浓度下，得到含有珠状物和纤维的混合材料；随着溶液浓度的增加，珠状物的形状由球形变成纺锤形；最后由于具有较高的黏滞阻力而形成直径较大的均匀纤维；在高浓度下无法维持溶液在针尖的流动，从而不能连线形成纤维。因此在静电纺丝过程中，应该有一个最佳的溶液浓度。

（2）分子量

聚合物的分子量对其黏度、表面张力、电导率、介电强度等流变性能和电学性能有显著影响，这是影响静电纺丝纤维形貌的另一个重要的溶液参数。通常高分子量的聚合物溶液为纤维的生成提供了所需的黏度。实验发现分子量过低的溶液倾向于形成珠状物而不是纤维，而高分子量的溶液形成较大直径的纤维。聚合物的分子量反映了溶液中聚合物链缠结的数量，从而反映溶液的黏度。链缠结在静电纺丝中起着重要作用，因此即使在聚合物浓度较低时，HM-PLLA（高分子量的聚-*L*-乳酸）仍能保持足够数量的聚合物链缠结，从而保证足够的溶液黏度，在静电纺丝过程中产生均匀的射流，抑制表面张力的影响，因此链缠结对静电纺丝纳米纤维是否成珠起重要作用。Gupta 等人合成了分子质量从 12.47～365.7kDa 变化的 PMMA[46]，他们发现随着分子量的增加，珠粒和液滴的数量减少。但有研究发现，如果分子间相互作用足以替代通过链缠结获得的链间链接，那么在静电纺丝过程中并不是必须使用高分子材料的。利用这一原理，研究者们通过静电纺丝将卵磷脂溶液中的低聚物磷脂制备成非织造膜。

(3) 黏度

溶液黏度在纺丝过程中对纤维的尺寸和形态起着重要作用。研究发现，在极低的黏度下，没有连续的纤维形成，而在极高的黏度下，射流很难从聚合物溶液中射出，因此静电纺丝有一个最佳的黏度要求。不同聚合物溶液纺丝时的黏度范围不同，据报道最大的纺丝黏度范围在 1～215P ($1P=10^{-1}Pa \cdot s$)。聚合物的黏度、聚合物浓度和分子量相互关联。溶液黏度与溶液浓度密切相关，聚合物黏度和/或浓度与通过静电纺丝获得的纤维之间的关系已经有了系统的研究。高黏度的聚合物溶液，通常表现出更长的应力松弛时间，可以防止静电纺丝过程中喷射射流的破裂。溶液黏度或浓度的增加会产生直径更大更均匀的纤维，溶液的黏度在是否能获得连续纤维中起着重要作用。对于低黏度的溶液，表面张力是主导因素，仅形成珠状物或珠状纤维；而高于临界浓度时，获得连续的纤维结构，其形态受溶液浓度的影响。综上所述，静电纺丝存在聚合物特定的最佳黏度值，对纤维的形态有显著影响。

(4) 表面张力

表面张力在静电纺丝过程中起着至关重要的作用，通过降低表面张力可以得到无珠的纳米纤维。不同的溶剂其表面张力不同，一般来说，高表面张力会因为射流的不稳定性和喷射液滴的产生而抑制静电纺丝过程。液滴、珠状物和纤维的形成依赖于溶液的表面张力，较低的纺丝溶液表面张力有助于静电纺丝在较低的电场下进行。然而，并不是说低表面张力的溶剂就一定更适合静电纺丝。基本上，如果所有其他变量保持不变，表面张力决定了静电纺丝电压窗口的上下边界。

(5) 电导率/表面电荷密度

聚合物大多是导电的，除极少数介电材料外，聚合物溶液中的带电离子对射流形成的影响很大。溶液电导率主要取决于聚合物类型、使用的溶剂和电离盐的可用性。研究发现，随着溶液电导率的增加，电纺纳米纤维的直径显著减小；溶液电导率较低时，射流在电力作用下伸长不足，可能产生均匀的纤维，也可能出现珠状物。一般情况下，在最高的电导率下可以得到最小直径的电纺纳米纤维，也就是说纤维尺寸的下降是由电导率增加导致的。另外射流半径与溶液电导率的立方根成反比。离子增加了射流的电荷承载能力，从而通过施加的电场使其承受更高的张力。据报道通过添加离子盐可以影响电纺纤维的形貌和直径，比如 KH_2PO_4、NaH_2PO_4、NaCl 等，随着离子盐的加入，纤维的均匀性增加，珠粒的形成减少。

3.5.2 工艺参数

(1) 电压

施加电压是电纺丝过程中一个至关重要的参数。只有在达到阈值电压后，才会形成纤维。电压在溶液中引入了必要的电荷和电场，启动了静电纺丝过程。实验已经证明，液滴的初始形状随纺丝条件（电压、黏度和进料速度）而改变。关于静电纺丝过程中施加电压的影响，目前尚有争议。Reneker 和 Chun（1996）表明静电纺丝聚乙烯时，电场对纤维直径的影响不大。有研究人员提出，当施加更高的电压时，有更多的聚合物喷射，这有利于形成更大直径的纤维。另一些研究者报道，增加电压（即增加电场强度），会增加流体射流上的静电斥力，最终有利于形成直径更小的纤维。在大多数情况下，较高的电压导致溶液更大拉伸，射流中存在更大的库仑力和更强的电场，导致溶剂从纤维中迅速蒸发，从而形成直径更小的纤维。在更高的电压下，更容易形成珠状物。拉朗多和曼利也观察到施加电压对纤维直径的类似影响。他们发现，通过增加一倍的外加电场，光纤直径大约减少了一半。因此，电压影响纤维直径，但随聚合物溶液浓度和尖端与收集器之间的距离而变化。

(2) 流量

聚合物从注射器流出的速度是一个重要的参数，因为它影响射流速度和材料的转化速度。进料速度较低使得溶剂有更足够的时间蒸发，纺丝溶液有一个最小的流速。对于聚苯乙烯（PS）纤维，随着聚合物流速的增加，纤维的直径和孔径增大，通过改变流速，可以略微改变其形态结构。少量工作系统地研究了溶液进料和流速对纤维形貌和尺寸的影响。高流速下材料在到达收集器之前没有足够的干燥时间，导致形成的纤维呈串珠状。

(3) 接收器的类型

静电纺丝工艺的另一个重要方面是接收器的类型。在这个过程中，接收器作为导电基底收集纳米纤维，通常使用铝箔作为接收器。但由于难以转移收集的纤维以及需要对纤维进行有序排列，各种接收器例如：导电纸、导电布、钢丝网、针、并行或网格线、旋转棒、旋转轮、非溶剂液体（如凝固状甲醇）等如今也得到应用。透明质酸的吹气辅助静电纺丝中，Wang 等采用了铝箔和铝丝网两种接收器。结果表明，导电面积较小的丝网对纤维收集有负面影响[47]。由于其比表面积小，导电面积较少时产生串珠状纤维。在另一项研究工作中，对比了具有铝箔的丝网和没有铝箔的丝网载体的同一导电区域，发现纯丝网是一种较好的纤维接收器。因为随着丝网的使用，纤维向其他衬底的转移变得更容易。纤维的排列方式由接收器的类型和转速决定。由于高电荷射流的弯曲不稳定性，所生成的纳米纤维以随机方式沉积在接收器上。几个研究小组证明了使用旋转的滚筒或旋转轮状的筒子或金属框架作为收集器，得到的静电纺纤维或多或少地相互平行。另外几种类型的分裂电极被用于制备有序排列的纳米纤维，这种典型的接收器由一个空隙隔开的两个导电基底组成，有序排列的纳米纤维沉积在该空隙中。

(4) 尖端到接收器的距离

尖端和接收器之间的距离也是控制纤维直径和形态的参数。在纤维到达接收器之前，需要有一个最小的距离，使纤维有足够的时间干燥，距离太近或太远，都会产生珠状结构。在聚乙烯醇、明胶、壳聚糖、聚偏氟乙烯等材料的静电纺丝过程中观察到针尖和接收器的距离对纤维形态的影响不像其他参数那样显著。据报道，使用具有纤维连接蛋白功能的类丝聚合物纺丝时，更近的距离得到扁平的纤维，随着距离的增加，纤维变得更圆。对于聚砜，尖端和接收器之间的距离越近，纤维越小。静电纺丝纳米纤维的一个重要物理特性是它们在溶解聚合物溶剂中的干燥度。因此，尖端与接收器之间的最佳距离应有利于溶剂从纳米纤维中蒸发。

3.5.3 环境参数

除了溶液和工艺参数外，还有环境参数，包括湿度、温度等。已有研究考察了环境参数对静电纺丝过程的影响。有人研究了在25～60℃范围内温度对聚酰胺-6纤维静电纺丝的影响，发现随着温度的升高，纤维直径减小，他们认为这是随着温度升高聚合物溶液的黏度降低造成的，因为溶液的黏度和温度成反比关系。另外随着纺丝湿度的增加，纤维表面会出现细小的圆形孔，进一步增加湿度会导致孔隙聚结。研究发现，在非常低的湿度下，随着溶剂的蒸发速度加快，挥发性溶剂可能会迅速干燥。但有时蒸发的速度比从针尖上去除溶剂的速度更快，这就会给静电纺丝造成问题：静电纺丝过程可能只进行几分钟针尖就会被堵塞。也有人指出，高湿度有利于静电纺丝纤维的放电。因此，除溶液和工艺参数外，环境参数也会对静电纺丝过程产生影响。

近几十年来，静电纺丝技术现已成为合成1D纳米材料的常用方法。通过改变喷嘴结构并控制实验条件，可以制备各种结构的1D纳米纤维。例如，固体纳米纤维可以通过传统的单喷嘴静电纺丝法来制备，包括金属化合物/碳纳米纤维和无碳纳米纤维。而且，通过同轴静电纺丝还可以制备具有特殊结构的纳米纤维，如核壳、空心以及多通道纳米结构。在电解水催化剂领域静电纺丝法也引起了广泛关注，因为：①纺丝过程的特点是聚合物材料在电纺过程中发生了强烈的变形，并且在毫秒内发生了非常迅速的结构形成过程；②制备组分可调、结构可控的纤维或管状纳米材料，静电纺丝是一种相对环保、经济的方法；③电纺纳米纤维具有较高的比表面积、长径比和孔隙率，可以提高电子的迁移率，增加活性位点的数量；④电纺纳米材料可以作为活性组分均匀分散的载体。由于这些优点，电纺纳米材料在电化学水分解过程中表现出了显著的催化效率和循环稳定性。电纺技术的发展为电化学水分解注入了新的能量。传统的纳米催化剂易于聚集成块状或大尺寸的颗粒，因此影响其分散性和利用率。1D电纺纳米材料可以用作载体，并使纳米颗粒均匀分散，从而提高电催化剂的效率。目前通过静电纺丝技术和后处理工艺相结合，已经制备了许多纳米纤维或纳米管，包括基于单金属、金属合金、金属氧化物、金属硫化物、金属磷化物以及金属碳化物的电催化剂等。比如Zhang等人采用静电纺丝和高温碳化相结合，合成了多孔的N掺杂的Co碳纳米纤维（Co-CNFs）作为HER电催化剂，并具有优异的性能，在0.5mol/L H_2SO_4溶液中，过电位为159mV。Zhao等人也报道，电纺丝合成的Co-NCNFs在碱性溶液中分解水具有优异的电化学活性，电流密度10mA/cm^{2}时电压为1.66V。这些材料具有优异的性能主要取决于：①独特的包覆和多孔结构加速了表面反应；②氮和钴的引入使CNF产生了更多的活性位点；③非金属掺杂金属对提高电催化活性具有明显的协同作用。另外，Wang等人通过电纺丝技术合成了N、P共掺杂的Co-CNFs，得益于N/P协同效应和大量的吡啶N和Co-N_x簇的活性位点，Co-N-P-CNFs展现出了优异的电催化活性。因此，将非金属元素引入CNF是增强一维电纺丝纳米催化剂电化学活性的良好选择。尽管电纺丝制备的单金属基催化剂已经取得了一定的成绩，但是就电解水催化剂的活性和稳定性而言，仍然不太理想。

3.6 材料的表征技术

为了适应纳米技术的发展，已经开发了许多先进的技术，包括表征和计算方法，用于探索纳米材料的形态、分布、缺陷、原子排列以及电子和电催化特性[48,49]。目前广泛采用的技术有扫描电子显微镜（SEM）、透射电子显微镜（TEM）、高分辨透射电子显微镜（HRTEM）、高角度环形暗场扫描透射电子显微镜（HAADF-STEM）、X 射线光电子能谱（XPS）和 X 射线衍射（XRD）等，通过对其尺寸、晶格、缺陷、暴露面和异质结构的观察来揭示活性位点。X 射线吸收光谱（XAS）作为应用最广泛的 X 射线工具之一，具有以下优点：对短程有序灵敏但对较轻元素、元素特异性和化学状态不敏感，在原子尺度上提供材料的定量结构信息，如氧化态、配位数、键长和原子种类[49-52]。考虑到每种技术都有其自身的优点和局限性，综合运用多种技术，对于认识材料结构和催化性能之间的强相关性是非常有用的。例如，用 HRTEM 很难观察到催化活性位点的缺陷和掺杂剂的存在，但采用 STEM 则很容易。此外，当材料的浓度极低时，也很难使用 XRD 和 XPS 检测材料的上述特征。幸运的是，XAS 的发展可以解决这个问题，它分为 X 射线吸收近边结构（XANES）和扩展 X 射线吸收精细结构（EXAFS）。然而，一些电催化材料可能会随着电化学反应过程中物理和化学条件的变化（例如温度和施加的电位）而改变其形态和电子结构。因此，发展可操作的表征技术来揭示催化反应的重组现象和识别真实的活性位点是非常必要的，这对于理解催化机理和开发新材料至关重要。在这种情况下，原位 XAS 对于揭示材料的真实反应状态非常重要，因为 XANES 提供了关于氧化态、电子结构和未占据态密度的元素相关信息，而 EXAFS 揭示了局部配位环境（键距和配位数）的影响[53-55]。

参 考 文 献

[1] Yang J, Ji Y, Shao Q, et al. A universalstrategy tometal wavy nanowires for efficient electrochemical water splitting at pH-universal conditions. Advanced Functional Materials, 2018, 28 (41): 1803722.

[2] Yao Q, Huang B, Zhang N, et al. Channel-rich RuCu nanosheets for pH-universal overall Water splitting electrocatalysis. Angewandte Chemie International Edition, 2019, 58 (39): 13983-13988.

[3] Fu L, Yang F, Cheng G, et al. Ultrathin Ir nanowires as high-performance electrocatalysts for efficient water splitting in acidic media. Nanoscale, 2018, 10 (4): 1892-1897.

[4] Bu L, Ding J, Guo S, et al. A general method for multimetallic platinum alloy nanowires as highly active and stable oxygen reduction catalysts. Advanced Materials, 2015, 27 (44): 7204-7212.

[5] Bu L, Guo S, Zhang X, et al. Surface engineering of hierarchical platinum-cobalt nanowires for efficient electrocatalysis. Nature Communications, 2016, 7 (1): 11850.

[6] Jiang K, Shao Q, Zhao D, et al. Phase and composition tuning of 1D platinum-nickel nanostructures for highly efficient electrocatalysis. Advanced Functional Materials, 2017, 27 (28): 1700830.

[7] Jiang K, Zhao D, Guo S, et al. Efficient oxygen reduction catalysis by subnanometer Pt alloy nanowires. Science Advances, 2017, 3 (2): 1601706.

[8] Zhang N, Feng Y, Zhu X, et al. Superior bifunctional liquid fuel oxidation and oxygen reduction electrocatalysis enabled by PtNiPd core-shell nanowires. Advanced Materials, 2017, 29 (7): 1603774.

[9] Yang J, Shao Q, Huang B, et al. pH-universal water splitting catalyst: Ru-Ni nanosheet assemblies. Science, 2019, 11: 492-504.

[10] Jang K, Kim H J, Son S U. Low-temperature synthesis of ultrathin rhodium nanoplates via molecular orbital symmetry interaction between rhodium precursors. Chemistry of Materials, 2010, 22 (4): 1273-1275.

[11] Bu L, Zhang N, Guo S, et al. Biaxially strained PtPb/Pt core/shell nanoplate boosts oxygen reduction catalysis. Science, 2016, 354 (6318): 1410-1414.

[12] Zhang N, Shao Q, Pi Y, et al. Solvent-mediated shape tuning of well-defined rhodium nanocrystals for efficient electrochemical water splitting. Chemistry of Materials, 2017, 29 (11): 5009-5015.

[13] Huang X, Chen Y, Chiu C Y, et al. A versatile strategy to the selective synthesis of Cu nanocrystals and the in situ conversion to CuRu nanotubes. Nanoscale, 2013, 5 (14): 6284-6290.

[14] Duan H, Yan N, Yu R, et al. Ultrathin rhodium nanosheets. Nature Communications, 2014, 5 (1): 3093.

[15] Dai L, Zhao Y, Qin Q, et al. Carbon-monoxide-assisted synthesis of ultrathin PtCu alloy nanosheets and their enhanced catalysis. ChemNanoMat, 2016, 2 (8): 776-780.

[16] Lv H, Chen X, Xu D, et al. Ultrathin PdPt bimetallic nanowires with enhanced electrocatalytic performance for hydrogen evolution reaction. Applied Catalysis B: Environmental, 2018, 238: 525-532.

[17] Stelter M, Bombach H. Process optimization in copper electrorefining. Advanced Engineering Materials, 2004, 6 (7): 558-562.

[18] Li C, Iqbal M, Lin J, et al. Electrochemical deposition: an advanced approach for templated synthesis of nanoporous metal architectures. Accounts of Chemical Research, 2018, 51 (8): 1764-1773.

[19] Lu F, Zhou M, Zhou Y, et al. First-row transition metal based catalysts for the oxygen evolution reaction under alkaline conditions: basic principles and recent advances. Small, 2017, 13 (45): 1701931.

[20] Zhang Y, Xiao J, Lv Q, et al. Self-supported transition metal phosphide based electrodes as high-efficient water splitting cathodes. Frontiers of Chemical Science and Engineering, 2018, 12 (3): 494-508.

[21] Yadav J B, Park J W, Cho Y J, et al. Intermediate hydroxide enforced electrodeposited platinum film for hydrogen evolution reaction. International Journal of Hydrogen Energy, 2010, 35 (19): 10067-10072.
[22] Mabayoje O, Liu Y, Wang M, et al. Electrodeposition of MoS_x hydrogen evolution catalysts from sulfur-rich precursors. ACS Applied Materials & Interfaces, 2019, 11 (36): 32879-32886.
[23] Hu X, Liu S, Chen Y, et al. One-step electrodeposition fabrication of iron cobalt sulfide nanosheet arrays on Ni foam for high-performance asymmetric supercapacitors. Ionics, 2020, 26 (4): 2095-2106.
[24] Wu R, Xiao B, Gao Q, et al. A janus nickel cobalt phosphide catalyst for high-efficiency neutral-pH water splitting. Angewandte Chemie International Edition, 2018, 57 (47): 15445-15449.
[25] Chi J Q, Yan K L, Xiao Z, et al. Trimetallic NiFeCo selenides nanoparticles supported on carbon fiber cloth as efficient electrocatalyst for oxygen evolution reaction. International Journal of Hydrogen Energy, 2017, 42 (32): 20599-20607.
[26] Wang D, Yang L, Liu H, et al. Polyaniline-coated $Ru/Ni(OH)_2$ nanosheets for hydrogen evolution reaction over a wide pH range. Journal of Catalysis, 2019, 375: 249-256.
[27] Simka W, Puszczyk D, Nawrat G. Electrodeposition of metals from non-aqueous solutions. Electrochimica Acta, 2009, 54 (23): 5307-5319.
[28] Zhao Y, Vandernoot T J. Electrodeposition of aluminium from nonaqueous organic electrolytic systems and room temperature molten salts. Electrochimica Acta, 1997, 42 (1): 3-13.
[29] Srivastava M, Yoganandan G, Grips V K W. Electrodeposition of Ni and Co coatings from ionic liquid. Surface Engineering, 2012, 28 (6): 424-429.
[30] Zhang D, Mu Y, Li H, et al. A new method for electrodeposition of Al coatings from ionic liquids on AZ91D Mg alloy in air. RSC Advances, 2018, 8 (68): 39170-39176.
[31] Zhou F, Izgorodin A, Hocking R K, et al. Electrodeposited MnO_x films from ionic liquid for electrocatalytic water oxidation. Advanced Energy Materials, 2012, 2 (8): 1013-1021.
[32] Wang A L, Xu H, Li G R. NiCoFe layered triple hydroxides with porous structures as high-performance electrocatalysts for overall water splitting. ACS Energy Letters, 2016, 1 (2): 445-453.
[33] Pasquale M A, Gassa L M, Arvia A J. Copper electrodeposition from an acidic plating bath containing accelerating and inhibiting organic additives. Electrochimica Acta, 2008, 53 (20): 5891-5904.
[34] Duan Y, Yu Z Y, Hu S J, et al. Scaled-up synthesis of amorphous NiFeMo oxides and their rapid surface reconstruction for superior oxygen evolution catalysis. Angewandte Chemie International Edition, 2019, 58 (44): 15772-15777.
[35] Gao D, Guo J, Cui X, et al. Three-dimensional dendritic structures of NiCoMo as efficient electrocatalysts for the hydrogen evolution reaction. ACS Applied Materials & Interfaces, 2017, 9 (27): 22420-22431.
[36] Cao X, Jia D, Li D, et al. One-step co-electrodeposition of hierarchical radial Ni_xP nanospheres on Ni foam as highly active flexible electrodes for hydrogen evolution reaction and supercapacitor. Chemical Engineering Journal, 2018, 348: 310-318.
[37] Chen X, Liu B, Zhong C, et al. Ultrathin Co_3O_4 layers with large contact area on carbon fibers as high-performance electrode for flexible zinc-air Battery integrated with flexible display. Advanced Energy Materials, 2017, 7 (18): 1700779.

[38] Li T, Liu J, Song Y, et al. Photochemical solid-phase synthesis of platinum single atoms on nitrogen-doped carbon with high loading as bifunctional catalysts for hydrogen Eevolution and oxygen reduction reactions. ACS Catalysis, 2018, 8 (9): 8450-8458.
[39] Skompska M, Zarębska K. Electrodeposition of ZnO nanorod arrays on transparent conducting Substrates-a review. Electrochimica Acta, 2014, 127: 467-488.
[40] Dong Z, Lin F, Yao Y, et al. Crystalline Ni(OH)$_2$/amorphous NiMoO$_x$ mixed-catalyst with Pt-like performance for hydrogen production. Advanced Energy Materials, 2019, 9 (46): 1902703.
[41] Wang B, Zhao K, Yu Z, et al. In situ structural evolution of the multi-site alloy electrocatalyst to manipulate the intermediate for enhanced water oxidation reaction. Energy & Environmental Science, 2020, 13 (7): 2200-2208.
[42] Lee J K, Yi Y, Lee H J, et al. Electrocatalytic activity of Ni nanowires prepared by galvanic electrodeposition for hydrogen evolution reaction. Catalysis Today, 2009, 146 (1): 188-191.
[43] Matijević E. Preparation and properties of monodispersed colloidal metal hydrous oxides. Pure and Applied Chemistry, 1978, 50, 1193-1210.
[44] Bhardwaj N, Kundu S C. Electrospinning: a fascinating fiber fabrication technique. Biotechnology Advances, 2010, 28 (3): 325-347.
[45] Liang D, Hsiao B S, Chu B. Functional electrospun nanofibrous scaffolds for biomedical applications. Advanced Drug Delivery Reviews, 2007, 59 (14): 1392-1412.
[46] Gupta P, Elkins C, Long T E, et al. Electrospinning of linear homopolymers of poly (methyl methacrylate): exploring relationships between fiber formation, viscosity, molecular weight and concentration in a good solvent. Polymer, 2005, 46 (13): 4799-4810.
[47] Wang X, Um I C, Fang D, et al. Formation of water-resistant hyaluronic acid nanofibers by blowing-assisted electro-spinning and non-toxic post treatments. Polymer, 2005, 46 (13): 4853-4867.
[48] Guan Y, Feng Y, Wan J, et al. Ganoderma-like MoS$_2$/NiS$_2$ with single platinum atoms doping as an efficient and stable hydrogen evolution reaction catalyst. Small, 2018, 14 (27): 1800697.
[49] Fabbri E, Abbott D F, Nachtegaal M, et al. Operando X-ray absorption spectroscopy: A powerful tool toward water splitting catalyst development. Current Opinion in Electrochemistry, 2017, 5 (1): 20-26.
[50] Conder J, Bouchet R, Trabesinger S, et al. Direct observation of lithium polysulfides in lithium-sulfur batteries using operando X-ray diffraction. Nature Energy, 2017, 2 (6): 17069.
[51] Dou J, Sun Z, Opalade A A, et al. Operando chemistry of catalyst surfaces during catalysis. Chemical Society Reviews, 2017, 46 (7): 2001-2027.
[52] Handoko A D, Wei F, Jenndy, et al. Understanding heterogeneous electrocatalytic carbon dioxide reduction through operando techniques. Nature Catalysis, 2018, 1 (12): 922-934.
[53] Zheng Y, Jiao Y, Zhu Y, et al. High electrocatalytic hydrogen evolution activity of an anomalous ruthenium catalyst. Journal of the American Chemical Society, 2016, 138 (49): 16174-16181.
[54] Man I C, Su H Y, Calle Vallejo F, et al. Universality in oxygen evolution electrocatalysis on oxide surfaces. Chem Cat Chem, 2011, 3 (7): 1159-1165.
[55] Yu P, Wang F, Shifa T A, et al. Earth abundant materials beyond transition metal dichalcogenides: A focus on electrocatalyzing hydrogen evolution reaction. Nano Energy, 2019, 58: 244-276.

第四章

高效电催化剂的设计

由于电解水能耗高，目前在工业电流密度下，工业电解水通常在1.8V以上的高压下运行，远大于理论电压1.23V，原因是电解水的析氢反应动力学和析氧反应动力学缓慢导致过电位较大[1-3]。电化学水分解的高能量转化效率需要低成本高效率的电催化剂。因此，开发基于非贵金属和丰富过渡金属的高性能电极材料对于其工业化应用至关重要。工业用电解水的电催化剂一般应满足以下要求：①具有与贵金属电催化剂性能相当甚至更优的高催化活性；②电解槽运行数千小时以上稳定性优良；③具有良好的可扩展性和较低的商业制造成本。本章重点介绍了高效电催化剂的设计策略，例如化学掺杂、引入缺陷（空位）、物相工程、晶面工程、结构工程以及采用自支撑电极等。

4.1 化学掺杂

将杂原子引入地球储量丰富的过渡金属（TM）材料中，包括过渡氧化物、硫属元素化物、氮化物、硼化物和磷化物（TMX，X＝O、OH、S、Se、Te、N、P、C和B），以此通过掺杂剂和催化剂本体之间的能级杂化来提高材料的电催化性能[4]。近年来，出现了几种有效的方法来改善TMX表面的电催化剂活性位点的密度。通常化学掺杂涉及对能带结构、d带中心、活性位点的价态和电荷重新分布进行工程设计，以优化中间体的形成能，从而促进H_2和O_2的释放。对材料进行非金属掺杂，可以促进TM基电催化剂的催化水分解性能[5-8]。

Zhang等人用数值模拟的方法研究了氮掺杂Ni-Co磷化物对费米能级和电子电导率的控制作用，发现了能级匹配和富活性中心的协同作用使材料在碱性介质中持续催化100h仍具有优异的电催化水分解性能[5]。Wu等报道了一种以聚钼酸和硝酸镍为金属源制备P掺杂$NiMo_4N_5$晶体阵列的有效方法，得到了具有高比表面积的活性物质，在1.59V和1.66V的低电压下，电解水的电流密度分别接近$50mA/cm^2$和$100mA/cm^2$[6]。此外，通过与其单掺杂对应物协同耦合来引入多个非金属（B、N、S）杂原子，是杂化复合材料内部产生快速界面电荷转移的良好方法[9]。

除了非金属掺杂以外，将过渡金属原子引入TMX也能改善电催化性能。Wu等报道了以过渡金属氧化物（Fe_3O_2、NiO、Cr_2O_3）为前驱体，

在双氰胺和 $(NH_4)_3PO_4$ 的混合物中，通过一步加热法大规模制备 Cr 掺杂 FeNi-P 纳米颗粒（图 4.1）。经过优化的 Cr 掺杂 FeNi-P/NCN 纳米片在 HER 和 OER 中均具有优异的催化活性，全电解水体系电流密度 $10mA/cm^2$ 时电池电压为 1.5V，电解 20h 性能依然很稳定，与贵金属 Pt/C 和 RuO_2 组成的全水解体系相当。该体系电催化性能的提高可以归因于 Cr 的掺杂增加了活性位点的数量且适当改善了表面电子结构。理论计算证实，掺杂金属确实可以优化相对吉布斯吸附能，降低理论过电位，与实验得到的结果一致[9]。

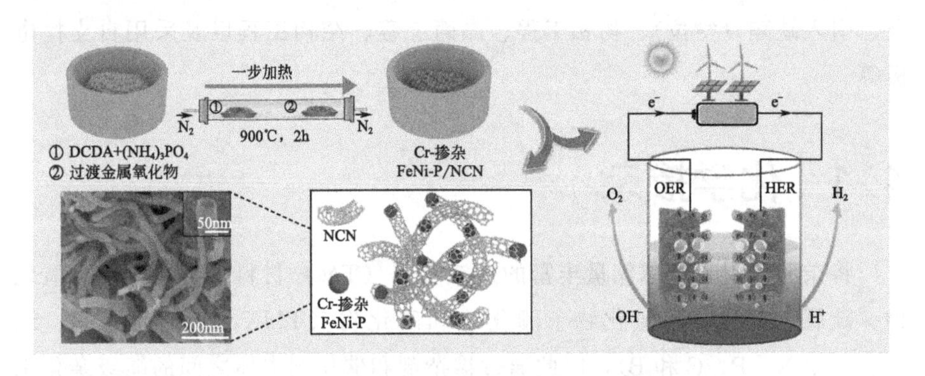

图 4.1　Cr 掺杂 FeNi-P/NCN 的合成路线示意图[9]

2019 年，Li 等人在 Ir 掺杂的 $NiV(OH)_2$ 表面分别设计了 HER 和 OER 的催化活性位点。由于这种化学掺杂，合成了具有较高催化活性的电解水催化剂（$10mA/cm^2$ 时电压为 1.49V）。在该催化剂中 Ir 离子的加入降低了相邻桥氧上的电荷密度来捕获 H^* 中间体，同时增加 V 离子上的电荷密度来吸引 O^* 中间体[10]。Sun 等人采用气相盐渍化技术，通过拓扑转变处理，将 Co 和 Fe 共掺杂到 $NiSe_2$ 多孔纳米片中[11]。在此过程中，Fe 和 Co 原子被引入到 $NiSe_2$ 晶格中，产生畸变原子排列。$NiSe_2$ 纳米片中 Fe 和 Co 原子的均匀分布不仅对电子结构有较大影响，同时还暴露了更多的活性位点，改善了电荷转移性能。因此，优化后的 $Fe_{0.09}Co_{0.13}$-$NiSe_2$/CFC 催化剂在碱性电解液中具有优异的 OER 和 HER 催化性能，在 $10mA/cm^2$ 下催化 OER 过电位为 251mV，催化 HER 过电位为 92mV，全水解电池电压为 1.52V。此外，Zhang 课题组以 $Ni(NO_3)_2 \cdot 6H_2O$、$Fe(NO_3)_2 \cdot 9H_2O$ 和 $RhCl_3 \cdot xH_2$ 为前驱体和阳离子掺杂剂，采用水热法合成了铑（Rh）掺杂的 NiFe-LDH 杂化物[12]。通过 X 射线吸收光谱（XAS）对 Rh/NiFeRh-

LDH 的结构进行了表征，证明存在 Rh 金属团簇，并得到了相应的杂化物。因此，设计 Rh 掺杂并取代催化位点，通过从本质上促进水的吸附和解离动力学来平衡 $NiFeO_xH_y$ 上的强 O_{Had}吸附，这对于碱性条件下全电解水体系催化至关重要。

4.2 缺陷（空位）工程

可以通过引入缺陷或空位改变电子结构来实现 TMX 的插入。为了降低过电位以改善反应效率，提高 TMX 的全电解水催化本征性能，最节能的缺陷是双功能电催化剂上的活性位点，即催化反应发生的地方[13,14]。例如，TMX 双功能电催化剂的催化性能可以通过界面耦合、酸蚀[14]和电子结构调控 TMX 材料晶格而得到显著增强[15]。钴铁氧化物能带内的氧空位缺陷态可以用来调节 H_2O 的吸附能，并且其 H 吸附最小值接近 0eV[16]。最早采用一步水热法对泡沫镍表面的超薄层状二氧化锰（δ-MnO_2）纳米片进行实验[13]。合成的样品在碱性溶液中显示出优异的 HER 和 OER 性能。理论和实验研究证实，不饱和配位的 Mn^{3+} 位点周围过多的电荷密度与大量的 O 空位有关，这些空位具有半金属性，促进了 H_2O 的吸附。另外，通过 Se 诱导的溶剂热处理泡沫镍，制备了富缺陷的 Se-$(NiCo)S_x/(OH)_x$ 纳米片 3D 杂化体。结果表明，Se 的掺入可有效地调控（NiCo)S/OH，形成结构缺陷和晶格畸变，大大增强电催化活性：电流密度为 $10mA/cm^2$，OER 和 HER 的过电位分别为 103mV 和 155mV，在 1.0mol/L KOH 电解液中，全电解水分解可稳定运行约 66h[17]。

Li 等人采用两步共沉淀法和退火工艺制备了具有丰富 S 空位的 FeS_2/CoS_2 界面纳米片[18]，以此研究了界面相互作用。他们发现产物具有强的电子顺磁共振（EPR）信号（$g=2.007$），表明 S 空位比前驱体多，这揭示了异质结构中层间相互作用导致硫空位缺陷。该材料具有良好的全电解水催化性能，电流密度 $10mA/cm^2$ 时只需要 1.47V 的电压。这归因于有缺陷的 MoS_2/NiS_2 异质多孔纳米片具有更多的电活性位点，更高的电子电导率，更有利于气体释放的纳米片表面。在电解水分解过程中，施加 1.59V 的电压就可获得 $10mA/cm^2$ 催化电流密度[19]。双功能电催化剂一般直接用作自支撑电极，不需要黏结剂和导电碳材料，因此在碱性环境中可以有效降低活性表面的损失，降低阻抗及改善溶液与电极之间的传质。另外，采用两步水热法合成了超薄的 Ni_3S_2/MnS_2 银耳状纳米片，再利用电氧化方式制备成富含氧空位的金属氧化物（图 4.2），增加了活性位点数和每个活性位点的本征活性[20]，在碱性条件下全电解水催化性能优异，当电流密度为 10mA/cm^2 时，只需 1.54V 的电压。

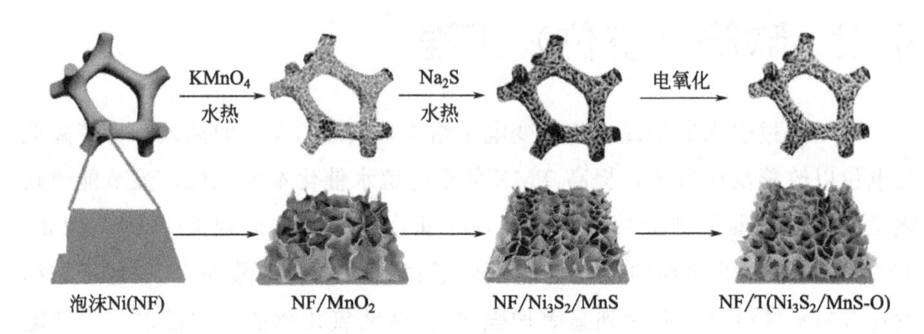

图 4.2 在泡沫镍上原位合成 NF/T（Ni_3S_2/MnS-O）的路线图[20]

下面简单介绍构建电催化剂中的阴离子空位的方法。

4.2.1 化学还原和电化学还原法

化学还原是通过还原剂将阴离子空位引入电催化剂的一种较为简便的方法，这些还原剂通常分为三类：气体（H_2、NH_3）、液体（$NaBH_4$、N_2H_4）和固体（Al、Zn）。例如，Chi 等人在 400℃下通过 H_2 和 NH_3 还原合成了富含 O 空位的 NF/$HCoMoO_4$[21]。Wei 等人利用 $NaBH_4$ 作为还原剂制备了富含 S 空位的 MoS_2[22]。在热退火过程中，MoS_2 中的 S 原子以 H_2S 和 Na_2S 形式被去除，从而形成 S 空位。Song 等人采用 Al 还原方法，在真空双区炉中用 Al 粉对 TiO_2 进行退火，在 TiO_2 中生成 O 空位[23]。与化学还原法相比，电化学还原法可以在温和的环境生成阴离子空位，并且可以通过调节电化学还原电流和时间来控制阴离子空位的数量[24]。例如，Zhang 课题组采用线性扫描伏安法（LSV）对 In_2O_3 进行了重构，合成了含有少量 O 空位的 In/In_2O_{3-x}[25]。

4.2.2 等离子处理

等离子体可以与电催化剂表面的原子发生相互作用，能量从高能离子转移到表面原子，打破共价键，导致表面原子射出从而形成空位。例如用 Ar 等离子体合成了含 S 空位的缺陷 Co_3S_4。在 Ar 等离子体处理过程中，高能电离的 Ar^+ 攻击 Co_3S_4 表面，使表面的 Co—S 键断裂，进而释放 S 原子产生 S 空位。除了 Ar 等离子体，H_2 和 NH_3 等离子体处理可以吸引气体自由基与表面原子的反应，有效地产生空位。如 Yan 等人利用 H_2 等离子体在 TiO_2 中获得了丰富的 O 空位[26]。

4.2.3 其他方法

除了上述合成方法外，还有一些方法可以在电催化剂中产生阴离子空位，如热处理和低价金属掺杂。例如，Bao 等人通过 Ni-Co/Co 氢氧化物在空气和 O_2 气氛中热转化，分别合成了富 O 空位和贫 O 空位的超薄 $NiCo_2O_4$ 纳米片[27]。Jain 等人通过掺杂过渡金属原子将 Se 空位引入 $MoSe_2$[28]。由于与富电子过渡金属掺杂剂相邻的 Se 原子的电子不稳定与金属—Se 键对氢吸附的削弱共同作用，使得在热力学上更容易形成 Se 空位。

阴离子空位的表征对研究阴离子空位在电解水催化剂中的作用很重要，表征手段一般有：显微镜表征，包括高分辨球差校正透射电子显微镜（AC-TEM）、光元素敏感环形亮场扫描透射电镜（ABF-STEM）和原子分辨环形暗场扫描透射电子显微镜（ADF-STEM）；光谱表征包括 X 射线光电子能谱（XPS）、X 射线吸收光谱（XAS）以及正电子湮没光谱（PAS）、电子顺磁共振（EPR）和电子自旋共振（ESR）光谱等。

离子空位的存在改变了催化剂的内在化学性质和物理性质，在提高电解水催化剂的性能方面起到了重要作用，比如增强催化剂内在活性、增加催化剂的活性位点、改善电导率以及提升稳定性等。通过各种先进的表征技术与理论研究相结合，初步建立了空位与催化剂活性之间的关系，有助于设计和合成高效的电催化剂。尽管电催化剂的空位工程已经取得了重要的研究进展，但是仍然存在一系列的挑战。首先，因为空位可能随时间动态变化，因此在反应过程中应该利用原位光谱表征来揭示更多关于离子空位稳定性、重分布和迁移率的细节；其次，需要先进的表征技术来深入了解离子空位的数量和分布，这有助于阴离子空位电催化剂的可控设计；最后，将理论模拟与实验分析相结合，阐明构效关系，为合理设计离子空位工程来制备高性能电催化剂提供理论指导。

4.3 物相工程

晶体结构显著影响 TM 基电催化剂的催化性能。一般来说，由于 1T MoS_2 中 Mo 的 4d 轨道不完全填充，TMS 的金属 1T 或金属单斜 1T′相比 2H 相，具有更好的电荷传输能力。因此提高了电极的导电性，使电催化剂电极动力学反应更容易，并增加了催化活性位点，更有利于提升 HER 的催化效率[29-32]。但其热力学不稳定性阻碍了其在实际电解水中的应用，可以通过物相工程来解决这一问题。比如通过使用两步溶剂热法制备杂化纳米管阵列作为双功能电催化剂[30]，表征结果表明：其晶体结构主要由金属单斜 1T′相 MoS_2 和 Fe、Co、Ni 基硫化物组成。其全电解水电流密度为 $10mA/cm^2$ 时只需要施加 1.429V 的电压。基于异位和原位同步辐射的扩展 X 射线吸收精细结构（EXAFS）表征和 HRTEM 图像表明，$(Co、Fe、Ni)_9S_8$ 向 MoS_2 的电荷转移引起的相变和 OER 电催化反应中三金属离子之间本征相互作用均协同改善了电催化水分解性能。Wei 等通过对铱（Ir）的吸附研究了 MoS_2 纳米片表面的相变[31]。密度泛函理论（DFT）计算表明，当吸附 Ir 原子浓度大于 20%时，Ir/2H-Mo_2 转化为更稳定的 Ir/1T-Mo_2。

影响电解水催化剂效率的因素包括活化基面、表面亲水性、大量异质结界面以及耦合效应。Oh 等人合成了由 $La_{0.5}Sr_{0.5}CoO_{3-\delta}$和 $MoSe_2$ 组成的异质结双功能电催化水分解催化剂，发现相变过程改善了其催化活性；Co 和 Mo 之间的电子转移诱导 $MoSe_2$ 发生局部相变，增加了 $La_{0.5}Sr_{0.5}CoO_{3-\delta}$中 Co 阳离子的部分氧化。该催化剂在 $100mA/cm^2$ 的高电流密度下催化水电解 1000h 仍然具有良好的稳定性，甚至超过贵金属铂和氧化铱电对[32]。此外，Yang 等利用超分子凝胶的结构优点和可控热转化技术，采用两步法策略，相控合成了 P 掺杂碳（PC），与 P 掺杂石墨烯（PG）复合后，包裹混合相磷化钴纳米晶，得到 CoP-Co_2P@PC/PG 复合材料，该材料具有优异的电催化水分解性能[33]。此外，还报道了 TMS 的相控合成和组成控制合成，并系统地比较了它们电解水分解反应的催化性能[34,35]。最近，采用多元醇溶液法制备了一系列 NiS_x（即 NiS、Ni_3S_2、NiS_2）纳米晶[35]。采用密度泛函理论计算了六方 Ni_3S_2 纳米颗粒吸附氢的自由能，其 ΔG_{H^*} 最接近催化剂 H* 的理想值（0eV）。

4.4 晶面工程

过渡金属基材料的晶面工程已成为微调物理化学性质的重要策略，从而优化双功能电催化剂最具催化活性的晶面密度[36-43]。然而晶面效应对电解水催化剂性能影响的研究较少，因为高表面能和暴露的晶面会使结构不稳定，从而导致催化活性不理想。

为了提高双功能电催化剂的性能，最近实现了一种晶面工程化过渡金属材料催化电解水的策略。2015 年发表了第一篇以硫脲为前驱体，采用水热法直接硫化泡沫镍合成高指数晶面 Ni_3S_2 纳米片阵列作为 HER 和 OER 催化剂的报道，该催化剂催化电解水的法拉第效率接近 100%，且稳定性良好[36]。Li 等人用 DFT 计算了 Ni_3S_2 非对称锯齿结构上氢吸附的最佳吉布斯自由能，发现（$\bar{1}11$）面表现出金属行为；另外，Ni 位点具有比 S 位点更小的 $\Delta G(H^*)$ 值，因此具有更好的催化活性，是催化 HER 的活性位点。这一结果被进一步证实：在 Na_2S 溶液中通过一步水热法处理泡沫镍合成具有（$\bar{1}11$）面的 Ni_3S_2 3D 纳米结构，该材料同样具有优异的电催化水分解性能[37]。Dong 等人报道了在镍箔（NTFs）上原位生长（003）晶面暴露的 Ni_3S_2 纳米多孔薄膜，在碱性电解水体系中表现出优异的性能[38]。优化后的 Ni_3S_2-360 NTFs 具有亲水-疏气表面，在 10mA/cm^2 电流密度下保持 30h 仍具有良好的稳定性（图 4.3）。

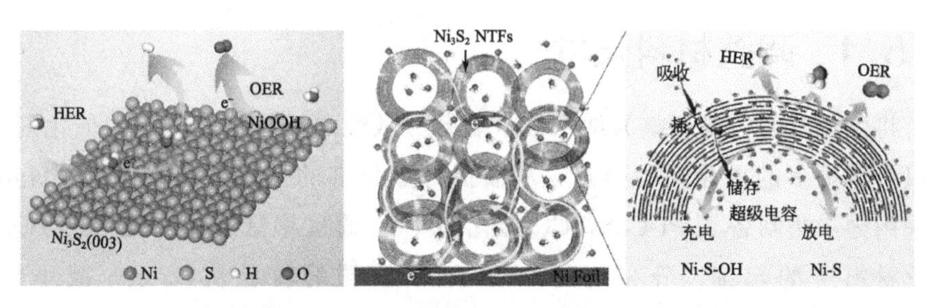

(a) Ni_3S_2(003)面的HER机理　(b) 自支撑Ni_3S_2 NTFs电极中电解质分子和电子传递的示意图(左)，以及纳米孔中3种能量转换和储存位点(右)

图 4.3

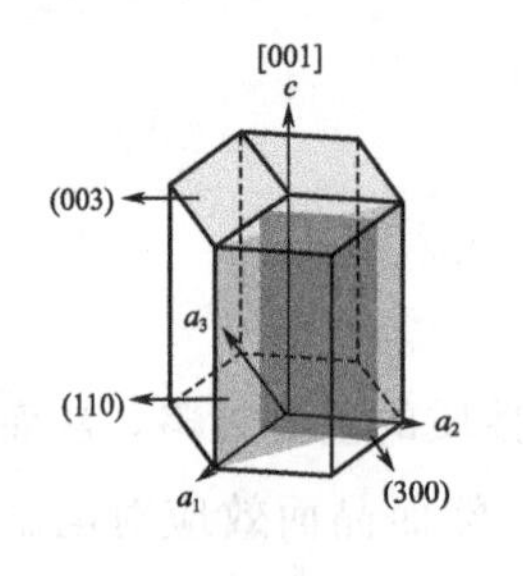

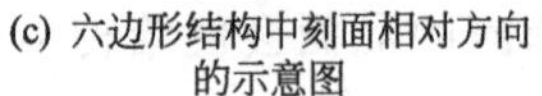
(c) 六边形结构中刻面相对方向的示意图

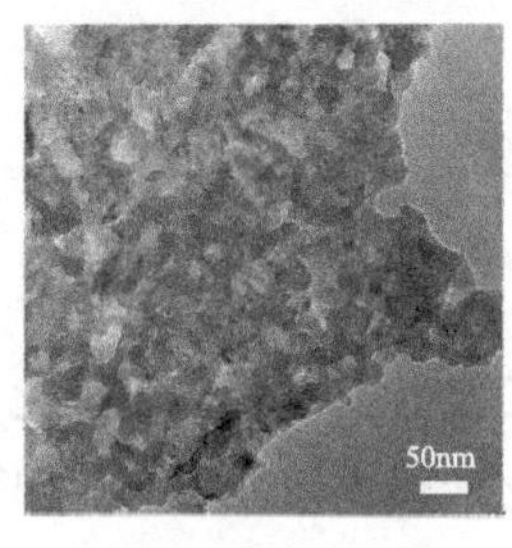

(d) Ni_3S_2-360的TEM图

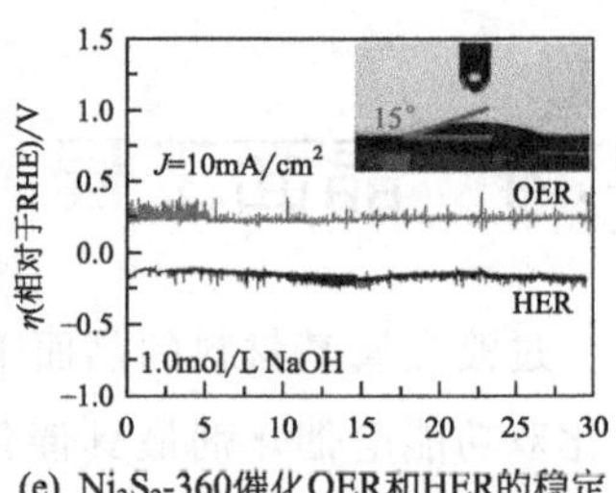

(e) Ni_3S_2-360催化OER和HER的稳定性测试插图为OER稳定性测试后的CA图

图 4.3 Ni_3S_2-360 NTFs 相关的机理图和性能测试图[38]

此外，Zhang 等人利用热分解法制备了四种不同晶面的 Co_3O_4 晶体，电催化活性的顺序为 {111}>{112}>{110}>{001}[39]，在 {111} 晶面上的有效电荷转移优于其他三个晶面。理论结果进一步表明，{111}||{111} 耦合电极的优异性能可归因于 {111} 晶面的协同作用，该晶面具有最高的表面能、较大的悬挂键密度和最小的 ΔG_{H^*} 绝对值。

4.5 结构工程

近年来对电解水催化剂的结构调控已有广泛的研究，出现了许多提高过渡金属基材料催化性能的可行方法[44-48]。总体可以分为四类：调整材料尺寸[49-51]、组分调制[52]、修饰表面/界面（杂交/异质结构和分级结构）[53-55]以及导电载体工程[56,57]。

4.5.1 调整材料尺寸

将催化剂的粒径调整为纳米级，制备分散性好、粒径分布窄的催化剂，可以暴露更多的催化活性位点，从而提升催化剂的催化性能。比如，Han 等采用分步法制备了 Pt-CoS_2/CC 催化剂[49]。通过将修饰在多孔 CoS_2/CC 杂化材料中的超细、分布均匀的 Pt 纳米颗粒的粒径从 2.8nm 减小到 1.7nm，增大了表面粗糙度，促进了气泡与电极表面的快速分离，有利于反应物的吸附和活化，提升了材料的电催化活性。在其他研究中也有类似的报道[50,51]。Li 等人通过简单的水热反应和后续磷化处理合成了分级结构的

CoP/CoP_2复合纳米颗粒。CoP/CoP_2复合纳米颗粒（12.9±1.6）nm 较 CoP 和 Co 纳米颗粒表现出更高的催化活性，在 $10mA/cm^2$ 处的电压低至 1.65V。结果表明，双活性 CoP/CoP_2 纳米粒子具有尺寸小、尺寸分布均匀等特点，显著促进了电荷转移过程，三维花状球状支撑体提供了合适的比表面积，阻止了粒子的团聚，进一步提高了对 HER 和 OER 的电催化性能。

4.5.2 组分调制

选择不同组成的前驱体，通过不同组分之间的调控作用使表面结构重新排列，进而影响电催化水分解催化剂的活性和选择性。2018 年，Menezes 等人首次使用亚磷酸镍合成了一类结构独特的磷基无机材料[52]。采用六水氯化镍（$NiCl_2 \cdot 6H_2O$）和磷酸二氢铵（$NH_4H_2PO_4$，作为参与反应的亚磷酸阴离子）为前驱体，通过温和的水热法合成了 $Ni_{11}(HPO_3)_8(OH)_6/NF$ 催化剂。该材料在碱性介质中仅需 1.6V 的电压就实现了 $10mA/cm^2$ 的电流密度，经过四天的电解水测试，其催化活性没有任何降低。亚磷酸镍之所以具有显著的活性和稳定性，是因为 OER 过程中在催化剂的近表面形成了氧化镍物种，并且在二价镍离子辅助下，亚磷酸根离子作为 HER 的催化活性位点。

4.5.3 修饰表面/界面

富含界面的过渡金属基复合材料中杂化或分级结构的表面/界面工程，通过活性中心的电子和几何变化以及组分间紧密界面相互作用的协同作用，对含氢和含氧活性中间体具有良好吸附能力，从而提升电解水催化剂的催化活性。例如，Xiong 等人通过简单的两步原位水热转化过程，在泡沫镍上合成了 2D 原子级的 Ni_3S_2/MnO_2 双相纳米阵列[53]。在碱性电解液中，与纯的 NF-Ni_3S_2 和 NF-$Ni(OH)_2/MnO_2$ 相比，NF-Ni_3S_2/MnO_2 表现出更好的电催化性能。采用 DFT 方法计算了 NF-Ni_3S_2/MnO_2 的杂化，发现分级非均相界面具有大量的活性位点，有利于水分子的吸附和裂解，进而提升了碱性条件下的催化活性。以 $CoSO_4 \cdot 7H_2O$、$Na_2MoO_4 \cdot 2H_2O$、CH_4N_2S 为原料，采用一锅水热法合成了 $MoS_2/Co_9S_8/Ni_3S_2/Ni$ 的分级纳米材料[54]。采用 DFT 计算 OER 时 MoS_2 层平行和垂直于 Co 端 Co_9S_8 的两种界面结构，发现在界面上形成了 Co—S 键，反应中间体的结合能发生变化，过电位降低，促使 Co_9S_8 电子转移到 MoS_2。研究人员还设计了多层 CoMo-NiS-NF-31 电催化剂，含有表面覆盖了 Ni_3S_2 纳米棒的三维多孔导电泡沫镍，然后再修饰 2D 超薄 MoS_2 和 Co_9S_8 纳米片（图 4.4，见文后彩插）。该材料具有丰富的活性位点，因此表现出优异的电催化性能，在酸性、中性和碱性条件下进行全电解水，电流密度为 10mA/cm² 时分别仅需 1.45V、1.80V 和 1.54V 的电压。

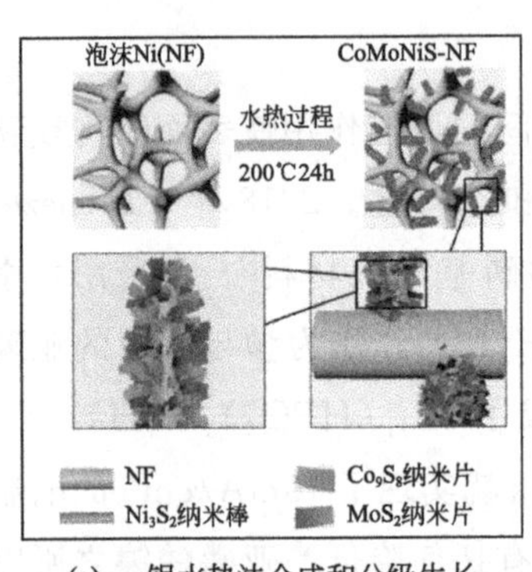

(a) 一锅水热法合成和分级生长 CoMoNiS-NF-*xy*复合材料的示意图

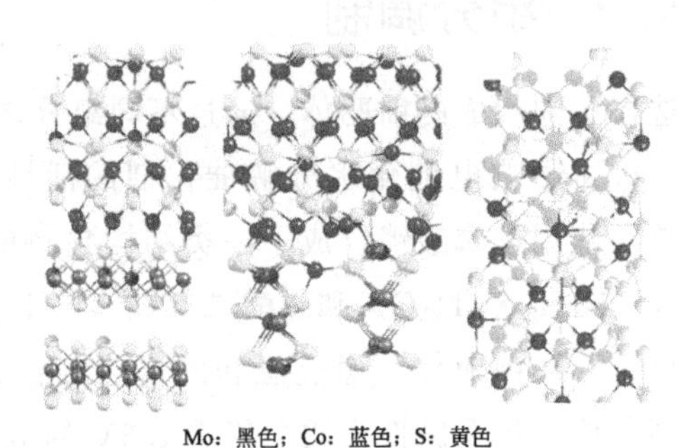

(b) DFT优化的MoS_2/Co_9S_8界面结构，左为MoS_2(001)平行于Co_9S_8(001)；中为MoS_2(001)垂直于Co_9S_8(001)；右为DFT计算吸附O*、OH*和OOH*的Co_9S_8(001)表面位点的顶视图

图 4.4 CoMoNiS-*xy* 的复合材料的合成及 DFT 计算结构[54]

4.5.4 导电载体工程

导电载体工程也是提高 HER/OER 催化性能的一个很有前景的方法。例如，Le 等人通过湿化学途径设计 2D 过渡金属碳化物 MXenes 的多孔导电支撑骨架，大大增强了电解水性能。催化性能的提升可归因于支撑材料的多孔表面与大量宏观/中/微孔的协同作用，不仅可以提供高密度分布的活性位点，而且可以有效缩短离子/传质迁移途径，电解质更易进入催化剂中[56]。

4.6 1D/2D 纳米结构自支撑电极

人们一直致力于开发涂覆型粉状电催化剂，这些催化剂需要用聚合物黏结剂固定在导电基底上，通常使用的黏结剂为 Nafion，导电基底一般有玻碳电极、碳纸、泡沫镍等。但黏结剂的引入会覆盖催化剂的活性位点，增加电子转移和质量传质阻力[1]。此外，工业电解水通常在高电流密度（约 500mA/cm^2）下长时间（数月甚至数年）运行，导致电催化剂受到大量气泡的冲刷。由于导电基底与电催化剂之间的结合力较低，在如此苛刻的条件下长时间运行后，涂层电催化剂往往会从导电集流体中脱落，导致性能下降[3]；另外由于结合力较低，涂层电催化剂的负载质量往往受到限制。而采用自支撑电极可以很好地解决粉末催化剂存在的缺点，将催化活性物质直接生长在导电基底上，不需要使用绝缘聚合物黏结剂和炭导电剂[2]。近年来，自支撑电催化剂的制备受到了广泛的关注，其具有以下优点：首先，该方法无需聚合物黏结剂和引入额外的导电剂，简化了电极制备过程，降低了生产成本[1,19]；其次，该方法使基底与活性物质紧密相连，提高了电子导电性，防止了在催化过程中催化材料的脱落。再次，电催化剂在基底上的负载质量可以通过合成条件来调节[2]。最后，通过对活性物质的形貌、组成和结构进行调整，更容易得到具有优异电化学性能和物理性能的催化剂[1]。因此自支撑电极催化剂在催化电解水的过程中活性和稳定性都有显著的提高。

调整电催化剂的形貌和结构是提高电催化剂催化活性的有效途径。通常，材料维度分为三类：零维（0D 纳米颗粒和纳米团簇）、一维（1D，纳米线、纳米纤维、纳米带、纳米棒和纳米管）和二维（2D，纳米片、纳米锥和纳米壁）[58]。一般来说，与 0D 纳米结构相比，1D 和 2D 纳米结构具有更好的电化学性能。因此，具有一维或二维纳米结构的活性材料作为高活性的电催化剂被广泛研究。研究证实，1D/2D 纳米结构通过暴露更多的活性位点和加强传质过程来提高电催化效率[58]。另外这些 1D/2D 纳米结构还具有高的比表面积和较大的表面缺陷，有利于表面反应。而且，相邻的 1D/2D 纳米结构间存在丰富的空隙，这也有利于电解液的渗透和气泡的释放。另外，1D/2D 纳米结构材料直接生长在导电基底表面较为容易，这种原位生长的材料对电荷转移至关重要。因此开发 1D/2D 纳米结构的自支撑电催化剂对于提高材料电解水的催化活性具有重要意义。1D/2D 纳米结构的电催化剂一般包括纳米阵列、分级纳米结构和中空纳米结构。

典型的 1D 纳米阵列主要包括原位生长在导电基底上的纳米棒、纳米线、纳米针等。典型的 2D 纳米阵列主要是原位生长在导电基底上的纳米片、纳米锥、纳米壁等。这些 1D/2D 纳米阵列为催化剂提供了许多优势，比如：①这些原位生长的纳米阵列可以提供更大的比表面积，从而提供更多的活性位点；②可以提供定向的电子传输途径；③阵列间的间隙有利于电解质的扩散和气泡的释放。

大多数分级结构的纳米阵列可以通过调整合成条件，由1D/2D纳米结构组装而成。对于1D纳米结构，其分级结构可以分为不同的种类：核/壳结构以及负载有纳米颗粒的1D结构。对于2D分级结构阵列，一般分为核/壳二维结构、负载有纳米颗粒的二维结构以及堆叠结构。分级结构的1D/2D纳米结构保持了1D/2D纳米结构阵列的优点，另外，一级结构和二级结构之间的强相互作将产生界面效应，可改善电化学反应动力学。

大量研究表明，中空的1D/2D纳米结构会显著提高催化活性。通常情况下，采用有针对牺牲纳米级材料模板的方法来制备中空纳米结构。合成方法可分为两类：①模板辅助法；②无模板法。对于自支撑空心1D/2D纳米结构电催化剂，其空心结构提供了更多可用的活性表面，有利于提高催化剂的活性。

4.6.1 自支撑电极作为HER催化剂

4.6.1.1 一般的1D/2D纳米阵列

目前，已经研发了很多自支撑电极的合成方法，包括：电沉积、水热/溶剂热、气相沉积、真空抽滤等。由于电催化水分解条件苛刻，常见的导电基底主要集中在耐碱或耐酸的金属和碳上，例如碳布（CC)、泡沫金属[例如泡沫镍（NF)]、金属网（例如钢网）、金属板/箔（如钛板）等，这些导电基底具有较高的机械强度和柔韧性，适合于工业应用。大量研究证实，地球上含量丰富的过渡金属，如磷化物、硫化物、硒化物等，在酸性和碱性溶液中均表现出较高的HER活性。2014年，Sun等通过对Co(OH)F/CC前驱体磷化处理，在碳布上原位生长磷化钴纳米线阵列（CoP/CC NWs)[59]。CoP/CC NWs阵列在pH 0～14范围内表现出高HER催化活性，并具有优异的稳定性。同样通过低温磷化α-$Co(OH)_2$前驱体在碳布上制备了CoP 2D纳米片阵列。CoP纳米片在酸碱性条件下同样具有优异的HER催化活性，并且其催化活性还比CoP纳米线略大。此外，采用类似的制备工艺可以在Ti板上原位生长CoP纳米片阵列以及其他过渡金属磷化物如FeP纳米线阵列[60]，在泡沫铜上原位生长Cu_3P纳米线等[61]。

阴阳离子的掺杂可以优化氢吸附的自由能，是一种调整电催化剂电子结构的有效方法[62,63]，目前有许多工作研究了杂原子掺杂对电催化剂性能的影响。2016 年，Sun 等人在碳布上制备了 $Fe_xCo_{1-x}P$ 纳米线阵列（$Fe_xCo_{1-x}P$/CC)[63]。优化后的 $Fe_{0.5}Co_{0.5}P$/CC 纳米线阵列的催化活性远远优于 CoP/CC 纳米线阵列，$Fe_{0.5}Co_{0.5}P$/CC 电极在 0.5mol/L H_2SO_4 中表现出类似 Pt 的 HER 催化性能，最低起始电位接近商用 Pt/C。密度泛函理论（DFT）计算表明，CoP 中 Fe 取代 Co 会削弱 H-Co 结合强度，产生有利的氢吸附自由能。另外还在 Ti 基底上生长了 Zn 掺杂的 CoP 纳米壁阵列[64]、Mn 掺杂的 CoP 纳米片阵列[65]，在泡沫镍上生长了 Mo 掺杂的 Ni_2P 纳米线阵列[66]。在电催化剂中引入金属和非金属都对催化剂的活性产生了显著的影响，有利于提高 HER 活性。在 H_2/PH_3 气氛下进行磷化得到 O 掺杂的 NiCoP 纳米线阵列，DFT 计算表明 O 的掺入确实减弱了 NiCoP 的表面 H 吸收，增强了其对水的吸附能力，从而获得了更好的 HER 活性。在碱性溶液中，电流密度 $10mA/cm^2$ 时 O 掺杂 NiCoP 纳米线阵列表现出极低的过电位 44mV，Tafel 斜率 38.5mV/dec。

一般来说，通过将一个组分与另一个组分结合形成异质结构，将形成界面独特的电子性质，促使界面电荷极化进而与反应中间体发生键合，大大增强催化动力学[67-69]。因此，设计具有丰富异质结构的 1D/2D 纳米材料以提高催化活性备受关注。采用湿化学-水热法结合随后再原位磷化反应，在碳毡上合成了自支撑的 $NiCo_2P_x$ 纳米线阵列，所得 $NiCo_2P_x$ 纳米线在 pH 为 0～14 范围内具有优异的 HER 电催化性能。而且，$NiCo_2P_x$ 在所有电解质中连续电解 30h 后均表现出优异的稳定性。实验研究表明，表面 Ni 位点会促进水解离，而表面 Co 位点会促进氢气的生成和释放，这种协同作用显著提高了 HER 的催化活性。

提高催化活性的另一个有效途径是降低活性物质与基底之间的界面电阻，改善从活性位点到导电基底的电子转移。采用原位生长催化剂的方法，有效地避免了使用额外的导电剂和黏结剂，从而降低了电子传递电阻，提高了催化剂稳定性[70]。例如，通过一步溶剂热法直接磷化泡沫镍制备了自支撑的 Ni_2P 纳米棒，在 $10mA/cm^2$ 的电流密度下表现出了显著的 HER 性能，其过电位为 131mV[71]。此外，利用磷蒸气直接磷化泡沫镍制备了泡沫镍负载的双相 Ni_5P_4-Ni_2P 纳米片阵列，同样表现出优异的催化活性，但该方法受到相应金属基底的限制[72]。

除了过渡金属磷化物外，过渡金属硫化物、硒化物、氮化物也表现出与磷化物类似的催化活性，如 MoS_2 纳米片阵列/CC[73,74]、$MoSe_2$ 纳米片阵列/CC[75]、$CoSe_2$ 纳米针阵列/Ti[76]、WN 纳米线阵列/CC[77]等。举一个特别的例子，Hou 等以 Mo 网格为 Mo 源，采用简单的方法在金属 Mo 片上合成了 MoS_2 纳米片阵列[78]。得到的 MoS_2/Mo 纳米片中，Mo 基底与 MoS_2 纳米片之间的紧密接触，同时电解液的快速扩散等优势，使得 MoS_2/Mo 电极具有优异的催化活性和稳定性。Liu 和同事在泡沫镍上制备了 Mn 掺杂的 NiS_2 纳米片阵列，其性能优于纯 NiS_2[79]。Nagaraja 等人报道了金属硒和水合肼直接硒化泡沫镍制备 NiSe 纳米片的方法[80]，特别是 Zhang 等人通过 NiCo LDH 的拓扑转变和酸蚀，在泡沫镍上制备了具有介孔结构的自支撑 $Ni_{1-x}Co_xSe_2$ 纳米片阵列[81]。介孔结构的 $Ni_{1-x}Co_xSe_2$ 纳米片表面积增大，其 HER 性能远远高于非介孔结构的 $Ni_{1-x}Co_xSe_2$ 纳米片阵列。另外电催化剂的效率还受材料润湿性的影响，为此，制备了具有良好润湿性的 MoS_2 纳米片[82]。实验结果表明，“超疏气”表面的构建能够使电极表面形成的气泡按一个数量级释放，并快速去除气泡以恒定材料的工作面积。

4.6.1.2 分级结构 1D/2D 纳米结构

分级结构 1D/2D 纳米结构主要包括以下几种：核/壳 1D 纳米结构、带颗粒的 1D 纳米结构、核/壳 2D 纳米结构、带颗粒的 2D 纳米结构、堆叠 2D 纳米结构等。

用于 HER 催化的核/壳 1D 纳米结构的各种材料得到了广泛的研究。2015 年，Chen 等人通过两步水热反应制备了 MoS_2 纳米片包覆的 CoS_2 纳米线阵列[83]，该材料的高电化学活性面积，以及 CoS_2 与 MoS_2 之间的协同作用，使其具有优异的 HER 活性和电化学稳定性。另外以 MoS_2 纳米片为壳、CoS_2 纳米棒为核的核/壳结构也表现出了优异的催化活性[84]。Chen 等人报道，合成的 $Co(OH)_2/CoS_2$ 纳米线阵列的核/壳结构中，异质界面有助于提高催化活性[85]，DFT 计算表明，$Co(OH)_2$ 会促进水解离形成 H_{ads}，而 CoS_2 有利于 H_{ads}的吸附。所制备的电极只需要 99mV 的低过电位，电流密度即可达到 $10mA/cm^2$。

除了将纳米片原位生长在纳米线或纳米棒上外，还发展了多孔纳米层包裹的一维纳米结构。Cao 等人制备了 N、P 掺杂碳包覆的 CoP 纳米棒阵列作为 HER 电催化剂，其中多孔 N、P 掺杂的碳和 CoP 纳米棒的协同作用能显著提高 HER 活性[86]。Yin 等采用可控磷化在 Co 基金属有机骨架（Co-MOF）纳米棒内制备 CoP[62]，DFT 计算结果表明，电子通过 N-P/N-Co 从 CoP 转移到 Co-MOF 中，会优化氢和水的吸附能。所得电极在 0.5mol/L H_2SO_4、1mol/L KOH 和 1mol/L 磷酸盐缓冲溶液中，输出电流密度为 10mA/cm^2 时的过电位分别为 27mV、34mV 和 49mV，表现出类 Pt 的催化活性。

对用于 HER 催化的纳米颗粒修饰的一维纳米结构也进行了大量的研究。例如，磷化 NiCo 前驱体，合成了由 CoP 纳米颗粒修饰的 NiCoP 纳米线组成的自支撑电极[87]。在碱性介质中，电流密度为 10mA/cm^2 时，这种独特的纳米结构催化 HER 过电位仅有 73mV。DFT 计算和结构分析表明，与纯 NiCoP（－0.23eV）和 CoP（－0.75eV）相比，H^*（ΔG_{H^*}）在 NiCoP 和 CoP（－0.15eV）界面的吸附自由能更接近于零，具有更优异的催化 HER 活性。2016 年，Sun 等人在碳布上合成 Ni-WN 异质结构纳米线阵列，作为一种活性稳定的 HER 电催化剂。通过电沉积 Ni 颗粒在 WN 纳米线表面上制备了 Ni-WN 异质结构。与纯 Ni 或 WN 相比，Ni-WN 异质结构纳米线表现出更优异的 HER 性能[88]。因此，分级结构中的一级和二级结构之间的协同相互作用有利于增强催化剂的活性。

对分级 2D 纳米结构材料作为高效的 HER 催化剂也有广泛的研究，如核/壳结构和纳米颗粒修饰的 2D 结构。有研究报道在小电流密度下，通过在 MoS_2/CC 纳米片上电沉积 $Ni(OH)_2$，合成了分级结构材料 [$Ni(OH)_2$/MoS_2/CC][89]，该材料在碱性电解液中具有良好的 HER 催化活性。结构分析表明，优异的催化性能主要归因于两分级结构的协同效应：$Ni(OH)_2$ 作为水分解催进剂并产生 H_{ads}，然后吸附在相邻的 MoS_2 活性位点上产生 H_2。另外，在泡沫镍上原位生长负载有 NiCo 纳米颗粒的 Ni-Co-Mo-O 纳米片的 2D 分级结构（NF/NiCoMo-H_2）也具有优异的 HER 电催化剂[31]。实验研究表明，Ni-Co-Mo-O 和 NiCo 纳米颗粒之间的协同效应可以显著地改变 H 吸收的结合能，削弱水的 O—H 键，从而促进水的解离增强。同时，XPS 结果证实电催化剂中的氧空位进一步提高 HER 活性。在 10mA/cm^2 的电流密度下，Tafel 斜率为 33.1mV/dec，过电位为 15mV。另外还有 CuS 纳米片上负载 CoS_2 纳米颗粒[90]、MoS_2 纳米片上负载单原子 Ni[91] 等分级结构的 2D 纳米材料也具有优异的 HER 催化活性。

除了纳米颗粒负载于纳米片的 2D 结构，核/壳结构也被广泛研究，包括纳米片负载于纳米片上的2D结构、夹心状的纳米片等。出现了许多纳米片负载于纳米片上的2D结构的研究工作。比如少层MoS_2纳米片原位生长在碳布支撑的石墨烯纳米片表面上[92]，与单纯碳布上的垂直石墨烯和碳布上的MoS_2纳米片相比，分级结构电极表现出更低的过电位和Tafel斜率，因为石墨烯阵列可以作为有效的电子传输途径，并调整具有丰富活性位点的MoS_2的性质。此外，分级结构为电解质渗入提供了合适的通道，这些都有利于增强HER催化活性。另外，Ren等报道了一种制备三明治结构电催化剂的有效方法，其中CoP作为薄皮覆盖在Ni_5P_4纳米片阵列的两侧，形成$CoP/Ni_5P_4/CoP$阵列[93]。自支撑$CoP/Ni_5P_4/CoP$不仅在较宽的pH范围内表现出优异的HER性能，而且易于扩大规模生产，为工业水电解提供了可能。

堆叠结构也是分级2D纳米结构的一个重要类别。Chen等人通过两步化学气相沉积法在导电MoO_2层上制备了MoS_2纳米片（MoS_2/MoO_2），形成了堆叠结构[94]。这种独特的结构可以高度暴露MoS_2的活性位点，并形成大量的S_2^{2-}内部二硫化物的活性位点。此外，在HER过程中，MoS_2纳米片可以保护内部MoO_2免受酸性电解质的影响。分级结构的MoS_2/MoO_2表现出优异的HER性能，过电位为142mV时电流密度达到$85mA/cm^2$，具有优异的稳定性。此外，Wang等人在$(NH_4)_2MoS_4$中以$Co(OH)_2$为前驱体，在相对较低的温度下，通过水热法制备了负载在CoS_2纳米片侧面边缘的MoS_2纳米片[95]。实验研究表明：①CoS_2的高导电性有利于电子从MoS_2活性位点转移到基底上；②MoS_2的低结晶度可以提供更多的活性位点；③CoS_2和MoS_2之间的界面提供了更多的活性位点，产生了协同效应。因此该电极表现出较高的HER催化活性，在$100mA/cm^2$的电流密度下过电位为118mV，Tafel斜率为37mV/dec，且具有优异的循环稳定性。更令人印象深刻的是，Wang等人设计了一种新颖的线-片结构三元$Co_{0.5}Ni_{0.5}P$，由于线-片纳米结构具有更多的活性位点，催化HER在96mV的过电位下电流密度竟达到$100mA/cm^2$[96]。

4.6.1.3 具有中空1D/2D纳米结构的电极

空心纳米结构具有较高表面积、高原子利用率、高暴露的活性位点和高质量/电子传输速率而被广泛研究。对于自支撑电催化剂，硬模板和自模板法是制备具有优异催化性能的中空1D/2D纳米结构电催化剂的最常用方法。对于硬模板法，目标材料首先包裹在预先制备的模板表面，然后选择性去除模板，形成中空结构。例如，采用硬模板法制备了Cu纳米点修饰的Ni_3S_2纳米管，该方法以ZnO纳米棒为模板，在其表面形成Ni_3S_2后，用碱溶液将模板ZnO除去[97]，得到Cu NDs/Ni_3S_2 NTs的纳米管阵列结构。实验结果表明，Cu纳米点和Ni_3S_2之间的快速传质、高表面积和强电子相互作用，有助于提升在碱性介质中对HER电催化活性。此外，以ZnO纳米棒为模板，还在泡沫镍上制备了中空NiMO合金，在F掺杂氧化锡上制备了ZnSe/$MoSe_2$纳米管阵列[98,99]。

然而，硬模板法需要移除模板，制备过程复杂，有些模板还较难除去。因此，在过去的几十年中开发了一种“更智能”的策略（即“自模板”），它直接将模板转换为空心结构[100]。因此，自模板法被认为是一种简便而有效地制备中空纳米结构电催化剂的策略[101]。目前采用自模板法制备自支撑1D/2D纳米结构催化剂已经取得了优异的成绩。例如，Gong等人以Co-MOFs作自模板前驱体，合成了自支撑的MoS_2/CoS_2纳米管阵列，在宽pH条件下具有优异的HER催化性能[102]。和纳米棒相比较，空心结构能提供更多的与电解质相接触的位点。比如自支撑的多孔$CoMoS_4$纳米管阵列[103]，Ni_4N/Cu_3N纳米管阵列[104]，$Ni_xCo_{3-x}S$纳米管阵列[105]，$Fe(PO_3)_2$@Cu_3P纳米管阵列[106]都是良好的HER催化剂。除了纳米管结构，也研究了通过自模板法形成的中空2D结构。超薄MoS_2纳米片阵列修饰的中空CoP纳米片，在酸性和碱性溶液中都是一种高效的HER电催化剂[107]。这种独特的结构提供了丰富的活性位点，促进氢气泡的释放。

4.6.2 自支撑电极作为OER催化剂

4.6.2.1 一般的1D/2D纳米阵列电极

层状双氢氧化物（LDHs）作为OER电催化剂已经被广泛地研究[108]。LDHs由带正电的层状阳离子（例如Ni^{2+}、Fe^{2+}和Co^{2+}）和层间阴离子（例如CO_3^{2-}、NO_3^-、Cl^-和SO_4^{2-}）组成。但是LDHs的导电性较差，在很大程度上阻碍了其催化活性。为了提高其稳定性和催化活性，LDHs通常直接生长在导电基底上。例如，通过一步水热法在泡沫镍上原位生长NiFe

LDH 纳米片，其催化 OER 活性比 20%（质量分数）Ir/C 催化剂更好[109]。为了进一步提高材料催化的稳定性以及简化制备工艺，Wang 等人直接用泡沫镍作为 Ni 源，通过室温氧化还原和水解共沉淀法在泡沫镍上制备了 NiFe LDHs 纳米片阵列（NiFe-OH NS/NF）[110]。NiFe-OH NS/NF 表现出优异的 OER 活性，过电位为 292mV 时电流密度可达到 $500mA/cm^2$，低于工业标准（达到 $500mA/cm^2$ 电流密度需要 300mV 的过电位）[1]。该方法稳定性高、合成规模大，为其工业化应用提供了可能。同样，Wang 等人也以泡沫镍为镍源，在泡沫镍上制备了 Fe 掺杂的 $Ni(OH)_2$ 纳米片，表现出很高的 OER 活性[70]。

为进一步提高 NiFe LDHs 的 OER 性能，杂原子掺杂是最有效的方法之一，通过引入杂原子调节其电子电导率和结构来改善 NiFe LDHs 导电性。Sun 等人在泡沫镍上合成了钒掺杂的 NiFe LDHs 纳米片阵列，调整了正电层的组成[111]，实验结果和 DFT＋U 计算表明，钒的掺杂可以调整电子结构，缩小带隙，从而提升自支撑电极 OER 催化性能。此外，还研究了其他种类的金属基 LDHs，如 NiCo LDHs 纳米片阵列[112]、CoZn 纳米片[113]等。

除了 LDHs 外，过渡金属氢氧化物和氧化物也作为 OER 电催化剂被广泛研究。例如，通过原位阳极氧化 α-$Co(OH)_2$ 纳米片，在碳布上生长 Fe 掺杂 CoOOH 纳米片阵列作为高效 OER 电极[114]。X 射线吸收精细光谱表明，在阳极氧化过程中，CoOOH 中的部分 CoO_6 八面体结构被 FeO_6 八面体取代，DFT 计算表明 FeO_6 八面体是 OER 的高活性位点。优化的 Fe 取代 CoOOH 纳米片阵列表现出良好的 OER 活性，优于迄今为止报道的大多数 Co 基电催化剂。此外，通过简单的水热结合掺杂方法在碳布上合成了 N 掺杂的 CoO 纳米线阵列，相比 CoO 表现出更强的 OER 催化性能[115]。此外，通过一种新颖的水热电沉积工艺在 Ti 网格上制备了 $Ni_{0.8}Fe_{0.2}$ 纳米片[116]。该材料表现出优异的催化 OER 性能，催化电流密度达到 $10mA/cm^2$ 时过电位为 206mV，且具有长时间的稳定性。与用于 HER 的 1D/2D 纳米阵列类似，具有异质结构的纳米阵列也被开发用于 OER 催化。Sun 等人在 Ti 网上合成了异质结构的 MoO_2-CoP_3 纳米线阵列，在碱性条件下作为高效的 OER 电催化剂[117]。实验研究表明，这种独特的结构有更多的活性位点参与水氧化反应。因此，电流密度为 $10mA/cm^2$ 时具有较低的过电位 288mV，比 MoO_2/Ti 电极低 120mV。

4.6.2.2 分级结构1D/2D纳米结构

用于OER催化的分级1D纳米结构主要有纳米片复合纳米线、纳米颗粒复合纳米线、纳米层复合纳米线。例如，Feng等人通过两步水热法在泡沫镍上合成了NiMn LDHs包覆$NiCo_2O_4$核壳纳米线作为OER电催化剂(NiMn LDH/镍钴氧/NF)[118]。所得NiMn LDH/镍钴氧/NF在电流密度为$10mA/cm^2$时过电位为310mV，优于贵金属IrO_2催化剂。优异的OER活性归因于材料良好的电导率、NiMn LDHs与镍钴氧之间的协同作用以及更合适的活性位点。此外，Xing等人在泡沫镍上制备了CoFe LDHs修饰镍钴氧核壳型纳米材料[119]，通过SEM和TEM表征证实了CoFe LDHs纳米片成功生长在$NiCo_2O_4$纳米线上。实验研究表明，在$20mA/cm^2$时，所制备电极对OER的过电位为273mV，但具有半导体特性的$NiCo_2O_4$核在一定程度上阻碍了电子转移。Wu等人首次合成了CoO_x纳米层包裹的金属Co_4N核材料作为一种有效的OER电催化剂[120]。原位形成的CoO_x层作为OER催化必要的活性位点，保护Co_4N内核不被进一步氧化。得益于Co_4N金属特性的协同作用，核壳型CoO_x/Co_4N在碱性条件下表现出优异的OER催化活性。此外，纳米颗粒修饰的纳米线复合材料也被用于OER催化，如Hu等人在碳纤维表面制备了FeOOH纳米颗粒包覆CoO纳米线阵列。CoO与FeOOH之间的协同催化作用，以及良好排列的纳米线阵列，使FeOOH/CoO具有优异的OER催化活性和稳定性[121]。

分级的2D纳米结构，如壳/核型2D结构和纳米颗粒复合2D纳米结构，包括纳米片包覆纳米片、纳米片上负载纳米线等也得到了广泛地研究。例如，Hu等制备了一种纳米/微米片-片结构，在泡沫铁负载的FeS微米片表面生长了排列整齐的$NiFe(OH)_x$纳米片阵列[$NiFe(OH)_x$/FeS/IF，图4.5(a)、(b)，见文后彩插][122]。因此，高电导率的FeS微米片、具有丰富活性位点的$NiFe(OH)_x$纳米片，以及快速的电子/传质等有利的结构特征赋予了$NiFe(OH)_x$/FeS/IF电极优异的OER催化活性[图4.5(c)和图4.5(d)，见文后彩插]。同样在泡沫镍上负载Ni_3S_2纳米片，并在该纳米片上形成NiFe双金属氢氧化物纳米片组成分级纳米结构材料[123]，得到的分级纳米结构材料具有良好的OER催化性能。更重要的是，得益于良好的结构稳定性和电子导电性，Ni-Fe-OH@Ni_3S_2/NF在大电流密度下，对OER也具有高催化活性和稳定性，为其用于实际电解水分解提供了条件。

另外一个例子是在铜箔上原位生长了 CuO 纳米片覆盖 Co_3O_4 纳米线 2D 纳米材料作为 OER 高催化活性电极，性能比 Co_3O_4 和 CuO 都好[124]。

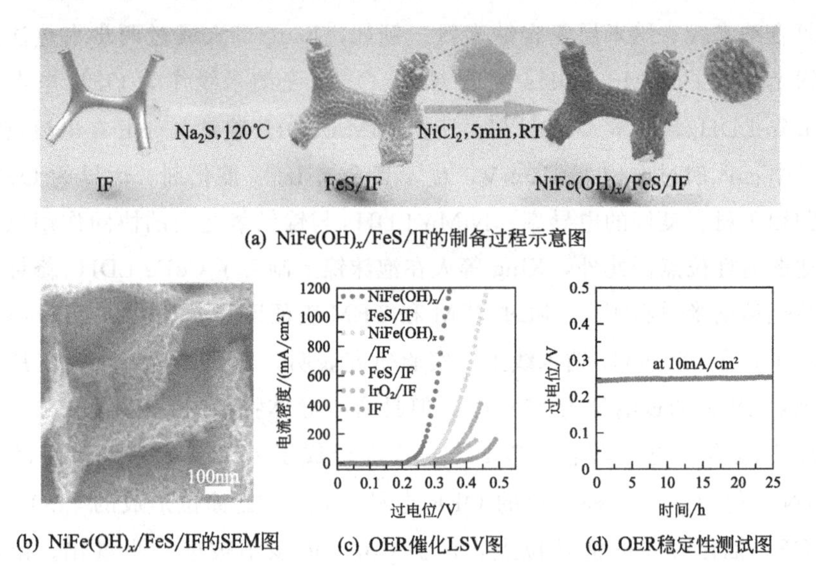

(a) NiFe(OH)$_x$/FeS/IF的制备过程示意图

(b) NiFe(OH)$_x$/FeS/IF的SEM图　(c) OER催化LSV图　(d) OER稳定性测试图

图 4.5　NiFe(OH)$_x$/FeS/IF 合成及 OER 催化性能测试[122]

Lu 等人采用一种简单的化学刻蚀方法在碳布上制备了 Fe-NiO 纳米颗粒覆盖 NiO 纳米片阵列分级结构[125]。实验结果表明，在 NiO 纳米片表面的 Fe-NiO 纳米颗粒平均粒径为 120nm。此外，Fe-NiO 纳米颗粒含有纳米多孔结构，提供了丰富的活性位点和电催化剂的表面通透性，因此 Fe-NiO/CC 具有优异的 OER 催化活性。

4.6.2.3　具有中空 1D/2D 纳米结构的电极

ZnO 纳米棒已被广泛用作制备中空结构 OER 电催化剂的硬模板。2014 年，Jin 等人以 ZnO 纳米棒为模板合成了 Ni-Fe 氧化物基纳米管阵列，与 Ni-Fe 氧化物基纳米棒阵列相比表现出更好的 OER 性能[126]。另外 2016 年，Li 等人报道了以泡沫镍为载体的 FeOOH/Co/FeOOH 杂化纳米管阵列用于 OER[127]。如图 4.6（见文后彩插）所示，首先在 ZnO 纳米棒上电沉积金属 Co，然后通过化学溶解除去 ZnO，形成 Co 纳米管阵列。最后在 Co 纳米管的内外表面电沉积 FeOOH 层，形成 FeOOH/Co/FeOOH 纳米管阵列。与 FeOOH 纳米管和 Co 纳米管相比，FeOOH/Co/FeOOH 复合纳米管表现出更好的 OER 催化活性，表明 Co 和 FeOOH 之间存在协同效应。

DFT 计算进一步证明了 Co 和 FeOOH 之间的强电子相互作用，可以降低中间体和产物的能垒，从而提高 OER 性能。与 FeOOH/Co/FeOOH 的制备类似，通过原位电化学氧化 NiFe 金属合金纳米管阵列制备 $Ni_xFe_{1-x}OOH$/NiFe/$Ni_xFe_{1-x}OOH$ 纳米管阵列[128]，该材料也具有优异的 OER 催化性能。

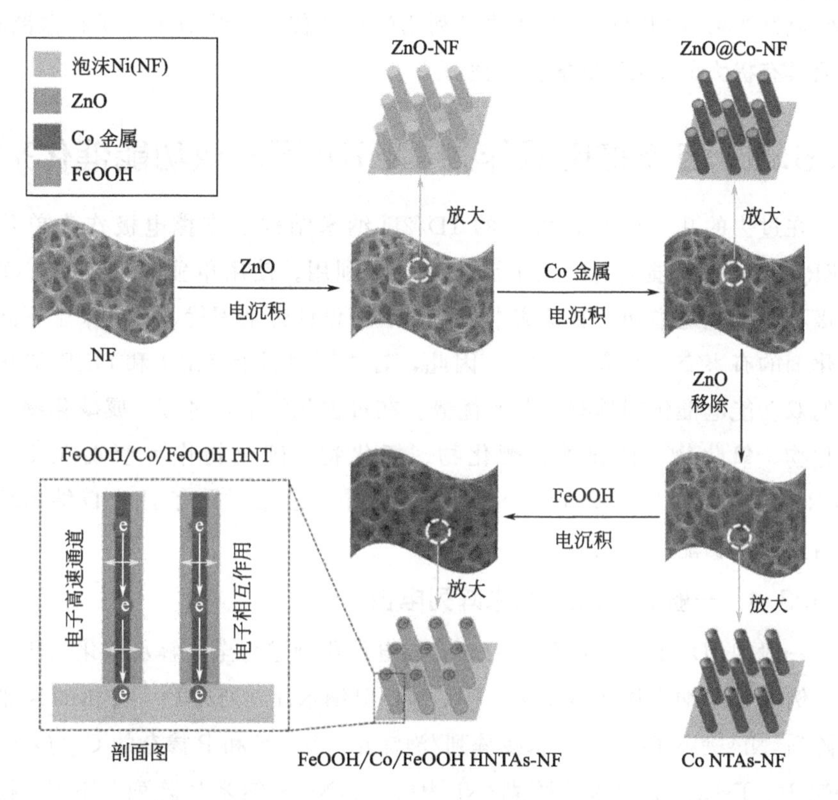

图 4.6 FeOOH/Co/FeOH-NF 的合成过程路线图[127]

自模板法可用于制备中空结构的 OER 电催化剂，MOF 和过渡金属氢氧化物/氧化物是制备中空结构最常见的前驱体。例如，以 $Cu(OH)_2$ 纳米线为前驱体形成了氢氧化镍纳米片修饰的 Fe-Cu 氢氧化物纳米管[129]。同样，以 NiCo 碳酸盐氢氧化物纳米线为前驱体，制备了 CeO_x 纳米颗粒覆盖的 $NiCo_2S_4$ 纳米管阵列[130]。以 Mn MOF 为前驱体，通过可控退火合成了 Mn_2O_3 纳米管阵列[131]。2017 年，Fang 和同事通过简单硒化 CoO 纳米锥，制备了一种新型的空芯枝状 $CoSe_2$ 阵列作为高效的 OER 电催化剂[132]。与 CoO 相比，中空 $CoSe_2$ 具有更大的比表面积和更高的电导率。对于中空的二维纳米结构，Wang 等人通过简单且高度可控的三步法设计了中空镍钴氧纳米壁阵列作为 OER 的电催化剂[133]。首先在碳布上原位生长了 Co-MOF 纳米壁阵列，然后，在 $Ni(NO_3)_2$/乙醇溶液中通过后续的离子交换/刻蚀工艺，将 Ni-Co LDHs 修饰在 Co-MOFs 纳米壁阵列表面，并在空气中退火。最终，在碳布上原位生长得到中空多孔的镍钴氧纳米壁阵列。所制备的中空镍钴氧纳米壁阵列提供了丰富的可利用的活性位点，缩短了离子扩散路径，从而具有极大的 OER 电催化活性。

4.6.3 自支撑电极作为 OER/HER 双功能催化剂

在过去的几十年里，大量的 1D/2D 纳米结构自支撑电极作为单独的 HER 或 OER 电催化剂得到了深入开发和利用。使用单独的 HER 和 OER 电极不可避免地增加了制备成本，给电解槽设计带来困扰，并带来了不同电催化剂的有害交叉效应[134,135]。因此，合成同时具有 HER 和 OER 催化活性的双功能电催化剂具有显著的优势。在过去几年中，过渡金属磷化物、氢氧化物、氧化物、硫化物、硒化物、氮化物、碳化物作为双功能 HER/OER 电催化剂被广泛研究和开发。在本节中，总结了具有 1D/2D 纳米结构的自支撑双功能电催化剂的研究进展。

4.6.3.1 一般的 1D/2D 纳米阵列电极

一般的 1D/2D 纳米阵列作为双功能电催化剂进行全电解水催化的研究很多，例如 CoP 纳米片阵列/碳布[136]、NiCoP 纳米片阵列/NF[137]、MnO_2 纳米片阵列/NF[138]、Co_3Se_4 纳米线阵列/泡沫钴[139]、Y 和 P 掺杂的 Co(OH)F/NF[140]、$FeCo_2S_4$ 纳米片阵列/Ni[141]、Fe_2Ni_2N 纳米片阵列/NF[142] 和 N 掺杂的 WC/碳纤维纸[143]。提高双功能催化活性的策略也与 HER 和 OER 相似，包括杂原子掺杂、表面工程、缺陷工程等。

由于 HER 和 OER 的催化机理不同，结合 HER 和 OER 电催化剂的优点，形成具有丰富界面的分级结构，增强 H_{ads} 和含氧中间体在表面的协同化学吸附，从而增强整体电化学水分解性能。整合单独的 HER 和 OER 电催化剂来形成异质结构，可促进不同活性位点和电子重整界面的动力学[144,145]。例如，Ni_3S_2 对 OER 具有很高的活性，但由于 HER 性能有限，其整体分解水性能较差。可以通过结合对 HER 具有很高活性的 MoS_2 来提高其 HER 活性和全电解水性能[146]。2017 年，采用两步水热法制备了 MoS_2 纳米片包覆 Ni_3S_2 纳米棒的分级纳米阵列（图 4.7，见文后彩插）[147]。

对于分级 MoS_2-Ni_3S_2 纳米棒阵列，XPS 结果表明，归属 Ni_3S_2 物种的峰出现在 852.9eV 和 871.4eV 处。相比之下，单纯的 Ni_3S_2 分别在 852.6eV 和 871.0eV 处出现峰值，与 MoS_2-Ni_3S_2 相比发生上移。这些结果表明 MoS_2 纳米片与 Ni_3S_2 纳米棒之间存在较强的电子相互作用，导致界面电荷重新分配。层状 MoS_2-Ni_3S_2 表现出优异的 HER、OER 活性，在 0.1mol/L KOH 全电解水时电流密度为 10mA/cm^2 仅需要 1.50V 的电压。实验研究表明，Ni_3S_2 和 MoS_2 之间丰富的界面有利于 MoS_2 的 H 化学吸附和 Ni_3S_2 的 HO 化学吸附。因此，相应中间体的吉布斯自由能显著降低，促进含氧中间体的 O—H 键的解离，从而提高了整体水分解动力学。此外，其他类型的异质结构双功能电催化剂，包括 $Mo_{1-x}W_xS_2$ 纳米球覆盖 Ni_3S_2 纳米棒/NF[148]，NiFe LDH 纳米片覆盖 NiCoP 纳米线/NF[149] 和 Ni_3S_2 与 MoS_2 纳米片覆盖 FeOOH 纳米线/NF[150]，也是 HER 和 OER 的有效电催化剂。

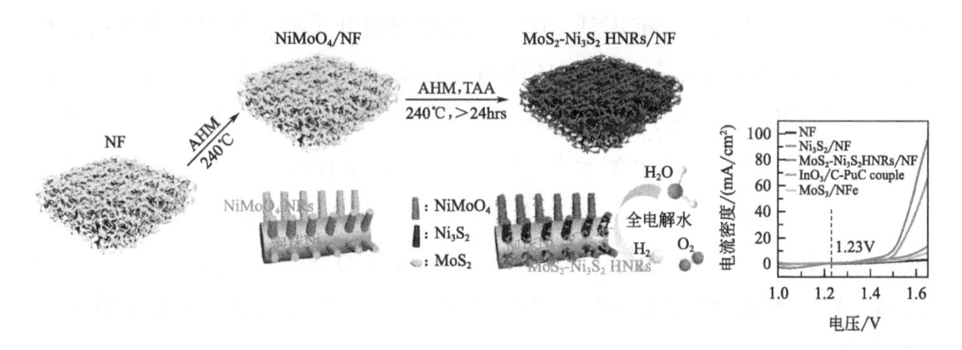

图 4.7 MoS_2-Ni_3S_2 HNRs/NF 的合成过程路线图及性能测试[147]

分级 1D/2D 纳米结构包括核壳型纳米线/纳米片、纳米线/纳米片中负载纳米颗粒也作为双功能电催化剂。例如，Sun 等人报道了在 Ti 网格上生长 Co-B 纳米粒子覆盖 CoO 纳米线阵列[151]；Dong 等人在泡沫镍上制备了 Ni_3S_2/VO_2 CSN 作为催化分解水的高效催化剂（图 4.8，见文后彩插）[152]。SEM 和 TEM 表征显示，该核/壳结构的平均直径大约 100nm，其中 VO_2 壳层厚度约为 5nm。实验研究和 DFT 计算表明，Ni_3S_2 核和 VO_2 壳层之间的界面可以降低能级带中心，优化了含 H 和 O 中间体的吉布斯自由能，从而提高对 HER 和 OER 的催化性能，全电解体系中只需 1.42V 的电压即可达到 10mA/cm^2 的电流密度。

纳米片包覆纳米线/纳米棒的分级纳米结构引起了极大的关注，如 NiFe LDH 纳米片包覆 Cu 纳米线的分级结构材料[153]，得益于高的表面积和快速

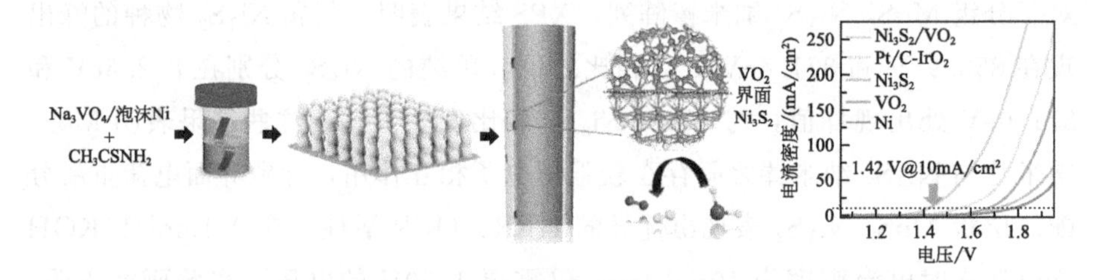

图 4.8　在泡沫镍上合成 Ni_3S_2/VO_2 CSN 的过程路线图及性能测试[152]

的电子/质量传输，所获得的电极在碱性溶液中表现出优异的 OER 和 HER 性能。同样，将 NiFe LDH 纳米片包覆在 Ni 纳米链上制备的分级结构纳米材料表现出比商业 RuO_2 和 IrO_2 更好的 OER 性能，在碱性介质中其 HER 活性与 Pt/C 相当。另外一个有趣的例子是，通过一步水热法，在泡沫镍上制备了 MoS_2 和 Co_9S_8 纳米片包覆 Ni_3S_2 纳米棒阵列分级结构纳米材料($MoS_2/Co_9S_8/Ni_3S_2$/NF)[54]。这种分层结构提升了材料的双功能性能，在电流密度为 10mA/cm^2 时，HER 和 OER 的过电位分别为 103mV 和 166mV，远优于 Ni_3S_2/NF、MoS_2/Ni_3S_2/NF 和 CoS_2/Ni_3S_2/NF。将该材料作为阴阳两极进行全电解水分解，电流密度为 10mA/cm^2 时，施加的电压为 1.54V。其优异的催化性能主要归因于 Co_9S_8 到 MoS_2 的快速电子转移。对于分级 2D 纳米结构，NiCoP 纳米棒覆盖的 NiCoP 纳米片/NF[154] 和 CoFe LDH 纳米片覆盖的 NiFe LDH 纳米片/NF[155] 也是高效的电催化双功能催化剂。

4.6.3.2　具有中空 1D/2D 纳米结构的电极

中空 1D/2D 纳米结构材料作为全电解水分解的高效双功能催化剂被广泛研究，主要通过硬模板法和自模板法来制备。2015 年，Asiri 等人通过硬模板辅助电沉积方法，在钛网上制备了 NiMo 合金空心纳米棒阵列，在碱性介质中表现出高 HER 和 OER 性能[156]。Zhao 等人通过简单的电化学脱合金方法，制备了 NiFe LDHs 覆盖的 Ni 纳米管作为 HER 和 OER 催化电极，在 1.51V 的槽电压下实现了 10mA/cm^2 的电流密度[157]。此外，通过自模板法制备了 MOF 衍生的中空 CoS_2 纳米管阵列，作为自支撑双功能电催化剂[158]。独特的中空纳米管阵列提供了很大的表面积，并增强了传质。同样，通过一种简单且易于放大的自模板法，制备了具有良好孔隙率的中空结构 Co_3O_4[159]。这种材料具有中空的内部和多孔的壳，可以增强传质并充分暴露活性位点，有利于提高材料的 HER 和 OER 催化性能。此外，以空心 CoN/Cu_3N 纳米管阵列为阳极和阴极组装的水电解槽，在 1.62V 电压下获得了 $10mA/cm^2$ 的电流密度[160]。还通过硬模板法和自模板法相结合，制备了 $Ni_3ZnC_{0.7}$纳米点修饰的氮掺杂碳纳米管（$Ni_3ZnC_{0.7}$/NCNT）阵列[161]，优化后的 $Ni_3ZnC_{0.7}$/NCNT 在低过电位下表现出优异的 HER 和 OER 活性。

中空的 2D 纳米结构也可作为高效的双功能电解水催化剂。2018 年，以 MOF 为前驱体在碳布上制备了一种新型的 Mo 掺杂 CoP 中空纳米盒子阵列[162]。Mo 掺杂的 CoP 经活化后转化为 Mo-CoOOH 用于 OER。结构分析和 DFT 计算表明，Mo 的掺杂会导致 H—P 键的增强，从而增强 P 位的 HER 活性。此外，Mo 向 H 的诱导转移可降低吸附 HO^* 中 H—O 键的强度，从而导致质子容易脱落。因此，Mo 掺杂可以提高 CoP 的 HER 活性和 CoOOH 的 OER 活性。

尽管在提高电催化性能方面取得了很大的成就，但在未来的研究中，开发高效分解水的自支撑电催化剂仍然存在一些挑战。

对于一般的 1D/2D 纳米阵列，未来在制备自支撑电催化剂应该更多地集中在精确控制具有理想孔隙率的 1D/2D 纳米结构上，形成具有更易接近的活性位点和传质通道的多孔 1D/2D 纳米结构。此外，为了进一步强化质量扩散，需要设计具有良好化学组成和形貌的 1D/2D 纳米结构，以获得对电解质具有超亲水性和对气泡具有超疏气性。此外，1D/2D 纳米结构与质量传输之间的关系研究得还相对较少，需要建立相应的质量传输机理和模型来指导设计自支撑纳米结构电催化剂。设计 1D/2D 异质纳米阵列的比例是提高一般 1D/2D 纳米阵列电催化活性的有效途径。然而，由于制备方法和材料的限制，精确控制具有理想组成的 1D/2D 纳米阵列仍然比较困难。活性材料和导电基底之间的界面，在电子转移和稳定性方面起着至关重要的作用，但很少有人对此进行研究。因此，为进一步了解它们的关系，需要进行更多的研究，包括活性材料与基底之间的黏附力（如静电引力、范德华力）和强相互作用（例如共价键）。

从材料制备的角度看，对于分级 1D/2D 纳米结构的制备通常涉及不同方法的结合，导致合成过程烦琐，严重限制了其商业应用。因此，发展反应时间短、高效简便的合成方法仍然是一个巨大的挑战。此外，不同成分之间的组合也值得进一步研究，尤其需要全面研究其中不同成分之间的相互作用机制。应通过建模来指导目标材料的结构和部件的设计，例如，应制定有效的策略来实现不同组成部分之间的强大结合力；应详细研究不同活性位点（例如，组分Ⅰ或组分Ⅱ的活性位点或组分Ⅰ和Ⅱ之间的界面位点）对整体反应的确切影响。此外，应更多关注分级结构对传质的影响，尤其是气泡的及时离开和电解质的渗入。

要制备中空的1D/2D纳米结构，模板起着决定性的作用。目前，ZnO和MOF前驱体是制备中空纳米结构的常用模板。因此，采用合适的策略研发更多的模板是推动中空纳米结构设计的最重要因素。例如，气泡或乳液可用作软模板。然而，有效策略的缺乏限制了它们在合成1D/2D中空纳米结构中的应用。此外，由于水分解的苛刻操作条件，应更多关注中空纳米结构的结构稳定性，应该开发有效的策略来保持中空纳米结构以获得高度稳定性。

工业上需要在电流密度500mA/cm^{2}以上，过电位在300mV以下，持续数千小时进行电解水，自支撑电极的催化活性和稳定性仍不能满足工业用电解水的需要[163]。因此需要解决苛刻操作条件下高电流密度材料稳定性。此外，非贵金属材料的降解机理还有待深入研究，一般可以从以下过程进行：电催化剂的溶解、基底的腐蚀、活性物质的脱落、催化剂表面的堵塞、催化剂尺寸的增长等。例如，阳离子从导电基底（例如Ni^{2+}）的溶解可能导致低活性组分在电催化剂表面成核和生长，从而减少电化学活性面积，甚至将活性组分从导电基底中排出。另外，应分别研究不同操作条件下的降解机理，需要发展原位表征技术和化学建模，以便更好地理解降解机理并提出有效的解决策略。

此外，大多数新型电催化剂的开发都以活性为导向的试错法为指导。对活性位点和反应机理的基本理解不足严重阻碍了新型电催化剂的合理设计[164]。例如，MoS_2的活性位点通过计算建模得到了验证，从而取得了巨大的进步[165,166]。然而，并没有研究材料合成与计算研究之间的关系以及大多数材料的基础认识。因此，需要研究具有可控形貌结构、电化学性能和晶体结构的更有效、简单和可扩展的制备方法。

另外，需要进一步阐明基底性质对电催化电极的影响。基底是自支撑电催化剂的关键部分，在制备过程中一般作为参与化学反应的原料。例如，Xu等人报道了在泡沫镍上原位生长带有MoS_2边缘的超薄Ni_3S_2纳米片，其中泡沫镍作为镍源和基底[167]。同时，部分泡沫镍在化学反应过程中溶解，会导致机械强度下降。因此，基底对电催化剂的活性组分和电极的实用性都起着重要的作用。了解催化活性物质和基底之间的界面行为，可以为修复自支撑电极的稳定性和催化活性提供有用的信息。

具有良好排列纳米结构的自支撑电极显著影响传质能力（例如 H^+、OH^-、O_2、H_2 和 H_2O）。特别是对于多相体系，纳米结构的表面性质对微尺度物质传质有着重要的影响。对 1D/2D 纳米结构电极进行适当表面设计，通过立即去除产生的气泡来促进析氢反应/析氧反应过程。然而，目前还没有系统地研究影响电化学响应的不同表面纳米结构和活性材料的传质效应和机理。因此，需要进一步研究纳米结构电极电化学响应与传质之间关系。然而，很难阐明传质和电子转移对电催化反应的影响。数值模拟与实验的结合有助于理解传质机理，尤其是传质与电化学响应之间的相互作用。因此，应更多地关注并发展有效的数值模拟。此外，如上所述，还需要考虑在激活、传质和欧姆过电位的情况下开发计算模型。

发展简单、可扩展的制备技术也很重要，而不是停留在探索制造尺寸小于 $10cm^2$ 的自支撑电极阶段。迄今为止，不同研究之间的催化性能仍然难以比较。为了避免催化性能的偏差，必须标准化测量和评价方法，因此，对性能评价和比较需要一个通用的标准。在今后的研究中，会更关注自支撑纳米结构电催化剂的经济性。

参 考 文 献

[1] Sun H，Yan Z，Liu F，et al. Self-supported transition-metal-based electrocatalysts for hydrogen and oxygen evolution. Advanced Materials，2020，32 (3)：1806326.

[2] Yan Z，Sun H，Chen X，et al. Anion insertion enhanced electrodeposition of robust metal hydroxide/oxide electrodes for oxygen evolution. Nature Communications，2018，9 (1)：2373.

[3] Sun H，Xu X，Yan Z，et al. Superhydrophilic amorphous Co-B-P nanosheet electrocatalysts with Pt-like activity and durability for the hydrogen evolution reaction. Journal of Materials Chemistry A，2018，6 (44)：22062-22069.

[4] Wu M，Liu Y，Zhu Y，et al. Supramolecular polymerization-assisted synthesis of nitrogen and sulfur dual-doped porous graphene networks from petroleum coke as efficient metal-free electrocatalysts for the oxygen reduction reaction. Journal of Materials Chemistry A，2017，5 (22)：11331-11339.

[5] Zhang R，Huang J，Chen G，et al. In situ engineering bi-metallic phospho-nitride bi-functional electrocatalysts for overall water splitting. Applied Catalysis B：Environmental，2019，254：414-423.

[6] Wu G，Zheng X，Cui P，et al. A general synthesis approach for amorphous noble metal nanosheets. Nature Communications，2019，10 (1)：4855.

[7] Yao N，Li P，Zhou Z，et al. Nitrogen engineering on 3D dandelion-flower-like CoS_2 for high-performance overall water splitting. Small，2019，15 (31)：1901993.

[8] Yu L, Xiao Y, Luan C, et al. Cobalt/molybdenum phosphide and oxide heterostructures encapsulated in N-doped carbon nanocomposite for overall water splitting in alkaline media. ACS Applied Materials & Interfaces, 2019, 11 (7): 6890-6899.

[9] Wu Y Q, Tao X, Qing Y, et al. Cr-doped FeNi – P nanoparticles encapsulated into N-doped carbon nanotube as a robust bifunctional catalyst for efficient overall water splitting, Advanced materials, 2019, 31 (15): 1900178.

[10] Li S, Xi C, Jin Y Z, et al. Ir-O-V catalytic group in Ir-doped $NiV(OH)_2$ for overall water splitting. ACS Energy Letters, 2019, 4 (8): 1823-1829.

[11] Sun Y, Xu K, Wei Z, et al. Strong electronic interaction in dual-cation-incorporated $NiSe_2$ nanosheets with lattice distortion for highly efficient overall water splitting. Advanced Materials, 2018, 30 (35): 1802121.

[12] Zhang B, Zhu C, Wu Z, et al. Integrating Rh species with NiFe-layered double hydroxide for overall water splitting. Nano Letters, 2020, 20 (1): 136-144.

[13] Porz L, Swamy T, Sheldon B W, et al. Mechanism of lithium metal penetration through inorganic solid electrolytes. Advanced Energy Materials, 2017, 7 (20): 1701003.

[14] Ji R, Zhang F, Liu Y, et al. Simple synthesis of a vacancy-rich NiO 2D/3D dendritic self-supported electrode for efficient overall water splitting. Nanoscale, 2019, 11 (47): 22734-22742.

[15] Liu H, He Q, Jiang H, et al. Electronic structure reconfiguration toward pyrite NiS_2 via engineered heteroatom defect boosting overall water splitting. ACS Nano, 2017, 11 (11): 11574-11583.

[16] Guo C, Liu X, Gao L, et al. Oxygen defect engineering in cobalt iron oxide nanosheets for promoted overall water splitting. Journal of Materials Chemistry A, 2019, 7 (38): 21704-21710.

[17] Hu C, Zhang L, Zhao Z J, et al. Water splitting: synergism of geometric construction and electronic regulation: 3D Se-$(NiCo)S_x/(OH)_x$ Nanosheets for Highly Efficient Overall Water Splitting. Advanced Materials, 2018, 30 (12): 1870085.

[18] Li Y, Yin J, An L, et al. FeS_2/CoS_2 Interface nanosheets as efficient bifunctional electrocatalyst for overall water splitting. Small, 2018, 14 (26): 1801070.

[19] Lin J, Wang P, Wang H, et al. Defect-rich heterogeneous MoS_2/NiS_2 nanosheets electrocatalysts for efficient overall water splitting. Advanced Science, 2019, 6 (14): 1900246.

[20] Wang Y, Zhang L, Yin K, et al. Nanoporous iridium-based alloy nanowires as highly efficient electrocatalysts toward acidic oxygen Eevolution reaction. ACS Applied Materials & Interfaces, 2019, 11 (43): 39728-39736.

[21] Chi K, Tian X, Wang Q, et al. Oxygen vacancies engineered $CoMoO_4$ nanosheet arrays as efficient bifunctional electrocatalysts for overall water splitting. Journal of Catalysis, 2020, 381: 44-52.

[22] Wei C, Wu W, Li H, et al. Atomic plane-vacancy engineering of transition-metal dichalcogenides with enhanced hydrogen evolution capability. ACS Applied Materials & Interfaces, 2019, 11 (28): 25264-25270.

[23] Song H, Li C, Lou Z, et al. Effective formation of oxygen vacancies in black TiO_2 nanostructures with efficient solar-driven water splitting. ACS Sustainable Chemistry & Engineering, 2017, 5 (10): 8982-8987.

[24] Zhang Z, Hedhili M N, Zhu H, et al. Electrochemical reduction induced self-doping of Ti^{3+} for efficient water splitting performance on TiO_2 based photoelectrodes. Physical Chemistry Chemical Physics, 2013, 15 (37): 15637-15644.

[25] Liang Y, Zhou W, Shi Y, et al. Unveiling in situ evolved In/In_2O_{3-x} heterostructure as the active phase of In_2O_3 toward efficient electroreduction of CO_2 to formate. Science Bulletin, 2020, 65 (18): 1547-1554.

[26] Yan Y, Cheng X, Zhang W, et al. Plasma hydrogenated TiO_2/nickel foam as an efficient bifunctional electrocatalyst for overall water splitting. ACS Sustainable Chemistry & Engineering, 2019, 7 (1): 885-894.

[27] Bao J, Zhang X, Fan B, et al. Ultrathin spinel-structured nanosheets rich in oxygen Deficiencies for enhanced electrocatalytic water oxidation. Angewandte Chemie International Edition, 2015, 54 (25): 7399-7404.

[28] Jain A, Sadan M B, Ramasubramaniam A. Promoting active sites for hydrogen evolution in $MoSe_2$ via transition-metal doping. The Journal of Physical Chemistry C, 2020, 124 (23): 12324-12336.

[29] Li H, Chen S, Jia X, et al. Amorphous nickel-cobalt complexes hybridized with 1T-phase molybdenum disulfide via hydrazine-induced phase transformation for water splitting. Nature Communications, 2017, 8 (1): 15377.

[30] Li H, Chen S, Zhang Y, et al. Systematic design of superaerophobic nanotube-array electrode comprised of transition-metal sulfides for overall water splitting. Nature Communications, 2018, 9 (1): 2452.

[31] Wei S, Cui X, Xu Y, et al. Iridium-triggered phase transition of MoS_2 nanosheets boosts overall water splitting in alkaline media. ACS Energy Letters, 2019, 4 (1): 368-374.

[32] Oh N K, Kim C, Lee J, et al. In-situ local phase-transitioned $MoSe_2$ in $La_{0.5}Sr_{0.5}CoO_{3-\delta}$ heterostructure and stable overall water electrolysis over 1000 hours. Nature Communications, 2019, 10 (1): 1723.

[33] Yang J, Guo D, Zhao S, et al. Overall water splitting: cobalt phosphides nanocrystals encapsulated by P-doped carbon and married with P-doped graphene for overall water splitting. Small, 2019, 15 (10): 1970052.

[34] Ma X, Zhang W, Deng Y, et al. Phase and composition controlled synthesis of cobalt sulfide hollow nanospheres for electrocatalytic water splitting. Nanoscale, 2018, 10 (10): 4816-4824.

[35] Zheng X, Han X, Zhang Y, et al. Controllable synthesis of nickel sulfide nanocatalysts and their phase-dependent performance for overall water splitting. Nanoscale, 2019, 11 (12): 5646-5654.

[36] Feng L L, Yu G, Wu Y, et al. High-index faceted Ni_3S_2 nanosheet arrays as highly active and ultrastable electrocatalysts for water splitting. Journal of the American Chemical Society, 2015, 137 (44): 14023-14026.

[37] Li L, Sun C, Shang B, et al. Tailoring the facets of Ni_3S_2 as a bifunctional electrocatalyst for high-performance overall water-splitting. Journal of Materials Chemistry A, 2019, 7 (30): 18003-18011.

[38] Dong J, Zhang F Q, Yang Y, et al. (003)-Facet-exposed Ni_3S_2 nanoporous thin films on nickel foil for efficient water splitting. Applied Catalysis B: Environmental, 2019, 243: 693-702.

[39] Liu L, Jiang Z, Fang L, et al. Probing the crystal plane effect of Co_3O_4 for enhanced electrocatalytic performance toward efficient overall water splitting. ACS Applied Materials & Interfaces, 2017, 9 (33): 27736-27744.

[40] Liang Q, Zhong L, Du C, et al. Interfacing epitaxial dinickel phosphide to 2D nickel thiophos-

phate nanosheets for boosting electrocatalytic water splitting. ACS Nano, 2019, 13 (7): 7975-7984.

[41] Hu J, Zhang C, Jiang L, et al. Nanohybridization of MoS_2 with layered double hydroxides efficiently synergizes the hydrogen evolution in alkaline media. Joule, 2017, 1 (2): 383-393.

[42] Yang H, Luo S, Li X, et al. Controllable orientation-dependent crystal growth of high-index faceted dendritic $NiC_{0.2}$ nanosheets as high-performance bifunctional electrocatalysts for overall water splitting. Journal of Materials Chemistry A, 2016, 4 (47): 18499-18508.

[43] Liu G, Yang H G, Pan J, et al. Titanium dioxide crystals with tailored facets. Chemical Reviews, 2014, 114 (19): 9559-9612.

[44] Mao H, Yu J, Li J, et al. A high-performance supercapacitor electrode based on nanoflower-shaped $CoTe_2$. Ceramics International, 2020, 46 (5): 6991-6994.

[45] Liu Y, Li P, Wang Y, et al. A green and template recyclable approach to prepare Fe_3O_4/porous carbon from petroleum asphalt for lithium-ion batteries. Journal of Alloys and Compounds, 2017, 695: 2612-2618.

[46] Menezes P W, Indra A, Zaharieva I, et al. Helical cobalt borophosphates to master durable overall water-splitting. Energy & Environmental Science, 2019, 12 (3): 988-999.

[47] Hu E, Feng Y, Nai J, et al. Construction of hierarchical Ni-Co-P hollow nanobricks with oriented nanosheets for efficient overall water splitting. Energy & Environmental Science, 2018, 11 (4): 872-880.

[48] Qian Y, Kang D J. Poly (dimethylsiloxane)/ZnO nanoflakes/three-dimensional graphene heterostructures for high-performance flexible energy harvesters with simultaneous piezoelectric and triboelectric generation. ACS Applied Materials & Interfaces, 2018, 10 (38): 32281-32288.

[49] Han X, Wu X, Deng Y, et al. Ultrafine Pt nanoparticle-decorated pyrite-type CoS_2 nanosheet arrays coated on carbon cloth as a bifunctional electrode for overall water splitting. Advanced Energy Materials, 2018, 8 (24): 1800935.

[50] Li W, Zhang S, Fan Q, et al. Hierarchically scaffolded CoP/CoP_2 nanoparticles: controllable synthesis and their application as a well-matched bifunctional electrocatalyst for overall water splitting. Nanoscale, 2017, 9 (17): 5677-5685.

[51] Fang L, Li W, Guan Y, et al. Water splitting: Tuning unique peapod-like $Co(S_xSe_{1-x})_2$ nanoparticles for efficient overall water splitting. Advanced Functional Materials, 2017, 27 (24): 1701008.

[52] Menezes P W, Panda C, Loos S, et al. A structurally versatile nickel phosphite acting as a robust bifunctional electrocatalyst for overall water splitting. Energy & Environmental Science, 2018, 11 (5): 1287-1298.

[53] Xiong Y, Xu L, Jin C, et al. Interface-engineered atomically thin Ni_3S_2/MnO_2 heterogeneous nanoarrays for efficient overall water splitting in alkaline media. Applied Catalysis B: Environmental, 2019, 254: 329-338.

[54] Yang Y, Yao H, Yu Z, et al. Hierarchical nanoassembly of $MoS_2/Co_9S_8/Ni_3S_2$/Ni as a highly efficient electrocatalyst for overall water splitting in a wide pH range. Journal of the American Chemical Society, 2019, 141 (26): 10417-10430.

[55] Liu Y, Jiang S, Li S, et al. Interface engineering of (Ni, Fe)S_2@MoS_2 heterostructures for synergetic electrochemical water splitting. Applied Catalysis B: Environmental, 2019, 247: 107-114.

[56] Le T A, Tran N Q, Hong Y, et al. Porosity-engineering of MXene as a support material for a

highly efficient electrocatalyst toward overall water splitting. ChemSusChem, 2020, 13 (5): 945-955.

[57] Kumar A, Chaudhary D K, Parvin S, et al. High performance duckweed-derived carbon support to anchor NiFe electrocatalysts for efficient solar energy driven water splitting. Journal of Materials Chemistry A, 2018, 6 (39): 18948-18959.

[58] Böhm D, Beetz M, Schuster M, et al. Efficient OER catalyst with low Ir volume density obtained by homogeneous deposition of iridium oxide nanoparticles on macroporous antimony-doped tin oxide support. Advanced Functional Materials, 2020, 30 (1): 1906670.

[59] Tian J, Liu Q, Asiri A M, et al. Self-supported nanoporous cobalt phosphide nanowire arrays: an efficient 3D hydrogen-evolving cathode over the wide range of pH 0-14. Journal of the American Chemical Society, 2014, 136 (21): 7587-7590.

[60] Pu Z, Liu Q, Jiang P, et al. CoP nanosheet arrays supported on a Ti Plate: An efficient cathode for electrochemical hydrogen evolution. Chemistry of Materials, 2014, 26: 4326-4329.

[61] Tian J, Liu Q, Cheng N, et al. Self-Supported Cu_3P nanowire arrays as an integrated high-performance three-dimensional cathode for generating hydrogen from water. Angewandte Chemie International Edition, 2014, 53 (36): 9577-9581.

[62] Liu T, Li P, Yao N, et al. CoP-Doped MOF-based electrocatalyst for pH-universal hydrogen evolution reaction. Angewandte Chemie International Edition, 2019, 58 (14): 4679-4684.

[63] Tang C, Gan L F, Zhang R, et al. Ternary $Fe_xCo_{1-x}P$ nanowire array as a robust hydrogen evolution reaction electrocatalyst with Pt-like activity: experimental and theoretical insight. Nano letters, 2016, 16 (10): 6617-6621.

[64] Liu T, Liu D, Qu F, et al. Enhanced electrocatalysis for energy-efficient hydrogen production over CoP catalyst with nonelectroactive Zn as a promoter. Advanced Energy Materials, 2017, 7 (15): 1700020.

[65] Liu T, Ma X, Liu D, et al. Mn doping of CoP nanosheets array: An efficient electrocatalyst for hydrogen evolution reaction with enhanced activity at all pH values. ACS Catalysis, 2017, 7 (1): 98-102.

[66] Sun Y, Hang L, Shen Q, et al. Mo doped Ni_2P nanowire arrays: an efficient electrocatalyst for the hydrogen evolution reaction with enhanced activity at all pH values. Nanoscale, 2017, 9 (43): 16674-16679.

[67] Bai Y, Huang H, Wang C, et al. Engineering the surface charge states of nanostructures for enhanced catalytic performance. Materials Chemistry Frontiers, 2017, 1 (10): 1951-1964.

[68] Liu X, Ni K, Niu C, et al. Upraising the O 2p orbital by integrating Ni with MoO_2 for accelerating hydrogen evolution kinetics. ACS Catalysis, 2019, 9 (3): 2275-2285.

[69] Subbaraman R, Tripkovic D, Strmcnik D, et al. Enhancing hydrogen evolution activity in water splitting by Tailoring Li^+-Ni $(OH)_2$-Pt Interfaces. Science, 2011, 334 (6060): 1256-1260.

[70] Wang P, Lin Y, Wan L, et al. Autologous growth of Fe-doped $Ni(OH)_2$ nanosheets with low overpotential for oxygen evolution reaction. International Journal of Hydrogen Energy, 2020, 45, 6416-6424.

[71] Wang X, Kolen'ko Y V, Liu L. Direct solvothermal phosphorization of nickel foam to fabricate integrated Ni_2P-nanorods/Ni electrodes for efficient electrocatalytic hydrogen evolution. Chemical Communications, 2015, 51 (31): 6738-6741.

[72] Wang X, Kolen'ko Y V, Bao X Q, et al. One-step synthesis of self-supported nickel phosphide nanosheet array cathodes for efficient electrocatalytic hydrogen generation. Angewandte Chemie In-

ternational Edition, 2015, 54 (28): 8188-8192.

[73] Xiang Z, Zhang Z, Xu X, et al. MoS_2 nanosheets array on carbon cloth as a 3D electrode for highly efficient electrochemical hydrogen evolution. Carbon, 2016, 98: 84-89.

[74] Zhang N, Gan S, Wu T, et al. Growth Control of MoS_2 Nanosheets on Carbon Cloth for Maximum Active Edges Exposed: An Excellent Hydrogen Evolution 3D Cathode. ACS Applied Materials & Interfaces, 2015, 7 (22): 12193-12202.

[75] Qu B, Yu X, Chen Y, et al. Ultrathin $MoSe_2$ nanosheets decorated on carbon fiber cloth as binder-free and high-performance electrocatalyst for hydrogen evolution. ACS Applied Materials & Interfaces, 2015, 7 (26): 14170-14175.

[76] Lee C P, Chen W F, Billo T, et al. Beaded stream-like $CoSe_2$ nanoneedle array for efficient hydrogen evolution electrocatalysis. Journal of Materials Chemistry A, 2016, 4 (12): 4553-4561.

[77] Ren B, Li D, Jin Q, et al. A self-supported porous WN nanowire array: an efficient 3D electrocatalyst for the hydrogen evolution reaction. Journal of Materials Chemistry A, 2017, 5 (36): 19072-19078.

[78] Xu Y, Zheng C, Wang S, et al. 3D arrays of molybdenum sulphide nanosheets on Mo meshes: Efficient electrocatalysts for hydrogen evolution reaction. Electrochimica Acta, 2015, 174: 653-659.

[79] Zeng L, Liu Z, Sun K, et al. Multiple modulations of pyrite nickel sulfides via metal heteroatom doping engineering for boosting alkaline and neutral hydrogen evolution. Journal of Materials Chemistry A, 2019, 7 (44): 25628-25640.

[80] Bhat K S, Nagaraja H S. Nickel selenide nanostructures as an electrocatalyst for hydrogen evolution reaction. International Journal of Hydrogen Energy, 2018, 43 (43): 19851-19863.

[81] Liu B, Zhao Y F, Peng H Q, et al. Nickel-cobalt diselenide 3D mesoporous nanosheet networks supported on Ni foam: An all-pH highly efficient integrated electrocatalyst for hydrogen evolution. Advanced Materials, 2017, 29 (19): 1606521.

[82] Lu Z, Zhu W, Yu X, et al. Ultrahigh hydrogen evolution performance of under-water "superaerophobic" MoS_2 nanostructured electrodes. Advanced Materials, 2014, 26 (17): 2683-2687.

[83] Huang J, Hou D, Zhou Y, et al. MoS_2 nanosheet-coated CoS_2 nanowire arrays on carbon cloth as three-dimensional electrodes for efficient electrocatalytic hydrogen evolution. Journal of Materials Chemistry A, 2015, 3 (45): 22886-22891.

[84] Huang N, Ding Y, Yan S, et al. Ultrathin MoS_2 nanosheets vertically grown on CoS_2 acicular nanorod arrays: A synergistic three-dimensional shell/core heterostructure for high-efficiency hydrogen evolution at full pH. ACS Applied Energy Materials, 2019, 2 (9): 6751-6760.

[85] Chen L, Zhang J, Ren X, et al. A $Ni(OH)_2$-CoS_2 hybrid nanowire array: a superior non-noble-metal catalyst toward the hydrogen evolution reaction in alkaline media. Nanoscale, 2017, 9 (43): 16632-16637.

[86] Huang X, Xu X, Li C, et al. Vertical CoP nanoarray wrapped by N, P-doped carbon for hydrogen evolution reaction in both acidic and alkaline conditions. Advanced Energy Materials, 2019, 9 (22): 1803970.

[87] Lai WH, Zhang LF, Hua WB, et al. General π-electron-assisted strategy for Ir, Pt, Ru, Pd, Fe, Ni single-atom electrocatalysts with bifunctional active sites for highly efficient water splitting. Angewandte Chemie International Edition, 2019, 58 (34): 11868-11873.

[88] Xing Z, Wang D, Li Q, et al. Self-standing Ni-WN heterostructure nanowires array: A highly efficient catalytic cathode for hydrogen evolution reaction in alkaline solution. Electrochimica

Acta, 2016, 210: 729-733.

[89] Zhang B, Liu J, Wang J, et al. Interface engineering: The $Ni(OH)_2/MoS_2$ heterostructure for highly efficient alkaline hydrogen evolution. Nano Energy, 2017, 37: 74-80.

[90] Li M, Qian Y, Du J, et al. CuS nanosheets decorated with CoS_2 nanoparticles as an efficient electrocatalyst for enhanced hydrogen evolution at all pH values. ACS Sustainable Chemistry & Engineering, 2019, 7 (16): 14016-14022.

[91] Zhang H, Yu L, Chen T, et al. Surface modulation of hierarchical MoS_2 nanosheets by Ni single atoms for enhanced electrocatalytic hydrogen evolution. Advanced Functional Materials, 2018, 28 (51): 1807086.

[92] Zhang Z, Li W, Yuen M F, et al. Hierarchical composite structure of few-layers MoS_2 nanosheets supported by vertical graphene on carbon cloth for high-performance hydrogen evolution reaction. Nano Energy, 2015, 18: 196-204.

[93] Mishra I K, Zhou H, Sun J, et al. Hierarchical $CoP/Ni_5P_4/CoP$ microsheet arrays as a robust pH-universal electrocatalyst for efficient hydrogen generation. Energy & Environmental Science, 2018, 11 (8): 2246-2252.

[94] Nikam R D, Lu A Y, Sonawane P A, et al. Three-dimensional heterostructures of MoS_2 nanosheets on conducting MoO_2 as an efficient electrocatalyst to enhance hydrogen evolution reaction. ACS Applied Materials & Interfaces, 2015, 7 (41): 23328-23335.

[95] Sun Y, Huang B, Li Y, et al. Trifunctional fishbone-like PtCo/Ir enables high-performance Zinc-Air batteries to drive the water-splitting catalysis. Chemistry of Materials, 2019, 31 (19): 8136-8144.

[96] Zhang X, Gu W, Wang E. Wire-on-flake heterostructured ternary $Co_{0.5}Ni_{0.5}P/CC$: an efficient hydrogen evolution electrocatalyst. Journal of Materials Chemistry A, 2017, 5 (3): 982-987.

[97] Feng J X, Wu J Q, Tong Y, et al. Efficient hydrogen evolution on Cu nanodots-decorated Ni_3S_2 nanotubes by optimizing atomic hydrogen adsorption and desorption. Journal of the American Chemical Society, 2017, 140: 610-617

[98] Cao J, Li H, Pu J, et al. Hierarchical NiMo alloy microtubes on nickel foam as an efficient electrocatalyst for hydrogen evolution reaction. International Journal of Hydrogen Energy, 2019, 44 (45): 24712-24718.

[99] Wu M, Huang Y, Cheng X, et al. Arrays of $ZnSe/MoSe_2$ nanotubes with electronic modulation as efficient electrocatalysts for hydrogen evolution reaction. Advanced Materials Interfaces, 2017, 4 (23): 1700948.

[100] Yu L, Wu H B, Lou X W D. Self-templated formation of hollow structures for electrochemical energy applications. Accounts of Chemical Research, 2017, 50 (2): 293-301.

[101] Xin Y, Huang Y, Lin K, et al. Self-template synthesis of double-layered porous nanotubes with spatially separated photoredox surfaces for efficient photocatalytic hydrogen production. Science Bulletin, 2018, 63 (10): 601-608.

[102] Tang B, Yu Z G, Zhang Y, et al. Metal-organic framework-derived hierarchical MoS_2/CoS_2 nanotube arrays as pH-universal electrocatalysts for efficient hydrogen evolution. Journal of Materials Chemistry A, 2019, 7 (21): 13339-13346.

[103] Pi Y, Shao Q, Wang P, et al. General formation of monodisperse IrM (M=Ni, Co, Fe) bimetallic nanoclusters as bifunctional electrocatalysts for acidic overall water splitting. Advanced Functional Materials, 2017, 27 (27): 1700886.

[104] Si Z, Lv Z, Lu L, et al. Nitrogen-doped graphene chainmail wrapped IrCo alloy particles on nNitrogen-doped graphene nanosheet for highly active and stable full water splitting. ChemCat-

Chem, 2019, 11 (22): 5457-5465.
[105] Wang MQ, Ye C, Bao SJ, et al. Ternary $Ni_xCo_{3-x}S_4$ with a fine hollow nanostructure as a robust electrocatalyst forhydrogen evolution. ChemCatChem, 2017, 9 (22): 4169-4174.
[106] Dai D, Wei B, Li Y, et al. Self-supported hierarchical $Fe(PO_3)_2$@Cu_3P nanotube arrays for efficient hydrogen evolution in alkaline media. Journal of Alloys and Compounds, 2020, 820: 153185.
[107] Ge Y, Chu H, Chen J, et al. Ultrathin MoS_2 nanosheets decorated hollow CoP heterostructures for enhanced hydrogen evolution reaction, ACS Sustainable Chemistry & Engineering, 2019, 7 (11), 10105-10111.
[108] Wang P C, Wan L, Lin Y Q, et al. NiFe hydroxide supported on hierarchically porous nickel nesh as a high-performance bifunctional electrocatalyst for water splitting at large current density. ChemSusChem, 2019, 12 (17): 4038-4045.
[109] Lu Z, Xu W, Zhu W, et al. Three-dimensional NiFe layered double hydroxide film for high-efficiency oxygen evolution reaction. Chemical Communications, 2014, 50 (49): 6479-6482.
[110] Zhu W, Zhang T, Zhang Y, et al. A practical-oriented NiFe-based water-oxidation catalyst enabled by ambient redox and hydrolysis co-precipitation strategy. Applied Catalysis B: Environmental, 2019, 244: 844-852.
[111] Li P, Duan X, Kuang Y, et al. Tuning electronic structure of NiFe layered double hydroxides with vanadium doping toward high efficient electrocatalytic water oxidation. Advanced Energy Materials, 2018, 8 (15): 1703341.
[112] Jiang J, Zhang A, Li L, et al. Nickel-cobalt layered double hydroxide nanosheets as high-performance electrocatalyst for oxygen evolution reaction. Journal of Power Sources, 2015, 278: 445-451.
[113] Wang J, Tan C F, Zhu T, et al. Topotactic consolidation of monocrystalline CoZn hydroxides for advanced oxygen evolution electrodes. Angewandte Chemie International Edition, 2016, 55 (35): 10326-10330.
[114] Ye S H, Shi Z X, Feng J X, et al. Activating CoOOH porous nanosheet arrays by partial iron substitution for efficient oxygen evolution reaction. Angewandte Chemie International Edition, 2018, 57 (10): 2672-2676.
[115] Zhang K, Xia X, Deng S, et al. N-doped CoO nanowire arrays as efficient electrocatalysts for oxygen evolution reaction. Journal of Energy Chemistry, 2019, 37: 13-17.
[116] Yao M, Wang N, Hu W, et al. Novel hydrothermal electrodeposition to fabricate mesoporous film of $Ni_{0.8}Fe_{0.2}$ nanosheets for high performance oxygen evolution reaction. Applied Catalysis B: Environmental, 2018, 233: 226-233.
[117] Xiong X, Ji Y, Xie M, et al. MnO_2-CoP_3 nanowires array: An efficient electrocatalyst for alkaline oxygen evolution reaction with enhanced activity. Electrochemistry Communications, 2018, 86: 161-165.
[118] Chen Q Q, Hou C C, Wang C J, et al. Ir^{4+}-Doped NiFe LDH to expedite hydrogen evolution kinetics as a Pt-like electrocatalyst for water splitting. Chemical Communications, 2018, 54 (49): 6400-6403.
[119] Zhang Y, Yang M, Jiang X, et al. Self-supported hierarchical CoFe-LDH/$NiCo_2O_4$/NF core-shell nanowire arrays as an effective electrocatalyst for oxygen evolution reaction. Journal of Alloys and Compounds, 2020, 818: 153345.
[120] Chen P, Xu K, Fang Z, et al. Metallic Co_4N porous nanowire arrays activated by surface oxidation as electrocatalysts for the oxygen evolution reaction. Angewandte Chemie International

Edition, 2015, 54 (49): 14710-14714.
[121] Wang Y, Ni Y, Liu B, et al. Vertically oriented CoO@FeOOH nanowire arrays anchored on carbon cloth as a highly efficient electrode for oxygen evolution reaction. Electrochimica Acta, 2017, 257: 356-363.
[122] Niu S, Jiang WJ, Tang T, et al. Autogenous growth of hierarchical $NiFe(OH)_x/FeS$ nanosheet-on-microsheet arrays for synergistically enhanced high-output water oxidation. Advanced Functional Materials, 2019, 29 (36): 1902180.
[123] Zou X, Liu Y, Li G D, et al. Ultrafast formation of amorphous bimetallic hydroxide films on 3D conductive sulfide nanoarrays for large-current-density oxygen evolution electrocatalysis. Advanced Materials, 2017, 29 (22): 1700404.
[124] Zhang S Y, Zhu H L, Zheng Y Q. Surface modification of CuO nanoflake with Co_3O_4 nanowire for oxygen evolution reaction and electrocatalytic reduction of CO_2 in water to syngas. Electrochimica Acta, 2019, 299: 281-288.
[125] Cao L M, Hu Y W, Zhong D C, et al. Template-directed growth of bimetallic prussian blue-analogue nanosheet arrays and their derived porous metal oxides for oxygen evolution reaction. ChemSusChem, 2018, 11 (21): 3708-3713.
[126] Zhao Z, Wu H, He H, et al. Self-standing non-noble metal (Ni-Fe) oxide nanotube array anode catalysts with synergistic reactivity for high-performance water oxidation. Journal of Materials Chemistry A, 2015, 3 (13): 7179-7186.
[127] Feng J X, Xu H, Dong Y T, et al. FeOOH/Co/FeOOH Hybrid nanotube arrays as high-performance electrocatalysts for the oxygen evolution reaction. Angewandte Chemie International Edition, 2016, 55 (11): 3694-3698.
[128] Wang A L, Dong Y T, Li M, et al. In situ derived $Ni_xFe_{1-x}OOH/NiFe/Ni_xFe_{1-x}OOH$ nanotube arrays from NiFe alloys as efficient electrocatalysts for oxygen evolution. ACS Applied Materials & Interfaces, 2017, 9 (40): 34954-34960.
[129] Czioska S, Wang J, Teng X, et al. Hierarchically structured multi-shell nanotube arrays by self-assembly for efficient water oxidation. Nanoscale, 2018, 10 (6): 2887-2893.
[130] Wu X, Yang Y, Zhang T, et al. CeO_x Decorated hierarchical $NiCo_2S_4$ hollow nanotubes arrays for enhanced oxygen evolution reaction electrocatalysis. ACS Applied Materials & Interfaces, 2019, 11 (43): 39841-39847.
[131] Liu P P, Zheng Y Q, Zhu H L, et al. Mn_2O_3 Hollow nanotube arrays on Ni foam as efficient supercapacitors and electrocatalysts for oxygen evolution reaction. ACS Applied Nano Materials, 2019, 2 (2): 744-749.
[132] Chen T, Li S, Wen J, et al. Rational construction of hollow core-branch $CoSe_2$ nanoarrays for high-performance asymmetric supercapacitor and efficient oxygen evolution. Small, 2018, 14 (5): 1700979.
[133] Guan C, Liu X, Ren W, et al. Rational design of metal-organic framework derived hollow $NiCo_2O_4$ arrays for flexible supercapacitor and electrocatalysis. Advanced Energy Materials, 2017, 7 (12): 1602391.
[134] Xiong B, Chen L, Shi J. Anion-containing noble-metal-free bifunctional electrocatalysts for overall water splitting. ACS Catalysis, 2018, 8 (4): 3688-3707.
[135] Tang C, Cheng N, Pu Z, et al. NiSe Nanowire film supported on nickel foam: An efficient and stable 3D bifunctional electrode for full water splitting. Angewandte Chemie International Edition, 2015, 54 (32): 9351-9355.
[136] Liu T, Xie L, Yang J, et al. Self-standing CoP nanosheets array: A three-dimensional bi-

functional catalyst electrode for overall water splitting in both neutral and alkaline media. ChemElectroChem, 2017, 4 (8): 1840-1845.

[137] Liang H, Gandi A N, Anjum D, et al. Plasma-sssisted synthesis of NiCoP for efficient overall water splitting. Nano Letters, 2016, 16 (12): 7718-7725.

[138] Jin Y, Wang H, Li J, et al. Porous MoO_2 nanosheets as non-noble bifunctional electrocatalysts for overall water splitting. Advanced Materials, 2016, 28 (19): 3785-3790.

[139] Chao T, Luo X, Chen W, et al. Atomically dispersed copper-platinum dual sites alloyed with palladium nanorings catalyze the hydrogen evolution reaction. Angewandte Chemie International Edition, 2017, 56 (50): 16047-16051.

[140] Zhang G, Wang B, Li L, et al. Phosphorus and yttrium codoped Co(OH)F nanoarray as highly efficient and bifunctional electrocatalysts for overall water splitting. Small, 2019, 15 (42): 1904105.

[141] Hu J, Ou Y, Li Y, et al. $FeCo_2S_4$ Nanosheet arrays supported on Ni foam: An efficient and durable bifunctional electrocatalyst for overall water-splitting. ACS Sustainable Chemistry & Engineering, 2018.

[142] Jiang M, Li Y, Lu Z, et al. Binary nickel-iron nitride nanoarrays as bifunctional electrocatalysts for overall water splitting. Inorganic Chemistry Frontiers, 2016, 3 (5): 630-634.

[143] Han N, Yang K R, Lu Z, et al. Nitrogen-doped tungsten carbide nanoarray as an efficient bifunctional electrocatalyst for water splitting in acid. Nature Communications, 2018, 9 (1): 924.

[144] Miao J, Xiao F, Yang H, et al. Hierarchical Ni-Mo-S nanosheets on carbon fiber cloth: A flexible electrode for efficient hydrogen generation in neutral electrolyte. Science Advances, 2015, 1: e1500259-e1500259.

[145] Wang X D, Xu Y F, Rao H S, et al. Novel porous molybdenum tungsten phosphide hybrid nanosheets on carbon cloth for efficient hydrogen evolution. Energy & Environmental Science, 2016, 9 (4): 1468-1475.

[146] Zhang J, Wang T, Pohl D, et al. Interface Engineering of MoS_2/Ni_3S_2 heterostructures for highly enhanced electrochemical overall-water-splitting activity. Angewandte Chemie International Edition, 2016, 55 (23): 6702-6707.

[147] Yang Y, Zhang K, Lin H, et al. MoS_2-Ni_3S_2 Heteronanorods as efficient and stable bifunctional electrocatalysts for overall water splitting. ACS Catalysis, 2017, 7 (4): 2357-2366.

[148] Zheng M, Du J, Hou B, et al. Few-layered $Mo_{(1-x)}W_xS_2$ hollow nanospheres on Ni_3S_2 nanorod heterostructure as robust electrocatalysts for overall water splitting. ACS Applied Materials & Interfaces, 2017, 9 (31): 26066-26076.

[149] Zhang H, Li X, Hähnel A, et al. Bifunctional heterostructure assembly of NiFe LDH nanosheets on NiCoP nanowires for highly efficient and stable overall water splitting. Advanced Functional Materials, 2018, 28 (14): 1706847.

[150] Zheng M, Guo K, Jiang W J, et al. When MoS_2 meets FeOOH: A "one-stone-two-birds" heterostructure as a bifunctional electrocatalyst for efficient alkaline water splitting. Applied Catalysis B: Environmental, 2019, 244: 1004-1012.

[151] Lu W, Liu T, Xie L, et al. In situ derived CoB nanoarray: A high-efficiency and durable 3D bifunctional electrocatalyst for overall alkaline water splitting. Small, 2017, 13 (32): 1700805.

[152] Lv Q, Yang L, Wang W, et al. One-step construction of core/shell nanoarrays with a holey shell and exposed interfaces for overall water splitting. Journal of Materials Chemistry A, 2019, 7 (3): 1196-1205.

[153] Yu L, Zhou H, Sun J, et al. Cu nanowires shelled with NiFe layered double hydroxide nanosheets as bifunctional electrocatalysts for overall water splitting. Energy & Environmental Science, 2017, 10 (8): 1820-1827.

[154] Wang J G, Hua W, Li M, et al. Structurally engineered hyperbranched NiCoP arrays with superior electrocatalytic activities toward highly efficient overall water splitting. ACS Applied Materials & Interfaces, 2018, 10 (48): 41237-41245.

[155] Yang R, Zhou Y, Xing Y, et al. Synergistic coupling of CoFe-LDH arrays with NiFe-LDH nanosheet for highly efficient overall water splitting in alkaline media. Applied Catalysis B: Environmental, 2019, 253: 131-139.

[156] Tian J, Cheng N, Liu Q, et al. Self-supported NiMo hollow nanorod array: an efficient 3D bifunctional catalytic electrode for overall water splitting. Journal of Materials Chemistry A, 2015, 3 (40): 20056-20059.

[157] Li D, Hao G, Guo W, et al. Highly efficient Ni nanotube arrays and Ni nanotube arrays coupled with NiFe layered-double-hydroxide electrocatalysts for overall water splitting. Journal of Power Sources, 2020, 448: 227434.

[158] Guan C, Liu X, Elshahawy A M, et al. Metal-organic framework derived hollow CoS_2 nanotube arrays: an efficient bifunctional electrocatalyst for overall water splitting. Nanoscale Horizons, 2017, 2 (6): 342-348.

[159] Zhu Y P, Ma T Y, Jaroniec M, et al. Self-templating synthesis of Hollow Co_3O_4 microtube arrays for highly efficient water electrolysis. Angewandte Chemie International Edition, 2017, 56 (5): 1324-1328.

[160] Wang Z, Ren X, Luo Y, et al. An ultrafine platinum-cobalt alloy decorated cobalt nanowire array with superb activity toward alkaline hydrogen evolution. Nanoscale, 2018, 10 (26): 12302-12307.

[161] Li R, Li X, Yu D, et al. $Ni_3ZnC_{0.7}$ nanodots decorating nitrogen-doped carbon nanotube arrays as a self-standing bifunctional electrocatalyst for water splitting. Carbon, 2019, 148: 496-503.

[162] Guan C, Xiao W, Wu H, et al. Hollow Mo-doped CoP nanoarrays for efficient overall water splitting. Nano Energy, 2018, 48: 73-80.

[163] Zhou H, Yu F, Sun J, et al. Highly active catalyst derived from a 3D foam of $Fe(PO_3)_2/Ni_2P$ for extremely efficient water oxidation. Proceedings of the National Academy of Sciences, 2017, 114 (22): 5607-5611.

[164] Seh Z W, Kibsgaard J, Dickens C F, et al. Combining theory and experiment in electrocatalysis: Insights into materials design. Science, 2017, 355 (6321): eaad4998.

[165] Hinnemann B, Moses P G, Bonde J, et al. Biomimetic hydrogen evolution: MoS_2 nanoparticles as catalyst for hydrogen evolution. Journal of the American Chemical Society, 2005, 127 (15): 5308-5309.

[166] Ding Q, Song B, Xu P, et al. Efficient electrocatalytic and photoelectrochemical hydrogen generation using MoS_2 and related compounds. Chem, 2016, 1 (5): 699-726.

[167] Wang C, Shao X, Pan J, et al. Redox bifunctional activities with optical gain of Ni_3S_2 nanosheets edged with MoS_2 for overall water splitting. Applied Catalysis B: Environmental, 2020, 268: 118435.

第五章

贵金属基纳米催化剂

电催化剂的内在活性、选择性和稳定性决定了电解水的效率。目前已有大量的工作研究了高性能电催化剂和反应机理。迄今为止，Pt 基和 Ru/Ir 基纳米材料是 HER 和 OER 最有效的电催化剂。本章主要总结近年来开发的性能优异的贵金属基电解水催化剂，重点介绍铱（Ir）基纳米材料和铂（Pt）基纳米材料。但是由于存在以下两方面的问题，导致利用贵金属基作为商业电解水催化剂存在相当大的挑战：①贵金属的高成本和稀缺性是电解水大规模利用的主要阻碍[1]；②贵金属在电催化过程中发生溶解、团聚、中毒、耐受性差等问题[2]。如今，开发高活性、高稳定性和低成本的电解水催化剂是实现氢经济的关键一步。与碱性电解槽相比，质子交换膜电解水槽（PEMWEs）具有更高的电流密度、更低的电阻、更高的气体纯度和更简便的系统设计等优点，是一种绿色、经济有效的制氢方法[3]。此外，在酸性环境下商品化的阳离子交换膜能在保证离子导电性要求下电解水，如 Nafion[4]。在酸性环境中，Ru 基材料被认为是最具活性的 OER 催化剂，但由于在阳极电位下生成了可溶性的 RuO_4，导致其性能不稳定。Ir 基催化剂比 Ru 基催化剂更稳定，但活性较低[5]。一般采用 Ir 基化合物和 Ru 基化合物来提升材料的催化活性和稳定性[6,7]，从而降低 Ir 和 Ru 的含量，降低经济成本。目前虽然开发了许多性能优异的非贵金属催化剂诸如金属硫化物/氮化物/磷化物基材料用于碱性电解水体系[8-10]，但是在酸性条件下，非贵金属基材料不稳定以及其存在的形式表现出的低催化活性，对工业应用来说很困难[11]。因此，特别是在酸性环境下，大力发展贵金属基催化剂，重点降低贵金属含量，提高贵金属基材料的催化效率和稳定性是非常重要的。

目前，研究者们致力于通过纳米技术，探索提高贵金属基材料催化水分解的活性和稳定性的方法。这些方法主要集中在结构设计上，通过引入杂原子（配体效应）、界面结构工程（应变效应或混合效应）和构建单原子来实现。当杂原子掺入贵金属基化合物，引起电子在两个不同原子团之间转移时，就会发生配体效应，从而导致电子结构发生变化[12,13]。相/界面效应源于两个不同相的接触来调整电子结构，从而加速水的解离吸附或反应中间体的解吸[14-16]。单原子效应是一个迅速兴起的研究领域，通过将催化剂尺寸缩小到单原子尺度，来获得最大原子利用效率和催化活性[17-20]。例如，将特定的过渡金属与贵金属合金化以调整贵金属的 d 带中心，并优化小分子在贵金属表面的吸附能，是提高相应交换电流密度的有效方法。一种比较有效的方法是通过构建 Pt 与其他过渡金属化合物之间的界面来调节催化剂性能，但在碱性环境中 Pt 电极上 HER 的动力学通常比酸性环境下低几个数量级[21]。除了构建贵金属/过渡金属复合物外，还有其他方法，例如制备尺寸和形状可控的纳米结构材料[22]、在应变效应和配体效应的基础上引入第二种非贵金属元素形成合金、在特制基底上设计单原子贵金属等，来优化电子结构、降低贵金属负载量、增大比表面积，从而提升催化剂性能和稳定性[23,24]。总之，与结构相关的因素对控制催化剂电催化性能至关重要。

贵金属基化合物（如 Pt、Pd、Rh、Ru 和 Ir）仍然是 HER 和 OER 最有效的催化剂，尤其是在酸性介质下。开发具有低贵金属负载量、高活性和优异耐受性的贵金属基催化剂对于能量转换系统具有重要意义。研究人员为实现这一目标做出了巨大的努力，主要有以下几种途径：①控制贵金属基催化剂的形状；②引入其他元素；③在先进的载体材料上构建贵金属单原子；④将贵金属基纳米催化剂的尺寸缩小到单原子尺度；⑤构建贵金属和过渡金属化合物之间的工程界面。

5.1 铱基纳米催化剂用于电化学分解水

目前，与其他水分解技术相比，酸性质子交换膜（PEM）电解水系统具有显著的优势，如结构紧凑、电流密度高、效率高等[25]。为了实现高效的 PEM 电解水，有必要设计高效且耐酸的电催化剂。Ir 基催化剂具有有利的电子特性和耐酸性，是作为酸性电解水催化剂最好的选择[26,27]。因此，开发 Ir 基电催化剂，有望以可持续、经济的方式实现大规模的电解水产氢。铱（Ir）作为铂族金属，在电催化应用中具有固有的优点，包括防腐能力强、催化活性高、导电性好[28-31]。目前 Ir 基纳米材料已经成为各种电化学反应的有效催化剂，如 OER、HER、ORR、氢氧化反应、二氧化碳还原反应和氮还原反应[32-34]，Ir 基材料（如金属铱、合金、氧化物等）作为电催化剂已有大量的报道。这些报道主要集中在新型催化剂的设计和揭示其 OER 催化机理[27,35]。Ir 基材料析氢过程中的吉布斯自由能（$\Delta G_H=0.03eV$）接近零，因此其作为 HER 的催化材料受到越来越多的关注[36]。

5.1.1 Ir 基催化剂用于 HER

Ir 相关的纳米材料，包括 Ir 金属、Ir 合金和氧化物以及其他含 Ir 化合物，已广泛应用于 HER。本小节总结了 Ir 相关的 HER 催化剂的最新进展。从图 5.1 金属的 HER 火山图可以看出[37]，Ir 具有良好的催化 HER 性能。为了进一步提高 Ir 的催化性能和利用率，通过形貌控制、设计 Ir 基杂化物和合金，开发更多良好的催化剂。

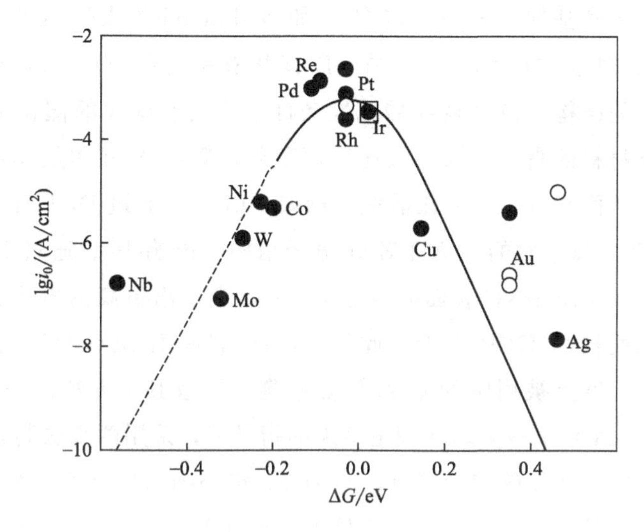

图 5.1　金属的 HER 火山图[37]

一般通过调整催化剂的形貌来调整暴露的活性位点数量，控制反应中间体/反应物的吸附/脱附行为来调节电催化性能[38]。目前，大多数文献报道的 Ir 金属催化剂都是以纳米颗粒的形式存在的，而 Yang 等人发现一维 (1D) 线状 Ir 纳米晶组装体 (ONAs) (η_{10}=47mV) 优于纳米颗粒 (NPs) 组装体 (η_{10}=79mV)[39]，这是因为 Ir ONAs 具有较高的质量/电荷转移率和长段的低指数晶面。因此，开发具有特定形貌的 Ir 催化剂 (例如维度、端面、孔隙率) 是改善其催化性能的比较合理的方式。

5.1.1.1　Ir 金属

Ir 活性位点的聚合是严重影响 Ir 纳米颗粒催化性能的一个关键问题，严重降低 Ir 的利用效率和催化性能。为了解决这一问题，开发了多种功能化支撑物/基底来锚定和分散 Ir 纳米颗粒。在某些情况下，Ir 和基底的相互作用也有利于提升 HER 催化性能，这些基底包括多孔碳[40,41]、碳纳米管 (CNT)、还原氧化石墨烯 (rGO) 和 NC[42,43] 等。He 等人采用了一种简便的原位碳化方法，将 Ir 有序分散在介孔碳上 (Ir-OMC) 作为优异的自支撑催化电极[41]。该材料中 Ir 纳米颗粒、有序介孔碳基体和三维导电基底（泡沫镍）之间的紧密接触，增强了电化学过程中的快速传质和电子传输，从而获得了良好的 HER 性能。

由于有机网络结构可以提供大的表面积和优异的电化学稳定性，最近被用作稳定金属基催化剂的载体[42]。如 Mahmood 等人采用了一种独特的方法，将 Ir 纳米颗粒均匀分散在 3D 笼状有机网络（CON）结构中[42]。由于其独特的结构，该材料在酸碱性条件下均具有优异的催化 HER 性能。除了纳米碳材料的高导电性和大比表面积外，碳-铱相互作用还调节了铱的电子结构，从而影响本征催化活性。例如，Li 等人通过 DFT 计算，研究了碳/氮（C/N）原子对的电负性对 Ir 电子结构的影响[43]。通过引入 C/N 配位，与纯 Ir 相比，IrNC 表面的空心 fcc 位点表现出更高的电子密度，有利于氢吸附从而提高 HER 性能。此外，IrNC 表面 Ir 位点的氢吸附/脱附行为，在很大程度上被周围的 C/N 位点平衡，导致其上升段斜率（0.04eV）比纯 Ir(0.25eV）小得多。在理论结果的指导下，采用两步法制备了 Ir 纳米颗粒锚定的中空氮化碳纳米球（Ir HNC)。空心纳米球厚度约为 14nm 且多孔。Ir 纳米颗粒均匀地分布在多孔壳上，尺寸范围为 0.9～2.5nm。如预期的那样，Ir 纳米颗粒锚定的中空氮化碳纳米球在 0.5mol/L H_2SO_4 中的 HER 性能优于 Ir 纳米颗粒[43]。

最近，硅基纳米材料作为一类很好的基底用于负载金属纳米颗粒，构建了优异的电催化剂。Sheng 等人制备了尺寸约为 2.2nm 的二元组分 Ir/Si 纳米线催化剂[44]。Si 除了控制 Ir 纳米颗粒的大小和防止其团聚外，在 HER 催化中也起着关键作用。在 Ir/Si 纳米线催化剂上，H_2 的生成过程分为三个步骤：①质子吸附在 Ir 原子表面并被还原为氢原子；②H 原子在 Ir-Si 界面由 Ir 原子迁移到 Si 原子；③吸附在 Si 原子表面的 H 原子相互结合或与电解液中的 H^+ 结合生成 H_2。Dai 等人研究报道称，从二维（2D）硅氧烷载体到 Ir 纳米颗粒的电子转移导致 Ir 纳米颗粒周围的电子密度增加，从而促进了 HER 中的氢吸附[45]。因此，通过调控电子结构，实现金属（Ir)-载体的相互作用，有利于提高 HER 的催化活性。因此应该更多地通过实验和计算的方法来探索载体与 Ir 金属的界面耦合来提升催化性能。

5.1.1.2 Ir 合金

通过合金化，异质金属元素可以调节主体金属元素的表面化学[46,47]，因此可以通过合金化来提高 Ir 的电化学性能。目前，已成功合成了许多 Ir 基合金，包括 Ir-NM 合金和 Ir-TM 合金用于 HER 催化。对于 Ir-NM 合金，如 Ir-Pt、Ir-Pd 和 Ir-Au 表现出优异的 HER 性能。以 Ir-Pd 合金为例，Wang 等人指出 Ir 掺杂的 Pd 纳米片组装纳米结构比纯 Pd 具有更高的 HER 性能，因为该合金具有可调节的电子结构，以及高度开放的 3D 纳米片组装结构。$Pd_{83.5}Ir_{16.5}$催化电流密度为 $10mA/cm^2$ 时仅需 73mV 的过电位[48]。

用 Co、Ni、Fe 等非贵金属 3d TMs 合金化 Ir，不仅保证了良好的催化性能，而且降低了材料的成本。具有不同纳米结构的 Ir-Ni 合金已被广泛用于 HER 催化，包括 Ir-Ni 薄膜、Ir-Ni 纳米颗粒[49]、GO 包覆的 Ir-Ni 合金(IrNi@OC)[50]和 $Ir_{25}Ni_{33}Ta_{42}$ 非晶合金[51]。例如：Gong 等人通过一步热解 Ir 掺杂 Ni 基 MOFs 制备了 O(吡喃 O，—OH 以及 C═O) 修饰的 IrNi@OC［图 5.2(a)］[50]。从图 5.2(b) 的高分辨 TEM（HRTEM）图像可以

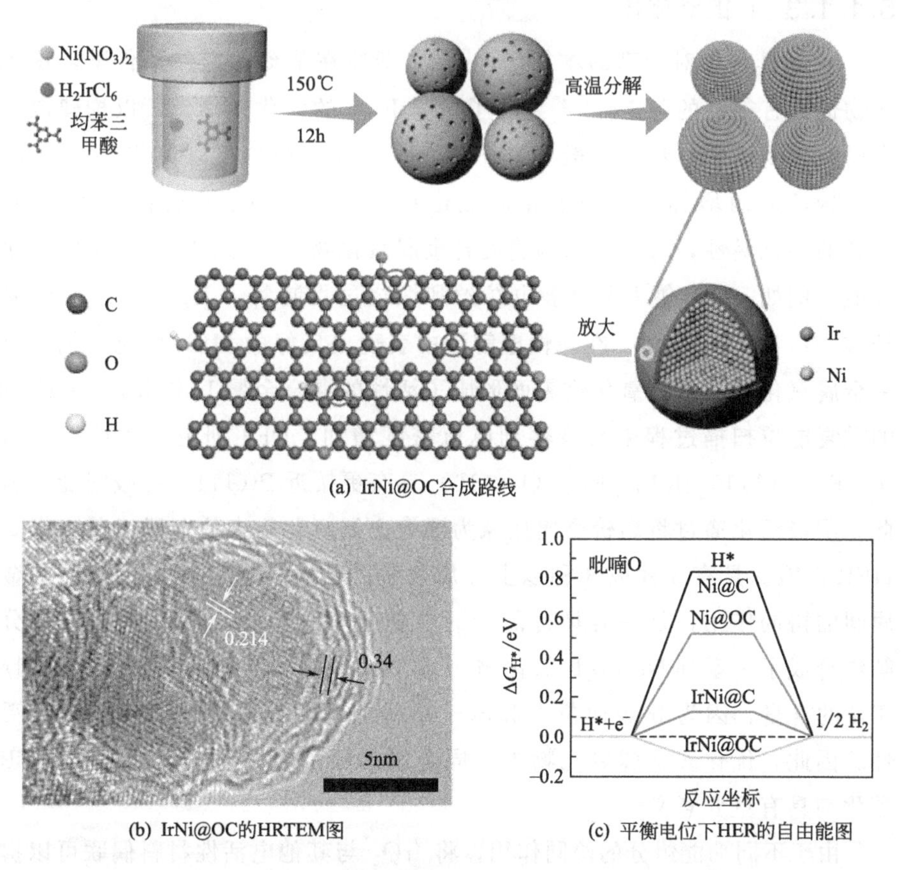

(a) IrNi@OC合成路线

(b) IrNi@OC的HRTEM图

(c) 平衡电位下HER的自由能图

图 5.2 IrNi@OC 的合成路线及表征（见文后彩插）[50]

看出，IrNi 合金被 3～4 层石墨烯包裹，具有典型的面间距 0.34nm。Ni 晶格中存在的 Ir 和石墨烯片中的吡喃 O 共同作用降低 ΔG_{H^*}［图 5.2(c)］。因此，H^* 在与氧相连的下一个最近的碳原子上的有利吸附，ΔG_{H^*} 也因 O 物种诱导的电子从合金核转移到石墨烯层中的碳原子而降低，从而提高了 HER 的性能。他们还研究了 N 掺杂石墨烯壳包裹 IrCo 纳米合金（IrCo@NC）的 HER 性能，接近商业 Pt/C 催化剂。更深入的理论研究表明，由于电荷从合金转移到石墨烯的 C 原子上，IrCo 合金可以显著地调节石墨烯的电子结构，氮掺杂和杂化结构协同作用导致 ΔG_{H^*} 值降低（−0.16eV）。因此，用碳覆盖的合金纳米颗粒构建纳米复合材料，是通过调节电子结构来提高催化剂本征催化活性的有效方法。另外，还可以设计多组分合金，进一步提升催化性能，同时降低 Ir 的含量。

5.1.1.3 Ir 的氧化物

纯 Ir 具有相对较高的本征活性[52]，即使在苛刻的操作条件下，Ir 氧化物仍具有很高的稳定性[53]。为了提高 IrO_2 的催化性能，可以构建混合金属氧化物复合材料。一般来说，掺入其他金属组分可以调节 IrO_2 的电子结构，从而增加或稳定催化活性位点，进而提高催化性能。Ru 具有较高的催化活性，可以用来构建混合金属氧化物[47,54]来提高 IrO_2 的催化性能。例如，Cho 等人通过静电纺丝和煅烧过程制备了 $Ir_xRu_{1-x}O_y$ 纳米纤维[54]。$Ir_{0.8}Ru_{0.2}O_y$ 在碱性电解液中表现出良好的 HER 性能，优于单一金属氧化物。由于氧化物表面形成了活性的混合金属 Ir 和 Ru，在 HER 的重复电位扫描过程中会观察到电流密度增加。DFT 研究还表明，金属 $Ir_{0.8}Ru_{0.2}$(111) 比 $Ir_{0.8}Ru_{0.2}O_2$(110) 具有更接近 Pt(111) 的吸附能。因此，阴极活化通过将高价金属还原为混合金属氧化物中的金属组分来提高 HER 性能。具有 3 种或 3 种以上金属物种的多金属氧化物，由于不同金属间的协同作用，是一类具有前景的电催化剂。Vidales 等人采用热盐分解法合成了一系列 Ni-Mo-Ir 氧化物[53]，Ni-Mo-Ir 氧化物的 HER 活性依赖于Ir 的含量，因为 Ir 表面位点和电子结构的调控可以提高材料的本征活性。因此，优化多金属氧化物中不同金属的添加量，对实现高性能 HER 催化剂具有重要意义。

由于不同功能组分的协同作用，将 IrO_2 与其他电活性材料偶联可以提高 HER 性能。例如，Zheng 等人通过两步电化学沉积法制备了 IrO_2-磷酸钴（CoPi)-CNTs复合物[55]。所制备的低Ir负载量［0.41%(质量分数)］的IrO_2-CoPi-CNTs催化剂表现出类Pt的HER性能，优于相应的单一组分。XPS测量和DFT计算进一步揭示了IrO_2和CoPi之间的协同效应。一方面，给电子的磷酸盐可以调节CoPi和IrO_2的电子结构，从而提供有利的ΔG_{H^*}(−0.13eV)，同时也防止了IrO_2团聚。另一方面，CoPi快速裂解HO—O键并将大量的H^*中间体转移到IrO_2相邻的催化位点上，促进H_2的形成和释放。这种协同作用也存在于其他化合物如Fe_2O_3[56]和$Er_2Si_2O_7$[57]中。目前，基于IrO_2的HER催化剂的研究仍然较少。

除了Ir和IrO_2外，其他的一些Ir基催化剂也表现出良好的HER性能，如Ir基磷化物[58]和氮化物[59]。2019年，Pu等人报道了一种二磷化铱（IrP_2）嵌入超薄氮掺杂碳层的电催化剂（IrP_2@NC)[58]。与商用的Pt/C相比，富P的IrP_2@NC催化剂具有更好的HER活性。DFT计算表明，P的引入显著降低了Ir/NC的H吸附能，从而有利于表面氢的形成。在另一项研究中，Kuttiyiel等人报道中空核壳结构的铱镍氮化物(IrNiN)，在酸性电解质中表现出与Pt/C催化剂相当的HER性能[59]。目前，只有几个实验揭示了Ir基材料在HER中的巨大潜力，更多的实验探索正在进行中。

5.1.2 Ir基催化剂用于OER

Ir基纳米材料是目前公认的极有前途的OER催化剂，因为Ir在苛刻的电化学条件下，特别是在酸性电解质中，具有良好的催化活性和稳定性。由于Ir价格高且储量低，因此设计高效、低Ir含量的电催化剂至关重要。目前，已经研发了多种Ir基纳米材料用于OER催化，包括Ir金属、Ir合金和Ir氧化物。

5.1.2.1 Ir金属

Ir金属纳米材料因其优异的催化活性、在恶劣环境中的稳定性，在电化学领域引起了极大的兴趣。为了设计具有最大Ir利用率和高催化活性的催化剂，已经研发了一些先进的策略，例如缩小尺寸和构建复合材料。

缩小尺寸是调节功能材料催化性能的可行方法，粒径决定了暴露的活性位点数[38,60]。通常催化剂的尺寸越小，比表面积越大，暴露的活性位点越多。因此，研究集中在纳米材料的设计上，尤其是新兴的单原子催化剂

(SACs)。2019 年 Yan 等人在 MnO_x/N-C 复合材料中掺入 Ir 单原子，该材料在碱性条件下表现出良好的 OER 活性[61]。Wang 等人采用另一种氧化物（NiO）构建 Ir-NiO 单原子催化剂，单个 Ir 原子不仅作为 OER 的催化活性中心，而且由于 Ir 原子引入了过量的电子调节了 NiO 附近的表面反应活性，这种协同效应提高了 OER 催化活性[62]。在另一个研究中，Luo 等人将 Ir 原子固定在 3D 非晶 NiFe 纳米线@纳米片上，得到的 $NiFeIr_x$/Ni NW@NSs 的 OER 活性显著提高，并研究了 Ir 和载体之间的协同效应对催化性能提升所起到的关键作用[63]。Jiang 等人通过可控沉积将 Ir 原子沉积在自支撑纳米多孔的（$Ni_{0.74}Fe_{0.26}$）$_3$P 材料上（np-Ir/NiFeP)，然后采用简单的自重构策略合成了 npIr-NiFeO［图 5.3（a)］[64]。随着在碱性条件下的电化学活化，单原子 Ir 转变为具有更高价态和更多氧配体的稳定状态，它们均匀地分散在无定形 Ni(Fe)（氧）氢氧化物上（np-Ir/NiFeO)，表现出优异的 OER 催化性能。XAS 表征得出 Ir 原子的—OH 配体经历了去质子化过程（Ir-OH 到 Ir-O*），并随着外加电位的增长转变为新的稳定氧位点。Ir 位点和活性氧物种组成的多个活性中心加速了 OER 过程。同时，Ni(Fe)—O 键的收缩形成了强健的表面结构，有助于锚定孤立的 Ir 原子并防止团聚。此外，从 Ni(Fe) 氧化物/氢氧化物到界面 Ir 原子的电荷转移，促进了 Ni(Fe) 羟基氧化物的形成并促进了 OER 反应过程［图 5.3（b)］。

另一种控制粒径的方法是将 Ir 纳米颗粒分散到具有高表面积的固体载体上。载体的孔隙率、电导率、比表面积、化学组成等性质对整体催化性能起着至关重要的调控作用。目前，被用作 Ir 纳米颗粒的载体有石墨烯[65-67]、碳材料[68]、二氧化硅[69]、蒙脱土[70]、锑掺杂氧化锡（ATO)[71,72] 等多种材料。在大多数情况下，这些载体不仅可以增强分散性和控制粒径，还可以通过强的催化剂-载体相互作用提高本征活性和结构稳定性[68,73]。除了 SACs 和 NPs 外，尺寸和形貌也显著影响 Ir 基催化剂的催化性能。例如，Liu 等人采用一锅水热法制备了多孔铱纳米线［Ir PNWs，图 5.4(a)］[74]。该 1D 纳米线结构具有特有的各向异性特征和丰富的微孔孔道，多孔 Ir 纳米线具有较大的 ECSA，从而表现出比商业 RuO_2 更好的 OER 催化活性和稳定性。Wu 等人使用一种简单的退火方法来制备 2D 无定形 Ir 纳米片［图 5.4(b)］[75]。无定形 Ir 纳米片的横向尺寸较小，厚度约为7.2nm，催化性能优于晶态对应物电催化剂。无定形 Ir 纳米片具有较高的 ECSA 和较

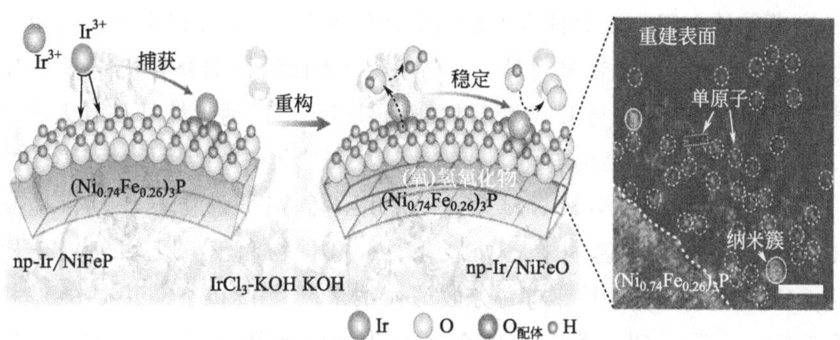

(a) np-Ir/NiFeO的制备路线示意图

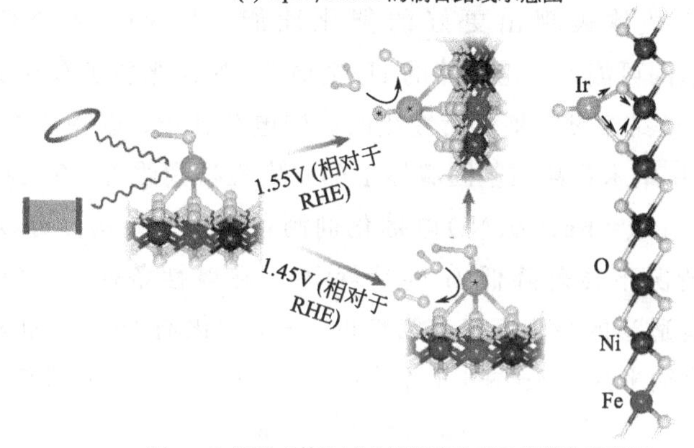

(b) np-Ir/NiFeO的XAS分析确定的OER机理的示意图

图 5.3 np-Ir/NiFeO 的合成路线及催化机理（见文后彩插）[64]

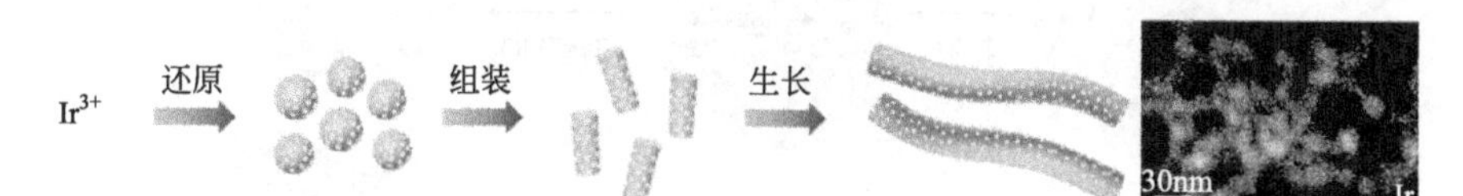

(a) Ir PNWs的合成机理示意图及EDX元素分析图谱

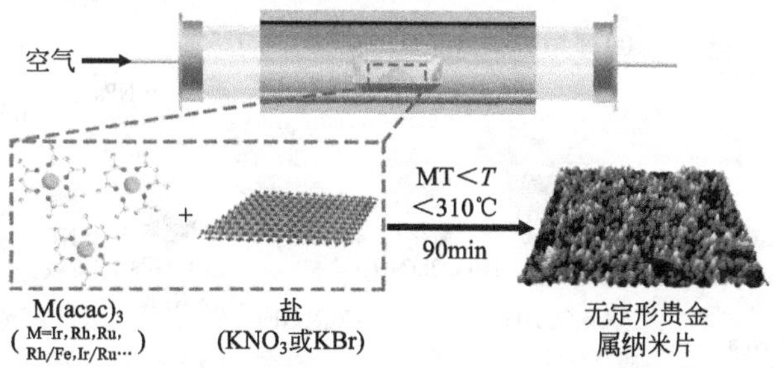

(b) 无定形贵金属纳米片的一般合成过程示意图

图 5.4 Ir PNWs[74]及无定形贵金属纳米片[75]的合成路线

MT 是金属乙酰丙酮化物的熔点

小的电荷转移电阻，因此具有丰富的活性位点和 OER 过程中快速的电荷转移。除了 1D 和 2D 纳米材料外，Oh 等人用表面活性剂介导法制备了 Ir 纳米枝晶（Ir-ND）[71]。与 Ir 纳米颗粒催化剂相比，Ir 纳米枝晶具有大的表面积、小的平均粒径和高的电化学孔隙率，从而催化剂的活性提高了两倍。

Ir 与其他电活性纳米材料杂化，利用不同组分的协同作用，提高其催化性能。过渡金属基材料，如氧化物、氮化物和氢氧化物已用于构建 Ir 基OER 杂化材料[76-81]。与高分散的亚纳米尺寸 Ir 掺杂介孔 Co_3O_4 相比［图 5.5(a)］[77]，Cho 等人发现金属氮化钴纳米纤维（Co_4N NFs）作为 Ir 纳米颗粒的载体表现出更好的催化性能［Ir@Co_4N NFs，图 5.5(b)］[76]。其原因可能是：①独特的 1D 金属 Co_4N 纳米粒子有利于电荷快速转移；②保持 Ir 纳米颗粒催化活性相的协同电荷补偿；③由于金属-载体的强相互作用，Ir 纳米颗粒在坚固载体上的稳定性得到改善。Tackett 等人发现，将 Ir/M_4N(M=Fe、Co、Ni)电催化剂制备成核壳结构，可以在不破坏 OER 活性的情况下显著降低 Ir 的消耗[78]。在碱性条件下，氢氧化物对 OER 中间体有适当的吸附，是高活性的 OER 催化剂[80,81]。Zhang 等人将 Co 基氢氧化物纳米片与 Ir 纳米颗粒耦合，发现在中性和碱性溶液中 Ir-Co

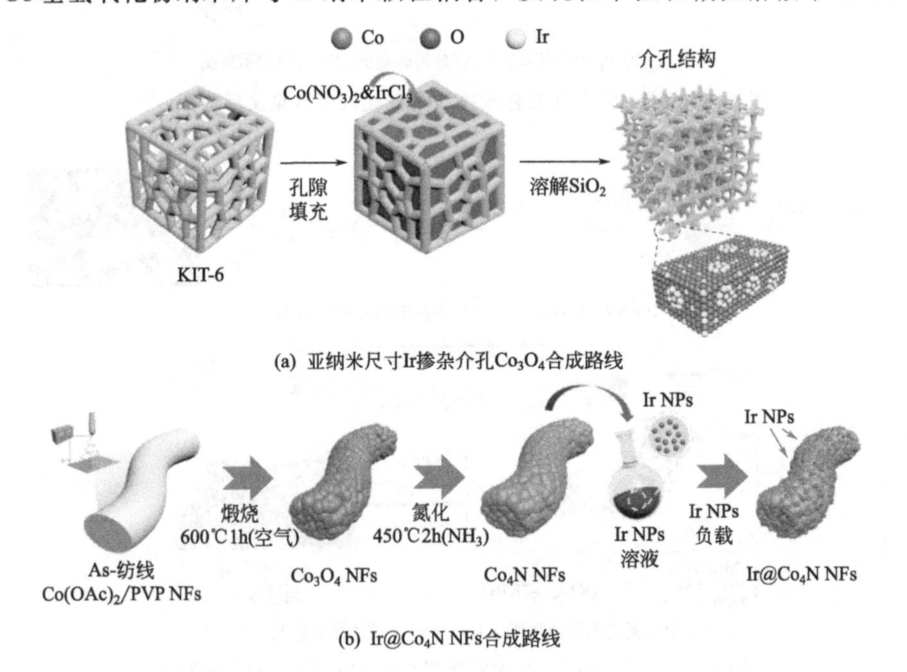

(a) 亚纳米尺寸Ir掺杂介孔Co_3O_4合成路线

(b) Ir@Co_4N NFs合成路线

图 5.5　Ir@Co_3O_4[77] 及 Ir@Co_4N NFs[76] 的合成路线（见文后彩插）

杂化物在 OER 中的性能优于商业 IrO_2 催化剂[79]。此外，Ir-Co 杂化物在 OER 过程中发生了不可逆的表面氧化和重构，α-Co(OH)$_2$ NSs 和原子 Ir 分别转化为 β-CoOOH 和高价 Ir，显著提升了 OER 活性。

5.1.2.2 Ir 合金

合金化可以调节 Ir 的电子结构和表面化学特性，通过调节反应中间体的吸附能来提高催化活性[82]。目前，许多贵金属（NM）（例如 Ru、Os、Pd、Pt）和非贵金属（TM）（例如 Fe、Co、Ni、W）被用来制备 Ir 基合金，它们不仅表现出优异的 OER 性能，而且降低了 Ir 的消耗，提高了 Ir 的利用效率。

最近，研究人员利用不同纳米材料之间的协同效应（例如集合效应、电子效应和应变效应）来设计结构稳定的 Ir-NM 合金作为 OER 催化剂。化学成分和纳米结构都显著影响 Ir NM 合金的催化性能。例如，Guo 等人报道了用微波辅助方法合成了一种可调成分的 RhIr 合金纳米粒子。该材料在 0.5mol/L H_2SO_4 中，具有优异的 OER 催化性能。通过将 Rh 与 Ir 合金化，整体效应和电子效应的协同作用以及—O 和—OOH 中间体的结合能差异降低，促进了 OER 活性。降低高成本金属含量，掺入 NM 合金来调节化学成分引起了越来越多的兴趣。一个成功的案例是采用湿化学方法合成 3D IrRuMn 纳米球[83]，在 IrRu 合金中掺杂 Mn 不但减少了贵金属的使用，且在酸性和碱性介质中都提供了很好的稳定性和高活性。

合金纳米结构的形成可以调节暴露的活性位点数量和表面结构，从而影响催化活性和结构稳定性。Zhang 等人通过一步溶剂热法制备了形状可调的 IrPd 合金纳米晶，包括纳米四面体（NTEs）、纳米线（NWs）和纳米空心球（NHSs）[84]。Ir-Pd 合金具有较高的表面微结构敏感性和催化活性，比活性与表面粗糙度呈正相关，其顺序为纳米线＞纳米空心球＞纳米四面体。用非贵金属 TM，特别是 3d TM(例如 Fe、Co、Ni、Cr、Cu) 合金化 Ir，可以有效地修饰 Ir 的电子结构，优化反应中间体（例如—OH、—O、—OH）的吸附能，从而提升催化性能，这种方法也能降低 Ir 的使用量[85-87]。Wang 等人认为 IrM 纳米线的催化活性和非贵金属相关，IrNi 纳米线具有比 IrFe 和 IrCo 纳米线更高的活性[85]。DFT 计算表明，配位环境决定了 3d TM 的活性；合金化后 Ir 的 d 带中心明显负移，减弱了反应中间体的吸附强度。Zhao 等人发现，Co 和 Ni 的掺入使纯 Ir 的态密度和双金属掺杂模型费米能级的电子密度变窄[88]。调谐的电子结构增强了催化剂的表面活性，并以快速的吸附/脱附过程进行电子-质子电荷交换。此外，与单金属催化剂相比，$(Ir_{0.7}Ni_{0.15}Co_{0.15})O_2$ 催化剂通过优化 OER 中间体的吸附强度显著降低了活化势垒，从而具有优异的催化性能。Ir-TM 合金的纳米结构也影响其催化性能，包括催化活性和结构稳定性。目前，各种复杂的 Ir-TM 纳米结构具有很高的 OER 催化活性和稳定性，如 IrNi 纳米笼[89]、IrFe 纳米线[85]、IrNi 纳米线[90]、多孔 IrCu 纳米晶[91]、IrNi 纳米颗粒[92]、CuNi@Ir 纳米框架[93]和 IrNi 纳米框架等[94]。值得注意的是，原子的排列对催化活性和稳定性起着至关重要的调控作用，由于 Ir 物种的抗腐蚀特性和较高的 OER 活性，富 Ir 的表面是合成良好催化剂的首选。例如，Park 等人合成了一种结构稳健的 Ir 基多金属双层纳米骨架（DNF）结构催化剂[93]。在长时间 OER 催化反应下，框架结构阻止了颗粒的生长和团聚，并原位生成了金刚石型 IrO_2 相，因此 IrNiCu 双层纳米骨架结构表现出优异的催化活性和稳定性。另外，Jin 等人采用一种镧系金属辅助合成方法，制备了具有合金成分且形貌可控的纳米框架材料（IrNi-RF）[94]。在整个纳米结构中包含许多晶界，由于局部区域原子排列无序，这些晶界被公认为是 OER 活性位点。因为，在 OER 催化过程中，IrNi 金属纳米框架材料表面氧化及其大的比表面积和丰富的晶界，共同促进了其在 0.1mol/L $HClO_4$ 中的 OER 性能。由于 IrNi 金属纳米框架具有独特的结构特点（内层为 IrNi 合金相，外层为 IrO_x 活性相）从而具有优异的稳定性。

5.1.2.3 Ir 氧化物

铱氧化物具有突出的催化性能，是最广泛研究的 OER 催化剂。由于铱氧化物在晶体结构、化学组成和纳米结构上的高度灵活性和复杂性，已报道的催化性能各不相同。金红石型铱氧化物（IrO_2）因其优异的催化性能和良好的晶体结构而成为 OER 的基准催化剂，也是 PEM 电解槽中最先进的 OER 电催化剂[95]。尽管金红石 IrO_2 的 OER 活性不如无定形 IrO_2，但在 Ir 溶解方面，结晶 IrO_2 在恶劣条件下更耐用[96]。为了解决这个问题并提高 IrO_2 的催化活性，通过成分优化、形貌控制等合理设计金红石型 IrO_2 来提升 OER 催化活性和降低 Ir 的使用量。将金属掺入 IrO_2 得到的混合金属氧化物提升了 OER 催化性能，Ir-Ru 氧化物同时具有 Ir 优异的耐久性和 Ru 的高 OER 活性（如多孔 $Ir_{0.7}Ru_{0.3}O_2$、IrO_2@RuO、$Ir_{0.7}Ru_{0.3}O_2$ 和 $Ru_{0.5}$

$Ir_{0.5}O_2$ NPs)[97]。Ir-Ru 氧化物的催化性能高度依赖于元素的组成和结构，比如 Faustini 等人通过与蒸发诱导自组装相关的喷雾干燥过程制备了多孔 $Ir_{0.7}Ru_{0.3}O_2$[98]。由于 Ir 和 Ru 的协同作用和材料的多孔结构，合成的 $Ir_{0.7}Ru_{0.3}O_2$ 的 OER 性能优于其类似物。此外，在 PEM 电解池中连续工作 48h，材料的孔结构依然保持良好。将稳定的 IrO_2 覆盖在其他材料上形成富 Ir 表面材料，也是一种开发优良 OER 催化性能的方法之一。一个典型的例子是核壳结构的 IrO_2@RuO_2[99]，其中活性 RuO_2 被稳定的 IrO_2 很好的保护。利用 IrO_2 的稳定性和 RuO_2 的本征活性，复合材料在催化 OER 活性和稳定性方面均表现出比纯的 IrO_2 和 RuO_2 更好的性能。相比较贵金属 Ru，低廉的非贵金属（Co[100,101]、Mo[102]、Cr[103]、Fe[104]、Mn[105]、Ce[106]和 W[107]）在研发性价比高的 Ir 基多金属氧化物作为 OER 高效催化剂更具优势。

掺杂是另一种常用来调节催化剂电子结构，从而优化材料催化性能的方法[46,105,108]。最近，Hao 等人制备了 Pt、La 共掺杂 IrO_2 纳米材料（$Pt_{0.1}La_{0.1}$-IrO_2@NC）[图 5.6(a)]，在酸性条件下表现出优异的 OER 催化活性和稳定性[108]。共掺杂的样品催化 OER 的性能高于单掺杂的样品 [图 5.6(b)]。La 和 Pt 共掺杂 IrO_2 除了 ECSA 增大和电荷转移电阻减小外，Ir 的 d 能带中心发生偏移，反应能垒从 1.83 eV 降低到 1.66eV。另外，向非贵金属和稳定的氧化物中掺杂 Ir 是构建低成本 OER 催化剂的另一种方法[105,109-111]。Ghadge 等人通过简单的水热和湿化学法制备了 Ir 和 F 共掺杂的 1D MnO_2 纳米棒（NRs）[图 5.6(c)][105]。由于纳米棒结构存在 1D 孔道以及独特电子结构，$(Mn_{0.8}Ir_{0.2})O_2$：10F 纳米棒表现出了优异的 OER 性能 [图 5.6(d)]。将 IrO_2 负载到稳定的基底上是开发稳健催化剂的有效方法，这种异质结构的好处包括：①稳定、小而活跃的纳米颗粒以防止其团聚；②通过提高 Ir 的利用效率来减少对 Ir 的需求；③调节 IrO_2 的电子结构，从而提高内在催化活性[112,113]。为了满足需求，有必要开发具有高比表面积和抗腐蚀能力的高导电性基底，比如 ATO、TiO_2、Nb-TiO_2、TiN、CNT、GCN 和 F 掺杂的 SnO_2(FTO)[113-119]。例如，Böhm 等人介绍了一种用小尺寸的 Ir 纳米颗粒均匀涂覆复杂多孔 ATO 新方法。扫描透射电子显微镜（STEM）的断层扫描表明，IrO_2 的分布与 ATO 支架的几何形状密切相关[120]。随后，形成的 ATO 负载 IrO_2 纳米颗粒具有约 89%的空隙体积，实现了极低的 Ir 堆积密度（约 0.08g/cm^3）的高质量活性（η=

300mV 时，63A/g_{Ir}）。TiN 和 GCN 除了增强电活性组分的分散性和抑制电活性组分的团聚外，还能有效的调制电子结构。例如，Li 等人发现，与纯 IrO_2 相比，TiN 能够很好地分散 IrO_2@Ir 纳米颗粒，并通过降低其 d 带中心 0.21eV 来调控 Ir 的电子结构[112]。此外，TiN 向 IrO_2@Ir 的电子转移抑制了 Ir 物种的进一步氧化溶解。与未负载 IrO_2@Ir 和市售 Ir 黑催化剂相比，IrO_2@Ir/TiN(60t%，质量分数）具有更好的稳定性和催化活性。Cheng 等人在 Ti 基纳米棒表面沉积了 IrO_x（Ir 和 IrO_2）纳米颗粒，所得到的 IrO_x-TiO_2-Ti(ITOT) 在 0.5mol/L H_2SO_4 中表现出良好的 OER 催化性能[119]。进一步的研究表明，TiO_2 载体有利于 Ir^{III} 和 $Ir^{III/IV}$ 混合价的无定形水合氧化物的形成，导致表面 OH^*_{ad} 的浓度很高，从而形成有效的 OER 过程。

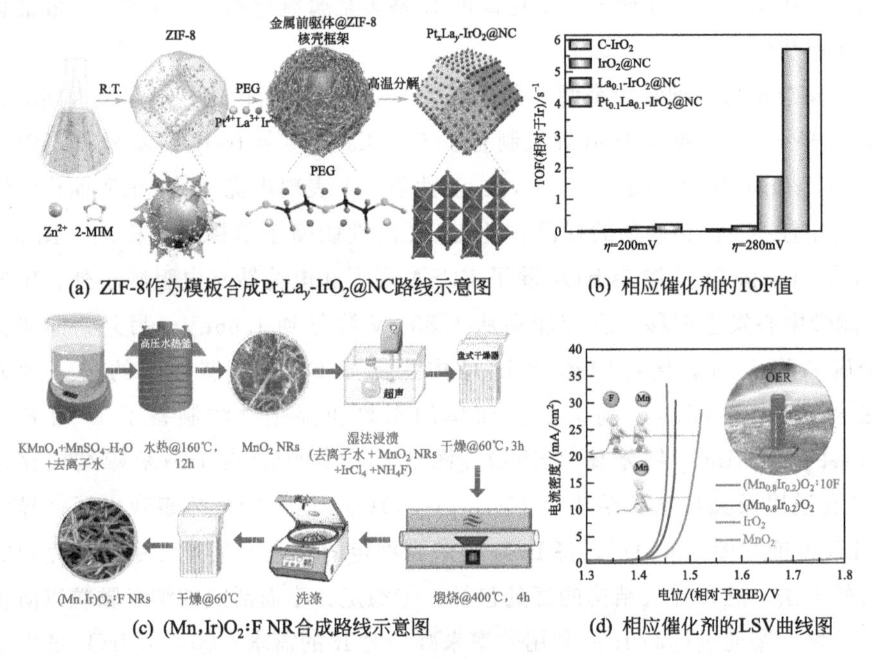

图 5.6 Pt_xLa_y-IrO_2@NC[108]和（Mn，Ir）O_2：F 纳米棒[105]的合成路线及性能测试（见文后彩插）

Ir-钙钛矿氧化物是一种良好的 OER 催化剂，通过在较便宜的框架材料中稀释 Ir 来减少 Ir 在合成活性高且稳定性好的催化剂中的使用量[121]。Ir 基钙钛矿的催化性能高度依赖于晶体结构，因此可以通过调节晶体学特征来促进催化剂的发展。例如，6H-$SrIrO_3$ 比 3C-$SrIrO_3$ 具有更好的 OER 催化性能[122]。DFT 计算和 XAS 结果表明，6H-$SrIrO_3$ 的优异性能来源于其共

面 IrO_6 八面体子单元，它削弱了表面 Ir—O 键，并促进了 OER 的决速步骤（比如 $O^* + H_2O \rightarrow HOO^* + H^+ + e^-$）。此外，3C-$SrIrO_3$ 在 OER 过程中发生了更显著的表面非晶化，故 6H-$SrIrO_3$ 比 3C 相对稳定。在另一种情况下，Chen 等人指出具有共角 IrO_6 八面体的赝立方 $SrCo_{0.9}Ir_{0.1}O_{3-\delta}$催化剂的 OER 性能超过了具有共角和共面 IrO_6 八面体混合的单斜 $SrIrO_3$（m-$SrIrO_3$）[123]。Ir 基钙钛矿的化学组成在决定其 OER 性能方面起着至关重要的作用。Diaz-Morales 制备了一组具有良好 OER 性能的 A_2BIrO_6 双钙钛矿（Ir-DPs），Ir-DPs 的催化活性由 B 位阳离子决定，其顺序为 Ce≈Tb≈Y<La≈Pr<Nd[121]。与未掺杂的 $SrIrO_3$ 相比，Co 掺杂 6H-$SrIrO_3$ 可以提高 OER 性能[124]。Co 的掺入不仅减小了材料的粒径，还促进了表面羟基的覆盖，调整了 IrO 键的共价键，并控制了 6H-$SrIrO_3$ 中 O 的 p 带中心。因此，6H-$SrIrO_3$ 中良好的表面调控和良好的电子结构显著加速了 OER 过程。这些研究结果为开发高性能 Ir 基钙钛矿型 OER 催化剂奠定了基础。

最近的研究表明，Ir 基焦绿石氧化物（$A_2Ir_2O_7$）因其独特的 IrO_6 八面体几何结构[125,126]，具有巨大的 OER 催化潜力。此外，通过改变 A 离子来改变电子关联，可以精确地调节它们的电子相关性，使焦绿石氧化物成为探索催化性质的一个极好的平台[127,128]。A 位阳离子主要控制 $A_2Ir_2O_7$ 的 OER 催化性能，随着 A 离子半径的增大，$A_2Ir_2O_7$（A＝Ho、Tb、Gd、Nd、Pr）的本征活性明显提高，这是因为增大的 A 离子半径削弱了混合氧化物中的电子关联[128]。这种效应诱导了绝缘体-金属的转变并提高了 IrO 键共价性，两者都有利于提高 OER 活性。然而，Abbott 等发现了不同的现象，即 $A_2Ir_2O_7$（A＝Yb、Gd、Nd）的 OER 性能随着 A 位阳离子尺寸的减小而提高[127]。产生这种矛盾结果的原因，可能源于催化剂的不同纳米结构（因为它们的合成方法完全不同）。因此，需要更深入的研究来揭示 $A_2Ir_2O_7$ 的 OER 性能与形貌/纳米结构、电子结构和表面化学等特性之间的结构-活性关系。

无定形纳米材料具有随机取向键、不饱和电子结构和柔性局域结构，是协调催化反应中氧化还原过程的理想催化剂[129,130]。无定形氧化铱（IrO_x）具有优异的 OER 催化性能受到大量的关注。前面讨论过电化学形成的 IrO_x，这里详细阐述未经电化学处理的无定形 IrO_x 材料（P-IrO_x），研究表明，P-IrO_x 的催化活性明显依赖于晶体结构[131]。Gao 等人采用一种简单的方法制备了 Li 掺杂的无定形 IrO_x，其催化 OER 活性比金红石型 IrO_2

更好[132]。虽然无定形 Li-IrO_x 和金红石型 IrO_2 具有相似的 IrO_6 八面体单元，但 IrO_6 八面体连接方式不同。与具有周期性互连的刚性 IrO_6 八面体的结晶 IrO_2 相比，具有无序 IrO_6 八面体的 Li-IrO_x 中的 Ir 在 OER 期间随着 IrO 键的收缩更容易被氧化，该键具有亲电性，能促进羟基氧化动力学和 O-O 键的形成，从而促进水的氧化。Willinger 等人还发现，类似碱硬锰矿结构的无定形 IrO_x 对 OER 的催化性能优于热力学稳定的金红石型结构[133]。这些结果表明，无定形 P-IrO_x 的性能优于晶型 IrO_x，P-IrO_x 的亚结构影响了受益于表面化学调控的催化性能。另外要在实际工程中应用，需确保无定形催化剂的稳定性。最好的方法是将电活性材料固定到合适的基底上。最近，Abbou 等人将无定形 IrO_x 纳米颗粒负载到掺杂 SnO_2 气凝胶上[113]。IrO_x/掺杂 SnO_2 气凝胶 OER 催化活性的稳定性受掺杂元素抗腐蚀性能的影响，SnO_2 基底中负载水平也决定了其性能。与 Sb 相比，适当的 Nb 或 Ta 掺杂 SnO_2 气凝胶在酸性条件下催化 OER 具有更好的稳定性，这对研发性能优异且稳定性好的电催化剂提供了更好的途径。

5.1.3 Ir 基催化剂用于全电解水

考虑到要简化电解槽、降低电解水的成本以及实现其在实际生产中的应用，在相同体系中耦合 HER 和 OER 电催化剂是非常有必要的[134,135]。因此，开发双功能催化剂对 HER 和 OER 都具有重要的意义。目前，已经有许多 Ir 基材料在各种 pH 条件下具有双功能催化水分解的性能，接下来对具有代表性的双功能催化剂进行简单的讨论。

5.1.3.1 Ir 金属基双功能催化剂

迄今为止，通过尺寸限制（例如单原子催化剂[30,136]和纳米颗粒[66,67]）和形态控制（例如纳米球[137]、纳米片[138]和纳米线[139,140]）已经研发出许多高效的 Ir 金属基双功能电催化剂。单原子催化剂（SACs）具有原子分散活性位点，其金属配位环境不饱和，有望最大限度地提高原子利用率并显著提升催化活性[30]。Lai 等人采用一种 π 电子辅助策略，将单个 M 原子（M=Ir、Pt、Ru、Pd、Fe、Ni）锚定在异质基底上［图 5.7(a)］[136]。Ir 原子可以同时锚定在两个不同的区域，Co 八面体中心（Ir@Co）和四重 N/C 原子（Ir@NC）上［图 5.7(b)］。与其他金属 SAC 相比，Ir_1@Co/NC 在 1.0mol/L KOH 电解液中对 OER 和 HER 均表现出较好的性能。

采用该材料（负载质量：0.2mg/cm^2）作为电解水的阳极和阴极催化剂，电位为1.603V时可获得10mA/cm^2的电流密度。通过DFT计算发现，优异的OER催化性能来源于Ir_1@CoO(Ir)位点，优异的HER催化性能来源于Ir_1@NC_3位点。这项研究还表明，基底-SA相互作用显著调节了单原子催化剂的几何结构和电子结构，从而改善了其催化性能。Luo等人的工作也证实了这一点[30]，他们通过将Ir原子分散在Fe纳米颗粒上，并将IrFe纳米颗粒嵌入到氮掺杂碳纳米管中（Ir-SA@Fe@NCNT），采用双重保护的方法来稳定原子分散的Ir SACs。在0.5mol/L H_2SO_4电解液中，Ir-SA@Fe@NCNT对HER和OER均表现出优异的催化性能和稳定性，优于商品化的Pt/C催化剂。

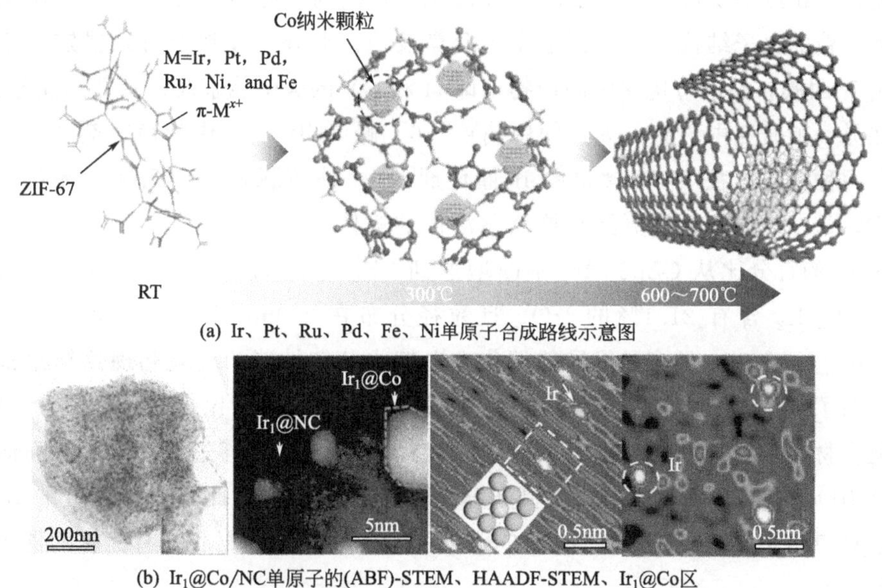

(a) Ir、Pt、Ru、Pd、Fe、Ni单原子合成路线示意图

(b) Ir_1@Co/NC单原子的(ABF)-STEM、HAADF-STEM、Ir_1@Co区域的HAADF-STEM以及Ir_1@NC区域的FFTI-HAADF

图5.7　Ir、Pt、Ru、Pd、Fe、Ni单原子的合成路线及表征[136]

不同尺寸的金属铱基纳米材料也表现出良好的全电解水催化性能。Fu等人通过简单的湿化学法成功合成了直径为1.7nm的超薄Ir波纹状1D纳米线[140]。与0D Ir纳米颗粒相比，1D纳米线具有独特的波浪形形貌和较大的催化活性面积，在0.1mol/L $HClO_4$溶液中表现出更优异的全电解水催化性能。最近，Cheng等人采用一种简单的湿化学法合成了厚度仅为5～6原子层的超薄Ir-纳米片[138]。Ir-纳米片部分羟基化表面在HER后保持不

变，OER 后主要被氧化为无定形的 IrO_xH_y，这些特性分别提高了 HER 和 OER 的催化活性。作者还认为，部分羟基化不仅增强了 Ir 位点与中间体（O）之间的表面电子转移，而且作为 H_2O 分解的第一步，保证了 HOH 的有效活化和键断裂。将 Ir 与其他材料（包括纳米碳材料和电活性材料）复合可以显著提高其催化性能，这类复合材料具有各组分的优点（如电活性高、导电性好、耐久性好等），表现出了良好的催化性能[38]。对于 Ir 纳米颗粒，采用高比表面积的载体来阻止颗粒团聚至关重要，而碳材料是最佳的载体。例如，Wu 等人合成了一种超细 Ir 纳米颗粒分散在氮掺杂石墨烯片载体上的催化材料（Ir@N-G-750）（直径＜6nm）[66]，该材料在酸性和碱性介质中均实现了高效的全电解水分解，性能优于 Ir@G-750 和商品化 Ir/C 催化剂。Ir@N-G-750 中丰富的 Ir-N 配位作用提高了电荷密度，改变了 Ir 纳米颗粒的电子结构，从而促进了电化学反应。此外，N 掺杂可以锚定和稳定 Ir 团簇，并通过优化中间产物（即 H 和 O）的吸附能来改善其催化性能。在另一项研究中，You 等人使用葫芦［6］脲（CB[6]）作为稳定剂，合成了一种具有可控 IrO_x 含量的 Ir 催化剂[141]。葫芦脲是一种刚性大环化合物，是一种比商业碳更具有抗氧化活性的载体。随着 CB[6] 含量的增加，IrO_x 的百分比从 CB[6]-Ir_3 不断增加到 CB[6]-Ir_1（从 11.5％到 32.8％）。CB[6]-Ir_2 含有 21.1％的 IrO_x 且粒径分布在 2.0nm 左右，在低负载量 $20ug_{Ir}/cm^2$ 时，具有最优异的催化全电解水分解性能。实验和理论结果都说明了 CB 和 Ir 之间的成键情况，与原始的 Ir(111) 相比，CB[6]-Ir(111) 复合材料中的富电子 Ir 位点更有利于捕获 O_2。电子转移过程促进了表面活性 IrO_x 物种的生成，从而提高了水分解的催化活性。此外，界面上的 Ir-O-C 键将 Ir NPs 紧密地锚定在 CB[6] 载体上，从而使材料具有优越的催化稳定性。

金属 Ir 与其他电活性材料的结合也改善了其催化性能，因为电活性材料的加入不仅增加了活性位点，而且还通过化学耦合作用调节了主体 Ir 基材料的本征活性[46]。最近，Zhao 等人通过多元醇还原过程将 Ir 纳米颗粒与 $Ni(OH)_2$ 纳米片杂交[80]。在 OER 催化过程中，Ir/$Ni(OH)_2$ 异质结构中的 Ir 和 $Ni(OH)_2$ 均发生氧化，分别转化为 IrO_x 和 NiOOH，显著地调节了 OER 中间体的吸附强度，促使 OER 动力学显著加快。Ir NPs 作为 HER 的活性位点，$Ni(OH)_2$ 大大加速了缓慢的水吸附/解离过程，Ir 与 $Ni(OH)_2$ 的协同作用提高了 OER 和 HER 的催化活性，因此该材料具有优异的催化电解水的双功能性能。在另一种情况下，Wei 等人注意到 Ir 原子在 MoS_2 上的吸附触发了 2H MoS_2 到 1T 的部分相变[142]。优化后的 Ir/MoS_2 复合物具有优异的全电解水催化活性，主要是因为：首先，Ir 物种的高分散性和超小尺寸提供了丰富的异质结界面和丰富的催化活性位点。其次，与 2H 类似物相比，Ir 物种诱导的 1T-MoS_2 具有良好的导电性和活化基面。再次，Ir/MoS_2 杂化物比原始 MoS_2 具有更高的亲水性（接触角分别为 76°、143°），促进了水的吸附和后续的电化学反应。最后，Ir 与 1T-MoS_2 之间的强协同和耦合作用促进了界面的水解离和电荷转移，从而促进了反应动力学，提高了催化活性。

5.1.3.2 Ir 合金基双功能催化剂

Ir-NM 合金：即将 Ir 与 NMs 如 Ag[143]、Pd[144]、Pt[82] 等合金化，可得到高效的双功能催化剂用于全电解水。例如，Zhu 等人研发的中空 Ir-Ag 纳米管（NTs）在 0.5mol/L H_2SO_4 中表现出优异的全电解水性能，这得益于该材料具有中空结构和表面富氧化 Ir[143]。Joo 等人研发了具有中空结构的仙人球状树枝结构的 $Cu_{2-x}S$@IrSy@IrRuNPs（CIS@IrRu NPs）[145]。由于垂直生长的尖峰状 IrRu 合金纳米结构所产生的活性位点密度较高，因此其催化活性较高，全电解水体系（CIS@$Ir_{48}Ru_{52}$ ‖ CIS@$Ir_{48}Ru_{52}$）仅在 1.47V 下便获得了 10mA/cm^2 的电流密度，其质量负载量仅为 12μg/cm^2。

将非贵金属 TM 引入贵金属合金中，可以在提高催化性能的同时降低催化剂成本。Sun 等人研发了具有可调整表面/界面结构的 PtCoIr 三金属鱼骨状纳米线（PtCo/Ir FB NWs），该材料具有优异的双功能催化性能[82]。在另一项研究中，Shan 等采用共还原多元醇合成了一系列非贵金属（TM）掺杂的 RuIr 纳米晶（TM＝Fe、Co、Ni）[146]。由于 Co 掺杂剂的双重作用，Co 掺杂 RuIr 表现出良好的双功能催化性能。在 OER 催化中，不可避免的 Co 浸出导致 O^{1-} 物种浓度增加，显著提高了 OER 催化活性。在 HER 催化中，Co 掺杂引起的电荷转移导致 Ru 表面价态的改变，优化了氢吸附，从而提高了 HER 活性。

如上所述，用低成本非贵金属合金化 Ir 不仅降低了 Ir 的使用量，而且还可以通过调控 Ir 的电子结构来提高其催化性能。目前，IrNi[86,147,148]、IrCo[86,147,149-151]、IrFe[147] 和 IrW[152] 等 Ir-TM 合金在 OER 和 HER 中均表现出良好的催化活性。Lv 等人合成了一系列具有3D 花状结构（Ir 基纳米花，NFs）的 Ir 基多金属纳米晶作为 OER 和 HER 的高效双功能催化剂[153]。在 HER 过程中，H 解吸/吸附峰的电位顺序为Ir/C≈Ir NFs＞IrCo 纳米花＞IrNi 纳米花，与 XPS 中 Ir 4f 峰的负移趋势一致。因此，在与 Co 或 Ni 元素合金化时，Ir 金属的电子结构受到调控，促使 Ir 上的氢结合能减弱，提高 HER 活性。Ir 与 Ni 或 Co 合金化引起的配体效应也促进了 Ir 基合金纳米纤维的 OER 催化活性，且这些合金的全电解水性能优于商用 Ir/C-Pt/C。同样，Feng 等人研发了一类 Ir 基多金属多孔空心纳米晶体（PHNCs）作为高效的全电解水双功能催化剂[86]。DFT 计算表明，与 Co 和 Ni 等 3d TMs 合金化后，Ir-TM 倾向于结合比纯 Ir 更弱的重要中间体（即—O、—OH、—OOH），从而降低活化能，促进了水分解过程。这些研究结果发现，Ir 与非贵金属合金化可以调节 Ir 的电子结构和中间体的吸附能，从而提高材料的催化性能。

5.1.3.3 Ir 氧化物/氢氧化物基双功能催化剂

Ir 的氧化物和类似物：利用 IrO_2 在 HER 和 OER 中良好的催化性能，一些研究试图将 IrO_2 基材料设计成电解水的双功能电催化剂。最近 Kundu 等人报道了一种 3D 分级多孔氧化铱/氮掺杂碳杂化物（3D-IrO_2/N@C），该杂化物在 0.5mol/L $HClO_4$ 溶液中具有良好的全电解水分解的催化性能[154]。与纯 IrO_2 相比，3D-IrO_2/N@C 复合物具有独特的多孔结构、高比表面积、快速的电子/质量传输以及 IrO_2 与 N@C 载体之间的强耦合作用，因此具有更优异的催化性能。Yu 等首次发现，由于单斜 $SrIrO_3$ 钙钛矿在 0.1mol/L KOH 溶液中 Sr 的浸出诱导表面重构，电化学活化后具有更高的 HER 催化性能（η_{10}=139mV，Tafel 斜率=49mV/dec）[155]，全电解水时其负载质量为 5mg/cm^2 表现出优异的催化性能和稳定性。与金属氧化物类似，金属硫族化物在全电解水领域也引起了巨大的研究兴趣[156,157]。Zheng 等人报道了一种锂插层的二硒化铱（Li-$IrSe_2$），作为 OER 和 HER 双功能催化剂[158]。通过向 $IrSe_2$ 中嵌锂活化产生丰富的孔隙和的 Se 空位，可用于高效的电化学反应。当 Li-$IrSe_2$ 组装成两电极水电解槽（负载量 3mg/cm^2）时，在 pH 为 0 和 7 时，10mA/cm^2 的槽电压分别仅为 1.44V 和 1.50V。

5.1.3.4 含 Ir 的氢氧化物

Ir 的层状双氢氧化物（LDHs）也具有优异的全电解水催化性能，如

NiVIr LDH[159,160]、NiFeIr LDH[161]、Ir-NiCo LDH/NF[162]等。这些化合物具有一个共同的特点：即在 TM LDHs 中掺入少量的 Ir 可以显著提高其电催化性能，特别是 HER 性能。例如，Chen 等发现 Ir 掺杂可以显著提高 NiFe LDH 在 HER 中的水溶解动力学，使其更有利于遵循有效的 Volmer-Tafel 过程[161]，因此该材料在 1mol/L KOH 溶液中具有高效且稳定的全电解水的催化性能。此外，Li 等发现 Ir 掺杂不仅降低了氢中间体在 NiV LDH 桥氧上的过强吸附，而且增加了 V 原子上的电荷密度，从而调节 OER 中间体的吸附能[160]。理论结果表明，Ir 的掺杂有望同时提高 NiVIr LDH 的 HER 和 OER 催化性能。实验结果与计算结果吻合得很好，NiVIr LDH 比 NiV LDH 和 Pt/C-Ir/C 具有更好的双功能催化性能。

将电解水体系和太阳能电池相结合，将太阳能直接转化为氢能，使可再生能源得到切实可行的转化、存储和再利用[163]。Fan 等人通过简单的大气溶剂热法和随后的自发电偶置换，制备了一种 Ir 原子掺杂的 NiCo LDH[162]。将 Ir 掺杂到 NiCo LDH 晶格中，不仅促进了它们在 HER/OER 中的协同作用，而且提供了更好的导电性和更大的电化学活性面积。因此，在 1.0mol/L KOH 溶液中，Ir 掺杂的 NiCo LDH 电对催化电解水时，在电流密度为 $10mA/cm^2$ 时仅需 1.45V 的低电压，表现出优异的全电解水性能。更有意思的是，通过将 Ir-NiCo LDH 与 Si 太阳能电池集成，太阳能驱动的水分解系统显示出 14.8%的高太阳能制氢效率。

在 HER 催化剂方面，金属 Ir 基催化剂是最佳的选择。通过其形貌控制、合金化和构建异质结来调节电子空间构型、提高电导率和增大 ECSA 是提高 Ir 基催化剂催化 HER 性能的有效策略。在 OER 催化剂方面，Ir 氧化物是一种较好的选择，氧化物的晶体结构（例如 IrO_6 八面体结构）影响 OER 活性。通过掺杂、组分调控和杂化等方法来提高 Ir 氧化物的 OER 催化性能。对于 Ir 金属和合金，通过调整尺度、控制纳米结构和杂化来改善 OER 活性。其中氧化阴离子引起的表面重构显著地调节了 Ir 基催化剂的表面化学，从而调节 OER 性能。利用 Ir 基催化剂对 HER 和 OER 均具有较高的催化活性，Ir 基金属、合金、氧化物和类似物作为双功能催化剂表现出良好的催化性能。与单一的 HER 或 OER 催化剂不同，双功能催化剂的设计表现出一定的特点：①将不同的 OER 活性材料与 HER 活性组分结合，如 Ir-LDH 复合材料；②利用原位阴离子氧化和阴极还原技术，对催化剂的表面活性物种进行调整，提高催化剂的固有活性，如 TM-Ir 合金；③探索提高催化性能的先进方法，如构建 SACs 和杂原子掺杂等。

尽管目前已经合成了许多优异的 Ir 基催化剂，但是仍然还需要进行继续的研究：

① 需要用先进的表征技术阐明反应机理。尽管 Ir 基纳米材料表现出良好的 OER 和/或 HER 性能，但对其潜在机理，尤其是酸性 OER 过程，包括真实的催化活性位点，表面演变/重构以及不同组分之间的协同作用，仍缺乏深入的研究。发展先进的原位操作技术（如 XPS、XAS、红外光谱、拉曼光谱等）对于监测催化剂的价态、化学组成、纳米结构、结构/表面溶解和原子水平的实时变化是十分必要的。这些信息有利于对表面演化、真实活性位点和反应中间体的探索，从而加强对反应过程的理解。此外，将实验结果与基于 DFT 计算预测紧密结合起来，将更好地揭示 Ir 基催化剂的电化学过程。

② 设计具有协同调制的催化剂。优异的电催化剂，需要具有高的内在活性、丰富的活性位点、有效的电子和传质以及高的化学和结构稳定性[164]。目前，大多数研究都集中在一个或几个方面，尽管这对于先进的水分解过程至关重要，但要同时实现所有这些方面的结合具有挑战性。因此，探索创新策略来协同内在活性位点、电子传输、传质和气体演化以及机械性和化学耐久性，将更好地满足 Ir 基电催化在实际水分解中的要求。

③ 规范电催化剂的评价。目前，在 $10mA/cm^2$ 下的过电位和 Tafel 斜率是评价催化剂活性最常用的参数。通常也经常使用交换电流密度、TOF 和稳定性等参数。对于 Ir 基催化剂，质量活性是另一个重要参数，但只有少数文章报道了这一参数。需要注意的是，各种工作中的评价方法不尽相同，如催化剂质量负载量、预处理工艺、电极制备方法等。这样的问题导致比较各种催化剂性能会不公平。因此，标准化的实验规程对于以公正的方式评价不同性质催化剂（例如化学成分、纳米结构等）的电催化活性至关重要。

④ 最大化 Ir 的利用效率。考虑到 Ir 价格昂贵、储量较低，在不牺牲性催化性能的前提下，应尽量减少 Ir 的消耗。在此背景下，良好分散的 Ir NPs、Ir 基多金属氧化物/合金以及 Ir 掺杂纳米材料是最有利的选择。尤其是 Ir 基单原子催化剂，表现出比其他类型催化剂更高的质量活性（例如 np-Ir/NiFeO 在 η 为 250mV 时，质量电流密度为 39900A/g），因此更值得关注。

⑤ 改善 pH 适用性。尽管 Ir 基作为 HER 或 OER 单催化剂适用于广泛 pH 值范围，但大多数报道的双功能 Ir 基纳米材料催化剂是在强酸性或强碱性条件下进行的。然而，在强酸、强碱电解质中存在一些严重问题，如需要具有高耐久性的特定离子交换膜、对碱/酸稳定的电催化剂以及耐腐蚀性环境的电池堆[165]。相比之下，在中性或接近中性条件下运行的设备更环保、经济。因此，应积极开发在中性电解质中高效且稳定的 Ir 基电催化剂的方法。

⑥ 推进实际应用。要将这些电催化剂应用于实际的电解水槽中，除上述问题外，还应考虑以下几个方面的问题：首先，有必要开发简单且可大规模合成的路线，以实现催化剂按需求进行大规模的生产。其次将无黏结剂催化剂构建成分级结构，消除黏结剂对电荷转移和稳定性的阻碍，提高可用活性位点密度。最后，将电解水系统与太阳能电池集成，直接将太阳能转化为氢能。因为此二元系统是实现可再生太阳能向清洁、高能量密度燃料可行转化、储存和后利用的最佳选择。

5.2 低 Pt 含量电催化剂用于 HER 催化

尽管 Pt 是最优异的 HER 电催化剂，但稀缺性和高价格限制了其大规模的应用。因此可以通过降低 Pt 含量来合成低 Pt 含量电催化剂（LPCEs），并保持其高催化活性来解决这一问题。本节概括了低 Pt 含量电催化剂的几种主要的合成方法，并总结了近年来一系列低 Pt 含量电催化剂在催化 HER 中的应用：载体负载的 Pt 纳米颗粒、载体负载的 Pt 纳米团簇、Pt 单原子催化剂、Pt 合金催化剂以及表面富 Pt 的低 Pt 含量电催化剂。Pt 催化 HER 反应过电位几乎为零，且在极端的 pH 条件下能保持优异的稳定性，被认为是 HER 电催化剂的“圣杯”（the Holy Grail）[166,167]。但是由于储量稀少、价格昂贵限制其在大规模商业化中的应用，目前很多科研者为规避这个问题，采用的普遍方法是使用储量丰富的替代品。尽管目前已经合成了许多性能优异、成本低廉的非贵金属基化合物用于 HER 催化，但其催化性能仍不如 Pt 电催化剂，还不能满足替代 Pt 的要求[168,169]。为了在性能和成本之间寻求平衡，研发低 Pt 含量电催化剂是一种可行的方法，在不降低催化活性的前提下降低 Pt 的负载量。大量的研究证实，电催化活性与暴露的活性位点的数量密切相关[170]。一般来说，暴露的活性位点数随着 Pt 含量的显著降低而降低。因此为了维持活性位点数，低 Pt 含量电催化剂应具有较高的 Pt 原子利用率（AUE）[171]。最直接方法是将 Pt 颗粒缩小到亚纳米团簇甚至单原子尺度上。因为催化剂尺寸的降低有利于增大材料的比表面积，从而具有更高的原子利用率[172,173]。当 Pt 活性中心在原子尺度上时，即所谓的 Pt 单原子催化剂，将获得最大的原子利用率[174,175]。然而，因为单个 Pt 原子具有极高的表面能，在高负载下容易团聚，因此进一步增加单原子催化剂的活性位点数是非常具有挑战性的[176,177]。材料是另一个实现高原子利用率的有效方法，在经济型材料表面包覆 Pt，构建原子级尺寸的 Pt 覆盖层/壳[178]。

另外各活性位点的本征活性，是直接决定电催化剂性能的关键因素。在这方面，优化低 Pt 含量电催化剂的结构有可能提高 Pt 活性中心的本征活性，有利于进一步降低催化剂负载量。普遍认为 Pt 基电催化剂可分为两类：①经济成本低的载体负载 Pt 基催化剂；②铂和其他非贵金属组成的合金或金属化合物。对于负载型 Pt 纳米颗粒，由于量子尺寸效应和表面几何效应，原子序数的减小促使反应活性大大增强[179,180]。在 Pt 原子分散的情况下，配位不饱和的 Pt 位点可以适度地与氢的反应中间体相互作用，从而表现出良好的 HER 活性[181-183]。此外，通过调节 Pt 位点中心的局部配位环境，进一步调控 Pt 单原子催化剂的活性，从而优化 HER 催化活性[184]。除了 Pt 物种的尺寸，还可以通过 Pt-基底相互作用来改变电子结构，进而提高 Pt 位点的本征活性[185,186]。此外，协同作用可以降低 Pt/载体界面周围氢物种的吸附/解吸障碍，从而提高 HER 催化性能。对于 Pt 基合金和金属间化合物，提高本征活性的途径主要包括：晶相工程、缺陷工程和成分优化[187,188]。由于在低 Pt 含量电催化剂中 Pt 的高原子利用率和强本征活性，促使 Pt 的使用量显著降低一个或两个数量级，从而控制了材料成本，使其在电化学大规模制氢气方面具有巨大的应用潜力。低 Pt 含量电催化剂的最新进展，为通过控制催化剂的组成、结构和形貌来提高电催化性能，提供了各种合成方法。目前，许多材料具有很大的潜力来取代商业 Pt/C 催化剂[186,189]。

5.2.1 低 Pt 含量电催化剂的合成

为了获得高 HER 催化性能，需要通过可控的合成策略赋予低 Pt 含量电催化剂合适的成分、形貌和结构。目前，合成低 Pt 含量电催化剂的方法主要有湿化学法、电化学法、退火法、原子层沉积法等。

湿化学法在大规模应用方面具有良好的潜力，已广泛用于制备低 Pt 含量电催化剂，通常采用 H_2PtCl_6 溶液作为 Pt 源。如图 5.8(a) 所示，Chi 等人通过 $NaBH_4$ 还原 $H_2PtCl_6 \cdot 6H_2O$ 溶液，在具有独特夹心结构界面的 Mo_2C@NC 表面上包覆并生长了 Pt 纳米颗粒[190]。获得的 Mo_2C@NC@Pt 纳米球具有典型的类似火龙果的异质结构，Pt 纳米颗粒均匀分布在 Mo_2C@NC 载体上。Pt 的负载量为 7.49%(质量分数) 时，其催化 HER 性能优于商业化的质量分数为 20%的 Pt/C。除了在低成本基底上外延生长低 Pt 含量电催化剂外，通过种子介导生长可以在 Pt 前驱体的镀液中得到具有核壳结构的低 Pt 含量电催化剂。在图 5.8(b) 中，向含有超小 Au 种子的水悬浮液中加入 $PtCl_6^{2-}$ 溶液，生长得到了 Pt 壳层[191]。超小的 Au@Pt 纳米颗粒可以暴露更多配位不饱和的 Pt 位点，提高 HER 催化活性。除了外延生长外，另一种合成低 Pt 含量电催化剂的方法是将载体与 Pt 纳米颗粒的水悬浮液进行杂化。例如，Hu 等人通过杂化过程在 CoS_2/CC 基底上负载了平均尺寸为 1.7nm 的 Pt 纳米颗粒，得到 Pt-CoS_2/CC 纳米片阵列［图 5.8(c)］[192]。超细 Pt 纳米颗粒均匀分散在多孔纳米片上，有助于提高 HER 的

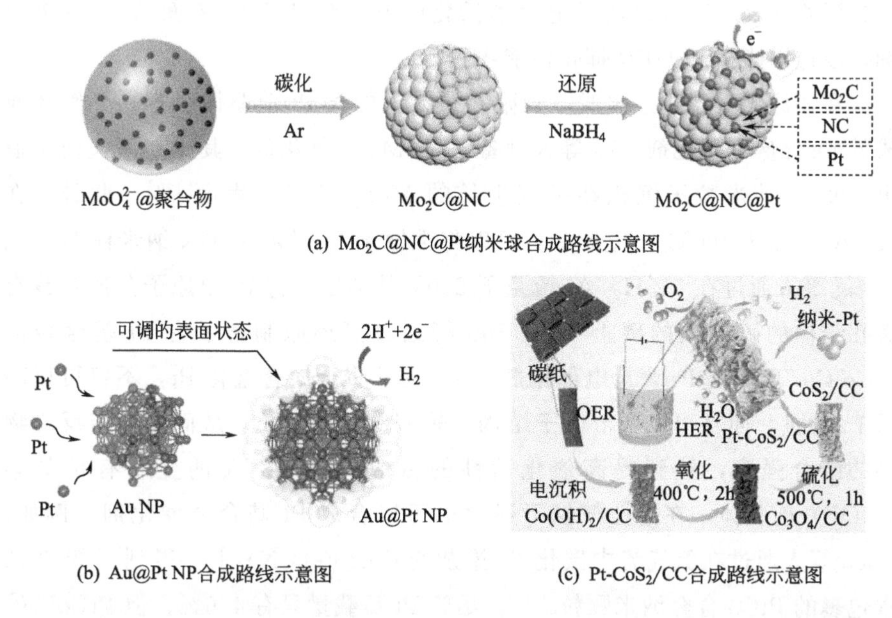

(a) Mo_2C@NC@Pt纳米球合成路线示意图

(b) Au@Pt NP合成路线示意图

(c) Pt-CoS_2/CC合成路线示意图

图 5.8　Mo_2C@NC@Pt[190]、Au@Pt NP[191] 以及 Pt-CoS_2/CC[192] 的合成路线图

催化活性和耐久性。此外，水热/溶剂热法作为湿化学法的一部分，可用于制备 Pt 纳米晶和 Pt 基合金催化剂，以获得较高的 Pt 原子利用率，并进一步降低成本[193]。例如 Huang 等人通过简单的水热路线合成了具有高密度孪晶缺陷的 Pt-Cu 分级的准十二面体[2]。

电化学方法具有几个独特的优势，包括通过电镀参数来精确控制 Pt 的数量和尺寸，对具有高 Pt 原子利用率的外层进行表面修饰，以及易于操作和扩展性，便于商业应用。合成低 Pt 含量电催化剂的电化学方法主要有电沉积和电化学活化两种。在没有离子交换膜的 HER 测量过程中，Pt 可以从 Pt 对电极溶解并重新沉积到 HER 催化剂上，从而显著改善 HER 测量结果。电解液中的 Pt 离子可以很容易地沉积在阴极电极上制备低 Pt 含量电催化剂。Chen 等人通过电沉积在蜂窝状 NiO@Ni 薄膜表面沉积了微小的 Pt 纳米团簇/纳米颗粒（Pt/NiO@Ni/NF）[194]。在静电吸附过程中，可以通过改变 H_2PtCl_6 溶液的 pH 来控制 Pt 在载体中的沉积量。此外，Yan 等人还报道了在原位电化学沉积 Pt 过程中，Pt 可以选择性地沉积在层状 WS_2 纳米片的边缘[195]。此外，电化学沉积可用于制备 Pt 单原子催化剂，采用电化学沉积方法在 CoP 基纳米管阵列/NF 上负载 Pt 单原子催化剂（PtSA-NT-NF），得到的 Pt 单原子催化剂具有比商用 Pt/C 催化剂更优异的 HER 催化活性[196]。还可以通过电化学活化的方法获得富 Pt 表面的 Pt 基催化剂，以进一步提高 HER 催化性能[197,198]。

在还原性气氛中退火时，吸附在载体上的含 Pt 前驱体很容易分解还原成低 Pt 含量电催化剂。Li 等人制备了 Pt-MoS_2 催化剂，其中在氢气流下退火，使 Pt 纳米粒子沉积在垂直生长的 MoS_2 纳米片表面[199]。此外，在 H_2/Ar 气氛中 100℃下退火 2h，可以将准原子 Pt 锚定在 TiN 纳米棒上[200]。Liang 等人通过在 5%H_2/Ar 流量下 250℃热还原，将 Pt 单原子分散在具有微孔结构的硫掺杂的碳上[201]。Sun 等通过氢还原制备了含 Pt 基合金的 PtCo-Co/TiM 低 Pt 含量电催化剂[202]。用非贵金属合金化 Pt，不仅可以降低 Pt 含量，而且可以调节电子结构、累积应变和键长，从而调节与反应物种的结合强度，达到提高催化活性的目的[203]。含 Pt 的金属有机骨架（MOF）前驱体，在惰性气氛下退火也可以合成 Pt 基合金催化剂。例如，Chen 等人通过在氮气流中碳化 Pt 掺杂的 $Co_3[Fe(CN)_6]_2$，得到了 N 掺杂碳包裹的 PtCo 合金纳米颗粒[197]。尽管 Pt 负载量只有 4.6%，但 PtCoFe@CN 的 HER 催化性能仍与商业 Pt/C 催化剂相当。此外，对含有 Pt 的 MOF

前驱体进行退火，通过刻蚀或蒸发消除金属纳米颗粒后可获得 Pt 单原子催化剂[204]。

原子层沉积（ALD）是在基底上实现材料共沉积的最新技术[205]。可以通过原子层沉积超薄 Pt 纳米颗粒在低成本基底上获得低 Pt 含量电催化剂。Nayak 等人使用原子层沉积，将 Pt 沉积在 2D 石墨烯片形成的 3D 互连网络上[206]。Saha 等人在 β-钼碳化物表面生长了亚纳米 Pt 颗粒[207]。在他们的工作中，沉积的 Pt 粒子的粒径可以通过原子层沉积循环次数来调节。通过调整原子层沉积工艺参数，可以精确控制在基底上分散的 Pt 原子。Sun 等人报道了一种利用原子层沉积技术合成 Pt 单原子和团簇的实用方法[208]。随着原子层沉积循环次数的增加，单个 Pt 原子和非常小的 Pt 团簇以更大的数量生长形成 Pt 纳米颗粒。

一般认为，形貌、结构、粒径以及 Pt 与基底的化学键对 Pt 基催化剂的 HER 活性有显著影响。形貌和结构决定了纳米晶催化剂的暴露晶面。高指数晶面通常比具有低米勒指数的晶面具有更高的活性，这是因为 Pt 原子在高指数晶面上配位不饱和。为了精确控制低 Pt 含量电催化剂的形貌和结构，较为理想的方法是水热/溶剂热合成，因为产物的特性与反应条件有关（如反应物浓度和反应温度）[2,188]。小粒径的粒子不仅有利于提高低 Pt 含量电催化剂的原子利用率，而且有利于协同不饱和活性位点的暴露[116]。通过原子层沉积方法，改变 Pt 沉积循环次数来控制 Pt 纳米颗粒尺寸[209]。Pt 与基底之间的化学键被广泛用于调节 Pt 活性位点的电子结构。一般使用湿化学法和电化学法构建具有强Pt-基底相互作用的纳米结构，因为 Pt 可以沉积在各种基底上，不同类型的基底使 Pt-基底界面的化学键存在差异。例如，电化学沉积 Pt 纳米粒子在 CoS_2 纳米片上，促进 Pt-S 键的形成，从而增加了 Pt 周围的电子密度[192]。另外，当 Pt 纳米粒子负载到 N、P 共掺杂碳纳米管上时，Pt 的 d 带发生了下移，这可能是由于电荷从 N、P 掺杂物向 Pt 转移所致。总的来说，不同方法制备的低 Pt 含量电催化剂具有明显不同的形貌、结构、颗粒尺寸和 Pt-基底相互作用，应针对所设计催化剂的特性选择合适的合成方法。

5.2.2 低 Pt 含量电催化剂在电催化 HER 的应用

研究表明，低 Pt 含量电催化剂可以保持甚至提高 HER 的催化活性。

一般来说，可以通过均匀分散Pt纳米材料改善Pt与催化剂载体之间的协同效应、减小Pt颗粒、改变电子结构和表面富集Pt纳米材料方式来提高HER性能。根据Pt在低Pt含量电催化剂中的不同存在形式，接下来简要描述负载Pt纳米颗粒、Pt纳米簇、Pt单原子催化剂、Pt合金以及具有富Pt表面的低Pt含量电催化剂在HER催化中的应用。

5.2.2.1 载体上的Pt纳米颗粒

众所周知，LPCEs主要以Pt纳米颗粒的形式负载在导电基底上[199]。碳基材料由于具有超高的比表面积、优良的导电性、低成本和多孔结构等特点，通常被用作载体。Yan等人在石墨烯纳米球上沉积了Pt纳米颗粒（Pt/GNs），Pt的负载量为14.7%(质量分数)，在0.5mol/L H_2SO_4 中具有优异的HER催化性能，电流密度10mA/cm^2 时的过电位为25mV[210]。正如他们所报道的，石墨烯存在的晶格缺陷和含氧基团可以增强Pt的均相锚定，增强材料的催化活性。除了尺寸均匀的Pt纳米颗粒外，碳基的功能化也有可能提高HER催化活性。例如，含氮官能团可能在酸性介质中捕获质子，在Pt活性位点周围创造富质子环境，从而提高HER催化活性[211]。同样，杂原子掺入碳基载体可以改善HER性能。N掺杂不仅可以通过促进Pt在N位成核的驱动力来增强Pt纳米颗粒的均匀分散，而且可以通过Pt纳米颗粒向N掺杂碳转移电子改变Pt的电子结构。Chen等人将Pt纳米颗粒高度分散在N掺杂的中空多面体多孔碳中制备了Pt@NHPCP催化剂[212]。与商用Pt/C催化剂相比，Pt@NHPCP具有更好的HER性能，这主要归因于材料中的N掺杂、Pt纳米颗粒均匀分散以及中空多面体多孔碳独特的纳米结构。过渡金属硫化物（TMS）在非贵金属电催化剂中表现出良好的HER性能和良好的导电性，为进一步提高HER催化活性，过渡金属硫化物可作为催化剂载体负载Pt纳米颗粒。Liu等人通过原位电沉积方法，在富含缺陷的 SnS_2 纳米片上沉积微量Pt（Pt-SnS_2）[213]。虽然Pt含量仅约为0.373%(质量分数)，但在电流密度为10mA/cm^2 时，催化HER过电位为117mV，远远优于富缺陷 SnS_2 纳米片的HER活性，HER性能显著改善的原因是Pt纳米颗粒与 SnS_2 纳米片富缺陷结构之间的协同作用。Tang等人选择性地将Pt纳米颗粒沉积在层状 WS_2 纳米片的边缘（er-WS_2-Pt），证明 WS_2 纳米片的边缘位点与Pt原子的相互作用比平台位点强得多[195]。这归因于边缘终止的 S_2^{2-} 或 S^{2-} 配体与Pt原子的结合能更低。er-WS_2-Pt在酸性和碱性介质中HER性能均表现出商业化的Pt性能。此外，Li等人将高分散的Pt纳米颗粒生长在边缘暴露的MoS_2纳米片上，表现出优异的HER催化活性，甚至优于商品化的Pt/C催化剂[199]。

过渡金属氢氧化物可以有效地裂解HO—H键，可作为载体制备在碱性介质中高活性的低Pt含量HER催化剂[214]。Tang等人将超薄Pt纳米线生长在单层氢氧化镍上组成低Pt含量电催化剂［Pt NWs/SL-$Ni(OH)_2$］[215]，该材料在0.1mol/L KOH和1mol/L KOH溶液中均展示了优于商业Pt/C良好的催化性能。Xing等人将Pt纳米颗粒沉积在$Co(OH)_2$纳米片上合成了低Pt含量电催化剂，具有优异的HER催化性能，在1mol/L KOH溶液中过电位为70mV时，展现出比商业Pt/C高4.8倍的电流密度[216]。此外，Anantharaj等将Pt纳米颗粒锚定在NiFe层状双氢氧化物晶片上，在1mol/L KOH中表现出较高的HER催化活性[217]。

MXene具有高电子电导率、亲水性和良好的稳定性，因此也可以作为制备低Pt含量电催化剂的基底[218]。例如，Jiang等人在Ti_3C_2纳米片上原位生长了$Pt_{3.21}Ni$超薄纳米线，在0.5mol/L H_2SO_4中具有优异的HER性能[186]。此外，Pt纳米颗粒还可以沉积在其他载体上，比如NiN纳米片[219]、镍单晶刺[220]、DNA分子自组装[221]。Lao等人将Pt纳米颗粒沉积在碳酸氢镍纳米片上合成的LPCE[Pt/$Ni(HCO_3)_2$]，在1mol/L KOH中显示出优异的电催化HER性能[189]。

5.2.2.2 载体上的Pt纳米簇

与Pt纳米粒子相比，Pt纳米簇（NCs）具有更小的尺寸、更高的比表面积、更高的AUE和更多的配位不饱和Pt位点。因此，科研者希望制备出基底负载Pt纳米簇的低Pt含量电催化剂，以获得优异的HER催化性能。Wang等人将超薄Pt纳米簇负载在三棱柱笼中制备了低Pt含量电催化剂(Pt@CIAC-121)[222]，因笼状CIAC-121的有限空间，形成的Pt纳米簇尺寸为1.4nm，该材料在0.5mol/L H_2SO_4溶液中，展现了比商业Pt/C更优异的催化性能。Chen等人将亚纳米氧化的Pt纳米簇负载在TiO_2载体上合成低Pt含量电催化剂（PtO_x/TiO_2）[223]。Pt纳米簇均匀分散在载体上，平均粒径为0.7nm。他们发现，Pt与TiO_2之间的Pt—O键增强了金属-载体相互作用，具有较高的稳定性。在0.5mol/L H_2SO_4中，过电位为

100mV 时，PtO_x/TiO_2 质量活性为 42.66A/mg，是 Pt/C 催化剂的 6.1 倍。Yan 等采用湿化学还原法，在二维 NiFe 层状双氢氧化物纳米片上沉积了平均尺寸为 0.59nm 的超薄 Pt 纳米簇（Pt-NiFe LDH/CC）[224]。在 1mol/L KOH 中，超低 Pt 含量（1.56%，质量分数）的 Pt-NiFe LDH/CC 表现出优异的 HER 催化活性，在电流密度为 $10mA/cm^2$ 的过电位为 28mV。Shi 等人将 Pt 纳米簇均匀分散在富缺陷 WO_3 上，在 0.5mol/L H_2SO_4 中电流密度为 $10mA/cm^2$ 下过电位为 42mV[225]。

5.2.2.3 Pt 单原子催化剂

Pt 进一步缩小到原子尺度，就可以得到单原子催化剂。Pt 单原子催化剂具有 Pt 的最大原子利用率和独特的 Pt 活性中心配位环境。从 Pt 纳米颗粒到 Pt 纳米簇再到 Pt 单原子催化剂，由于 Pt 中心较低的配位环境、量子尺寸效应和更强的金属载体相互作用，Pt 活性位点的本征活性随着尺寸的减小而提高[226]。然而，由于 Pt 的表面能越来越高，使其粒径从纳米量级减小到亚纳米量级和单原子量级较为困难[227]。

目前报道最多的用于 HER 的 Pt 单原子催化剂，是将原子级 Pt 高分散负载在碳基材料上的催化剂。杂原子，特别是 N 和 O，可以为 Pt 原子提供均匀的锚定位点，从而在碳基材料中形成活性部分[182]。Song 等人用原子层沉积将 Pt 原子均匀分散在洋葱状碳纳米球上（Pt1/OLC）[183]。根据扩展 X 射线吸收精细结构（EXAFS）测量和密度泛函理论（DFT）计算的结果，由一个碳原子稳定的 Pt 单原子和两个氧原子（$Pt_1O_2C_1$）组成高均一化活性组分。当 Pt 含量为 0.27%(质量分数）时，Pt_1/OLC 催化剂在 0.5mol/L H_2SO_4 中展现出优异的 HER 催化性能。DFT 计算结果表明，Pt_1/OLC 优异的 HER 催化性能，归因于−0.01eV 的低自由能变化和质子向 Pt 活性位点的有效传质。此外，Lu 等人通过构建 Pt 活性物种的配位环境，优化了基于石墨二炔（GDY）的 Pt SACs 的 HER 催化性能[184]。其中孤立的 Pt 原子以两种形式均匀分散在石墨二炔上，一种形式是以五配位 C_1-Pt-Cl_4 形式分散在 Pt-GDY1 中，另一种以四配位 C_2-Pt-Cl_2 形式分散在 Pt-GDY2 中。由于 Pt-GDY2 中 Pt 位点的配位数较低，在 100mV 过电位下，Pt-GDY2 催化剂的质量活性为 23.64A/mg，分别是 Pt-GDY1（7.26A/mg）和商业 Pt/C 催化剂（0.88A/mg）的 3.3 倍和 23.64 倍。此外，Zeng 等人通过桥氧分子，将 Pt 原子嫁接到 Fe 掺杂石墨化碳的 FeN_4 物种上，得到了具有新型 Pt_1-

O_2-Fe_1-N_4 单元的 Pt 单原子催化剂，在酸性介质中催化 HER，在电流密度为 $10mA/cm^2$ 下，过电位为 60mV[228]。

有趣的是，单个 Pt 原子也可以分散在其他金属中形成 Pt 的单原子合金。当引入合金时，改变了电子结构和几何构型，促使单原子 Pt 的化学性质被改变[227]。因此这类 Pt 单原子催化剂合金很有希望作为优异的 HER 催化剂。如 Bard 等人制备了 Bi 和 Pd 为基底的 Pt 单原子合金作为 HER 的催化电极[229]。Li 等人将原子级 Cu-Pt 二元合金分散在 Pd 纳米环上组成的 Pt 单原子催化剂[181]。Pd/Cu-Pt 纳米环具有六方环状结构，保持了 Pd 基底的晶相，Pt 含量为 1.5%(原子分数)。Pt 原子被 Cu 和 Pd 原子包围，其中 Pt@Cu 和 Pt@Pd 中 Pt 的配位数分别为 3.6 和 5.1，在 0.5mol/L H_2SO_4 溶液中展现出比 Pt/C 更优异的催化性能。此外，非贵金属基化合物可以稳定 Pt 单原子催化剂，Wang 等人将 Pt SACs 固定在双过渡金属 MXene 纳米片上（$Mo_2TiC_2T_x$-Pt_{SA}）[230]，他们发现 MXene 中的 Mo 空位为单个 Pt 原子提供锚定位点。与商用 Pt/C 催化剂相比，$Mo_2TiC_2T_x$-Pt_{SA} 催化剂表现出优异的 HER 性能，在 0.5mol/L H_2SO_4 中电流密度为 $10mA/cm^2$ 时过电位为 30mV。Lou 等人报道了将单个 Pt 原子锚定在 CoP 的纳米管阵列上得到 Pt 单原子催化剂，在磷酸缓冲液中具有优异的 HER 催化活性[196]。Tang 等人将单原子 Pt 修饰在纳米多孔的 $Co_{8.5}Se$ 基底上得到 Pt 单原子催化剂，该材料在中性介质中展现出优异的 HER 催化性能[231]。

5.2.2.4 Pt 合金催化剂

在 Pt 与其他金属合金化过程中，可以通过调整 Pt 的电子结构、累积应变和键长来提高 Pt 的催化活性[203]。可以通过金属成分和比例来调节合金的电子结构，从而优化合金的 HER 催化性能。Zheng 等人用湿化学法合成了六方密排的纳米多枝状 Pt-Ni 合金，在 0.1mol/L KOH 中，$10mA/cm^2$ 的电流密度下过电位为 65mV，具有良好的 HER 催化活性[232]。Zhang 等人制备了具有良好结晶的莲花-丘脑形 Pt-Ni 合金催化剂，在 1mol/L KOH 溶液中具有优异的 HER 催化性能[188]。除了二元 Pt 基合金外，三元 Pt 基合金也有报道。更复杂的三元 Pt 基合金提供更多的可能性来调整金属组成和比例，从而进一步提高电催化性能[233]。例如，Xiong 等人合成了 PtFeCo 三元合金催化剂，其 HER 催化性能与原子间电荷极化和 d 电子耦合均有很强的相关性[234]。研究表明，可以通过改变化学试剂的剂量和反应温度来调节 PtFeCo 三元合金的组成。如图 5.9(a) 所示，$Pt_{81}Fe_{28}Co_{10}$ 材料具有三星形状，分支长约 12nm，直径约 4nm。该材料在酸性条件下，展示了明显优于商业 Pt/C 的催化 HER 性能 [图 5.9(b)]。也有报道将 Pt 合金催化剂生长在基底上的[234]，比如 Qin 等人在多孔碳上合成了 PtCo 二元金属合金，在酸性条件下具有优异的 HER 催化性能[235]。

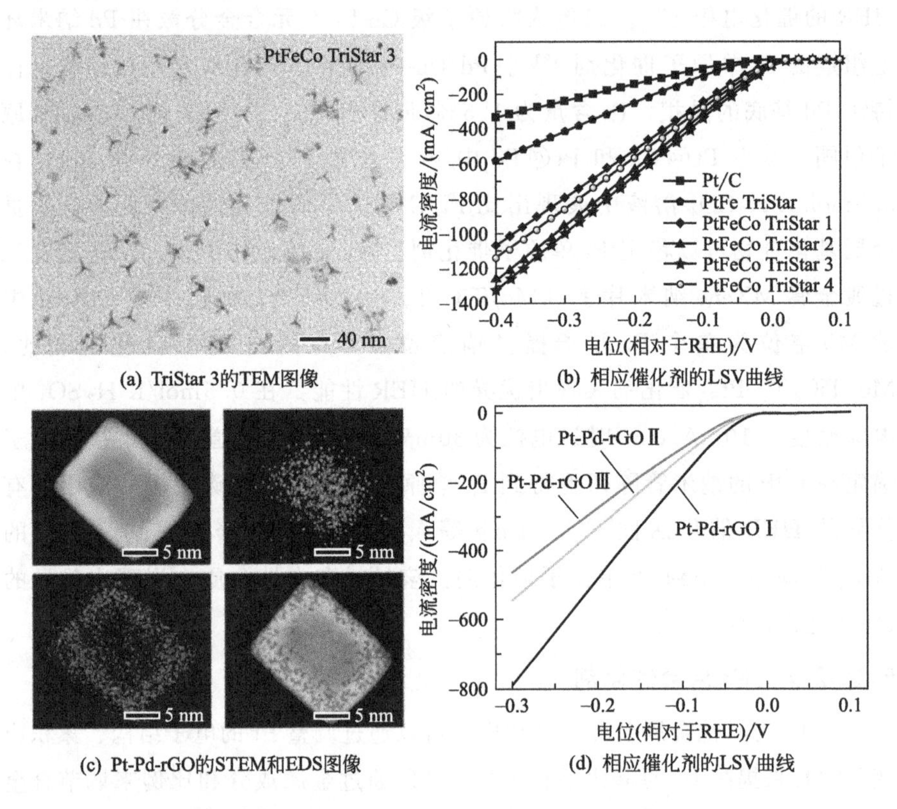

图 5.9 TriStar 3[234] 及 Pt-Pd-rGO[236] 的 TEM 图像及 LSV 测试

5.2.2.5 表面富 Pt 的低 Pt 含量电催化剂

为了提高 Pt 的原子利用率，通过将 Pt 涂覆在基底上来构建具有原子薄层的 Pt 表面，可以得到具有富 Pt 表面的低 Pt 含量电催化剂，在薄 Pt 壳层 Pt 使用量略有增加的情况下，可以有效地提高 HER 催化性能[237]。Xiong 等人报道了在还原的氧化石墨烯上涂覆了 Pt 壳层（0.8nm）的独特 Pd 堆结构 (Pt-Pd-rGO Ⅰ)[236]。在图 5.10(c) 中 Pt 富集在 Pd 核的壳层中，由图 5.9

(d) 可知，在所有 Pt-Pd-rGO 催化剂中，Pt 含量最低的 Pt-Pd-rGO Ⅰ 催化剂表现出最高的 HER 催化性能。值得注意的是，随着 Pt 富集表面壳厚度的减小，HER 活性增加。薄的富 Pt 表面可以通过表面极化使 Pt 和 Pd 之间的功函数不同，从而提高 HER 性能。Chen 等人研发了一种具有核壳结构的 Cu-Pt 纳米线，在中性介质中具有较好的 HER 活性[238]。

要实现高 HER 催化性能，需要对低 Pt 含量电催化剂的成分、形态、结构进行合理的设计和调谐。可以通过高比表面积、多孔结构、高密度缺陷和及/或高边含量的载体来负载高度均匀分散的 Pt，以此增加暴露的 Pt 活性位点数量。同时，Pt 与载体之间的协同作用可以提高 HER 催化活性，如构建 Pt—O 键、Pt—S 键、与官能团的相互作用等。减小 Pt 或低 Pt 含量电催化剂的尺寸可以有效提高 Pt 的 UAE，暴露更多的低配位 Pt 位点，这可以通过制备 Pt 纳米簇甚至 Pt 单原子催化剂来实现。另一方面，电子结构的修饰可以进一步提高 Pt 的本征活性，这可以通过调控 Pt 基合金或 Pt 基复合材料的成分和结构来实现。同样，Pt 表面的富集通过表面极化和提高 Pt 的原子利用率来提高 HER 性能，这可以通过在基底上包覆原子薄层的 Pt 膜来实现。

尽管低 Pt 含量电催化剂在 HER 方面取得了一些重大突破，但仍需进一步努力，可以向以下几方面进行发展：

① 为实现工业化大规模应用，应该采用易于规模化生产的方法来制备低 Pt 含量电催化剂。就目前的制备方法来看，可以通过湿化学和电化学方法来获得低 Pt 含量电催化剂。一般将 Pt 原子通过强相互作用，选择性的锚定在基底的缺陷或边缘上。因此，基底应具备这些特征，但这些特征很难大规模获得。故有必要开发易于规模化合成制备低 Pt 含量电催化剂的合适基底。

② 其他一些廉价的 Pt 族金属，例如 Ru，在化学惰性和电化学活性方面与 Pt 具有相似的性质。为了进一步降低低 Pt 含量电催化剂的成本，这些更廉价的 Pt 族金属可以用来替代 Pt 以获得低成本的电催化剂，例如低 Ru 含量的电催化剂。

③ 许多研究表明，Pt 与载体或其他金属之间的协同效应在提高 HER 活性方面起着关键作用。在很多情况下，科研人员都是采用 DFT 计算来支撑这一说法，并没有强有力的直接证据。因此在这方面，原位光谱测量可以用来研究电化学 HER 过程中 Pt 和基底之间的界面结构，有助于分析和证明协同效应。

参考文献

[1] Wu H B, Lou X W. Metal-organic frameworks and their derived materials for electrochemical energy storage and conversion: Promises and challenges. Science Advances, 2017, 3 (12): eaap9252.

[2] Huang R, Sun Z, Chen S, et al. Pt-Cu hierarchical quasi great dodecahedrons with abundant twinning defects for hydrogen evolution. Chemical Communications, 2017, 53 (51): 6922-6925.

[3] Oh A, Kim H Y, Baik H, et al. Topotactic transformations in an icosahedral nanocrystal to form efficient water-splitting catalysts. Advanced Materials, 2019, 31 (1): 1805546.

[4] Vielstich W, Lamm A, Gasteiger H. Handbook of fuel cells. Fundamentals, technology, applications. Chichester (United Kingdom), 2003.

[5] Reier T, Oezaslan M, Strasser P. Electrocatalytic oxygen evolution reaction (OER) on Ru, Ir, and Pt catalysts: A comparative study of nanoparticles and bulk materials. ACS Catalysis, 2012, 2: 1765-1772.

[6] Reier T, Nong H N, Teschner D, et al. Electrocatalytic oxygen evolution reaction in acidic environments-reaction mechanisms and catalysts. Advanced Energy Materials, 2017, 7 (1): 1601275.

[7] Sardar K, Petrucco E, Hiley C I, et al. Water-splitting electrocatalysis in acid conditions using ruthenate-iridate pyrochlores. Angewandte Chemie International Edition, 2014, 53 (41): 10960-10964.

[8] Chen P, Xu K, Fang Z, et al. Metallic Co_4N porous nanowire arrays activated by surface oxidation as electrocatalysts for the oxygen evolution reaction. Angewandte Chemie International Edition, 2015, 54 (49): 14710-14714.

[9] Anantharaj S, Ede S R, Sakthikumar K, et al. Recent trends and perspectives in electrochemical water splitting with an emphasis on sulfide, selenide, and phosphide catalysts of Fe, Co, and Ni: A review. ACS Catalysis, 2016, 6 (12): 8069-8097.

[10] Cabán-Acevedo M, Stone M L, Schmidt J R, et al. Efficient hydrogen evolution catalysis using ternary pyrite-type cobalt phosphosulphide. Nature Materials, 2015, 14 (12): 1245-1251.

[11] Sultan S, Tiwari J N, Singh A N, et al. Single atoms and clusters based nanomaterials for hydrogen evolution, oxygen evolution reactions, and full water splitting. Advanced Energy Materials, 2019, 9 (22): 1900624.

[12] Stamenkovic V R, Mun B S, Arenz M, et al. Trends in electrocatalysis on extended and nanoscale Pt-bimetallic alloy surfaces. Nature Materials, 2007, 6 (3): 241-247.

[13] Greeley J, Mavrikakis M. Alloy catalysts designed from first principles. Nature Materials, 2004, 3 (11): 810-815.

[14] Luo M, Guo S. Strain-controlled electrocatalysis on multimetallic nanomaterials. Nature Reviews Materials, 2017, 2 (11): 17059.

[15] Guo C, Jiao Y, Zheng Y, et al. Intermediate modulation on noble metal hybridized to 2d metal-organic framework for accelerated water electrocatalysis. Chem, 2019, 5 (9): 2429-2441.

[16] Lai J, Huang B, Tang Y, et al. Barrier-free interface electron transfer on PtFe-Fe_2C janus-like nanoparticles boosts oxygen catalysis. Chem, 2018, 4 (5): 1153-1166.

[17] Chen W, Pei J, He C T, et al. Rational design of single molybdenum atoms anchored on

N-doped carbon for effective hydrogen evolution reaction. Angewandte Chemie International Edition，2017，56 (50)：16086-16090.

[18] Zhang L，Fischer J M T A，Jia Y，et al. Coordination of atomic Co-Pt coupling species at carbon defects as active sites for oxygen reduction reaction. Journal of the American Chemical Society，2018，140 (34)：10757-10763.

[19] Yao Y，Hu S，Chen W，et al. Engineering the electronic structure of single atom Ru sites via compressive strain boosts acidic water oxidation electrocatalysis. Nature Catalysis，2019，2 (4)：304-313.

[20] Wang J，Li Z，Wu Y，et al. Fabrication of single-atom catalysts with precise structure and high metal loading. Advanced Materials，2018，30 (48)：1801649.

[21] Stamenkovic V R，Strmcnik D，Lopes P P，et al. Energy and fuels from electrochemical interfaces. Nature Materials，2017，16 (1)：57-69.

[22] Bu L，Zhang N，Guo S，et al. Biaxially strained PtPb/Pt core/shell nanoplate boosts oxygen reduction catalysis. Science，2016，354 (6318)：1410-1414.

[23] Xu J，Liu T，Li J，et al. Boosting the hydrogen evolution performance of ruthenium clusters through synergistic coupling with cobalt phosphide. Energy & Environmental Science，2018，11 (7)：1819-1827.

[24] Wang N，Sun Q，Bai R，et al. In situ confinement of ultrasmall Pd clusters within nanosized silicalite-1 zeolite for highly efficient catalysis of hydrogen generation. Journal of the American Chemical Society，2016，138 (24)：7484-7487.

[25] Shan J，Zheng Y，Shi B，et al. Regulating electrocatalysts via surface and interface engineering for acidic water electrooxidation. ACS Energy Letters，2019，4 (11)：2719-2730.

[26] Mccrory C C L，Jung S，Ferrer I M，et al. Benchmarking hydrogen evolving reaction and oxygen evolving reaction electrocatalysts for solar water splitting devices. Journal of the American Chemical Society，2015，137 (13)：4347-4357.

[27] Wang C，Lan F，He Z，et al. Iridium-based catalysts for solid polymer electrolyte electrocatalytic water splitting. ChemSusChem，2019，12 (8)：1576-1590.

[28] Li C，Baek J B. Recent advances in noble metal (Pt，Ru，and Ir)-based electrocatalysts for efficient hydrogen evolution reaction. ACS Omega，2020，5 (1)：31-40.

[29] Zhu J，Chen Z，Xie M，et al. Iridium-based cubic nanocages with 1. 1-nm-thick walls：A highly efficient and durable electrocatalyst for water oxidation in an acidic medium. Angewandte Chemie International Edition，2019，58 (22)：7244-7248.

[30] Luo F，Hu H，Zhao X，et al. Robust and stable acidic overall water splitting on Ir single atoms. Nano Letters，2020，20 (3)：2120-2128.

[31] Zhu J Y，Xue Q，Xue Y Y，et al. Iridium nanotubes as bifunctional electrocatalysts for oxygen evolution and nitrate reduction reactions. ACS Applied Materials & Interfaces，2020，12 (12)：14064-14070.

[32] Li Y，Sun Y，Qin Y，et al. Recent advances on water-splitting electrocatalysis mediated by noble-metal-based nanostructured materials. Advanced Energy Materials，2020，10 (11)：1903120.

[33] Sun X，Chen C，Liu S，et al. Aqueous co2 reduction with high efficiency using α-Co(OH)$_2$-supported atomic Ir electrocatalysts. Angewandte Chemie International Edition，2019，58 (14)：4669-4673.

[34] Yang X，Ling F，Zi X，et al. Low-coordinate step atoms via plasma-assisted calcinations to enhance electrochemical reduction of nitrogen to ammonia. Small，2020，16 (17)：2000421.

[35] Zhang R, Pearce P E, Duan Y, et al. Importance of water structure and catalyst-electrolyte interface on the design of water splitting catalysts. Chemistry of Materials, 2019, 31 (20): 8248-8259.

[36] Sheng M, Jiang B, Wu B, et al. Approaching the volcano top: Iridium/silicon nanocomposites as efficient electrocatalysts for the hydrogen evolution reaction. ACS Nano, 2019, 13 (3): 2786-2794.

[37] Skúlason E, Tripkovic V, Björketun M E, et al. Modeling the electrochemical hydrogen oxidation and evolution reactions on the basis of density functional theory calculations. The Journal of Physical Chemistry C, 2010, 114 (42): 18182-18197.

[38] Chen Z, Duan X, Wei W, et al. Boride-based electrocatalysts: Emerging candidates for water splitting. Nano Research, 2020, 13 (2): 293-314.

[39] Yang F, Fu L, Cheng G, et al. Ir-oriented nanocrystalline assemblies with high activity for hydrogen oxidation/evolution reactions in an alkaline electrolyte. Journal of Materials Chemistry A, 2017, 5 (44): 22959-22963.

[40] Ming M, Zhang Y, He C, et al. Room-temperature sustainable synthesis of selected platinum group metal (PGM=Ir, Rh, and Ru) nanocatalysts well-dispersed on porous carbon for efficient hydrogen evolution and oxidation. Small, 2019, 15 (49): 1903057.

[41] He Y, Xu J, Wang F, et al. In-situ carbonization approach for the binder-free ir-dispersed ordered mesoporous carbon hydrogen evolution electrode. Journal of Energy Chemistry, 2017, 26 (6): 1140-1146.

[42] Mahmood J, Anjum M a R, Shin SH, et al. Encapsulating iridium nanoparticles inside a 3d cage-like organic network as an efficient and durable catalyst for the hydrogen evolution reaction. Advanced Materials, 2018, 30 (52): 1805606.

[43] Li F, Han G F, Noh H J, et al. Balancing hydrogen adsorption/desorption by orbital modulation for efficient hydrogen evolution catalysis. Nature Communications, 2019, 10 (1): 4060.

[44] Sheng M, Jiang B, Wu B, et al. Approaching the volcano top: Iridium/silicon nanocomposites as efficient electrocatalysts for the hydrogen evolution reaction. ACS Nano, 2019, 13 (3): 2786-2794.

[45] Dai Q, Meng Q, Du C, et al. Spontaneous deposition of Ir nanoparticles on 2D siloxene as a high-performance her electrocatalyst with ultra-low Ir loading. Chemical Communications, 2020, 56 (35): 4824-4827.

[46] Chen Z, Duan X, Wei W, et al. Recent advances in transition metal-based electrocatalysts for alkaline hydrogen evolution. Journal of Materials Chemistry A, 2019, 7 (25): 14971-15005.

[47] Yu J, He Q, Yang G, et al. Recent advances and prospective in ruthenium-based materials for electrochemical water splitting. ACS Catalysis, 2019, 9 (11): 9973-10011.

[48] Wang C, Xu H, Shang H, et al. Ir-doped Pd nanosheet assemblies as bifunctional electrocatalysts for advanced hydrogen evolution reaction and liquid fuel electrocatalysis. Inorganic Chemistry, 2020, 59 (5): 3321-3329.

[49] Papaderakis A, Pliatsikas N, Patsalas P, et al. Hydrogen evolution at Ir-Ni bimetallic deposits prepared by galvanic replacement. Journal of Electroanalytical Chemistry, 2018, 80: 21-27.

[50] Gong S, Wang C, Jiang P, et al. O species-decorated graphene shell encapsulating iridium-nickel alloy as an efficient electrocatalyst towards hydrogen evolution reaction. Journal of Materials Chemistry A, 2019, 7 (25): 15079-15088.

[51] Wang Z J, Li M X, Yu J H, et al. Low-iridium-content IrNiTa metallic glass films as intrinsically active catalysts for hydrogen evolution reaction. Advanced Materials, 2020, 32

(4): 1906384.

[52] Kim S J, Jung H, Lee C, et al. Comparative study on hydrogen evolution reaction activity of electrospun nanofibers with diverse metallic Ir and IrO_2 composition ratios. ACS Sustainable Chemistry & Engineering, 2019, 7 (9): 8613-8620.

[53] Vidales A G, Dam-Quang L, Hong A, et al. The influence of addition of iridium-oxide to nickel-molybdenum-oxide cathodes on the electrocatalytic activity towards hydrogen evolution in acidic medium and on the cathode deactivation resistance. Electrochimica Acta, 2019, 302: 198-206.

[54] Cho Y B, Yu A, Lee C, et al. Fundamental study of facile and stable hydrogen evolution reaction at electrospun Ir and Ru mixed oxide nanofibers. ACS Applied Materials & Interfaces, 2018, 10 (1): 541-549.

[55] Zheng X, Nie H, Zhan Y, et al. Intermolecular electron modulation by P/P bridging in an IrO_2-CoPi catalyst to enhance the hydrogen evolution reaction. Journal of Materials Chemistry A, 2020, 8 (17): 8273-8280.

[56] Yang X, Li Y, Deng L, et al. Synthesis and characterization of an IrO_2-Fe_2O_3 electrocatalyst for the hydrogen evolution reaction in acidic water electrolysis. RSC Advances, 2017, 7 (33): 20252-20258.

[57] Karfa P, Majhi K C, Madhuri R. Shape-dependent electrocatalytic activity of iridium oxide decorated erbium pyrosilicate toward the hydrogen evolution reaction over the entire pH range. ACS Catalysis, 2018, 8 (9): 8830-8843.

[58] Pu Z, Zhao J, Amiinu I S, et al. A universal synthesis strategy for P-rich noble metal diphosphide-based electrocatalysts for the hydrogen evolution reaction. Energy & Environmental Science, 2019, 12 (3): 952-957.

[59] Kuttiyiel K A, Sasaki K, Chen W F, et al. Core-shell, hollow-structured iridium-nickel nitride nanoparticles for the hydrogen evolution reaction. Journal of Materials Chemistry A, 2014, 2 (3): 591-594.

[60] Chen Z, Liu Y, Wei W, et al. Recent advances in electrocatalysts for halogenated organic pollutant degradation. Environmental Science: Nano, 2019, 6 (8): 2332-2366.

[61] Yan N, Detz R J, Govindarajan N, et al. Selective surface functionalization generating site-isolated Ir on a MnO_x/N-doped carbon composite for robust electrocatalytic water oxidation. Journal of Materials Chemistry A, 2019, 7 (40): 23098-23104.

[62] Wang Q, Huang X, Zhao Z L, et al. Ultrahigh-loading of Ir single atoms on NiO matrix to dramatically enhance oxygen evolution reaction. Journal of the American Chemical Society, 2020, 142 (16): 7425-7433.

[63] Luo X, Wei X, Zhong H, et al. Single-atom Ir-anchored 3d amorphous NiFe nanowire@nanosheets for boosted oxygen evolution reaction. ACS Applied Materials & Interfaces, 2020, 12 (3): 3539-3546.

[64] Jiang K, Luo M, Peng M, et al. Dynamic active-site generation of atomic iridium stabilized on nanoporous metal phosphides for water oxidation. Nature Communications, 2020, 11 (1): 2701.

[65] Roy S B, Akbar K, Jeon J H, et al. Iridium on vertical graphene as an all-round catalyst for robust water splitting reactions. Journal of Materials Chemistry A, 2019, 7 (36): 20590-20596.

[66] Wu X, Feng B, Li W, et al. Metal-support interaction boosted electrocatalysis of ultrasmall iridium nanoparticles supported on nitrogen doped graphene for highly efficient water

electrolysis in acidic and alkaline media. Nano Energy, 2019, 62: 117-126.

[67] Jiang B, Wang T, Cheng Y, et al. Ir/g-C_3N_4/nitrogen-doped graphene nanocomposites as bifunctional electrocatalysts for overall water splitting in acidic electrolytes. ACS Applied Materials & Interfaces, 2018, 10 (45): 39161-39167.

[68] Xue Q, Gao W, Zhu J, et al. Carbon nanobowls supported ultrafine iridium nanocrystals: An active and stable electrocatalyst for the oxygen evolution reaction in acidic media. Journal of Colloid and Interface Science, 2018, 529: 325-331.

[69] Sugita Y, Tamaki T, Kuroki H, et al. Connected iridium nanoparticle catalysts coated onto silica with high density for oxygen evolution in polymer electrolyte water electrolysis. Nanoscale Advances, 2020, 2 (1): 171-175.

[70] Boshnakova I, Lefterova E, Slavcheva E. Investigation of montmorillonite as carrier for OER. International Journal of Hydrogen Energy, 2018, 43 (35): 16897-16904.

[71] Oh H S, Nong H N, Reier T, et al. Oxide-supported Ir nanodendrites with high activity and durability for the oxygen evolution reaction in acid PEM water electrolyzers. Chemical Science, 2015, 6 (6): 3321-3328.

[72] Hartig-Weiss A, Miller M, Beyer H, et al. Iridium oxide catalyst supported on antimony-doped tin oxide for high oxygen evolution reaction activity in acidic media. ACS Applied Nano Materials, 2020, 3 (3): 2185-2196.

[73] Saveleva V A, Wang L, Kasian O, et al. Insight into the mechanisms of high activity and stability of iridium supported on antimony-doped tin oxide aerogel for anodes of proton exchange membrane water electrolyzers. ACS Catalysis, 2020, 10 (4): 2508-2516.

[74] Liu Z, Li J, Zhang J, et al. Ultrafine Ir nanowires with microporous channels and superior electrocatalytic activity for oxygen evolution reaction. ChemCatChem, 2020, 12 (11): 3060-3067.

[75] Wu G, Zheng X, Cui P, et al. A general synthesis approach for amorphous noble metal nanosheets. Nature Communications, 2019, 10 (1): 4855.

[76] Cho S H, Yoon K R, Shin K, et al. Synergistic coupling of metallic cobalt nitride nanofibers and IrO_x nanoparticle catalysts for stable oxygen evolution. Chemistry of Materials, 2018, 30 (17): 5941-5950.

[77] Wang W, Xi S, Shao Y, et al. Sub-nanometer-sized iridium species decorated on mesoporous Co_3O_4 for electrocatalytic oxygen evolution. ChemElectroChem, 2019, 6 (6): 1846-1852.

[78] Tackett B M, Sheng W, Kattel S, et al. Reducing iridium loading in oxygen evolution reaction electrocatalysts using core-shell particles with nitride cores. ACS Catalysis, 2018, 8 (3): 2615-2621.

[79] Zhang Y, Wu C, Jiang H, et al. Atomic iridium incorporated in cobalt hydroxide for efficient oxygen evolution catalysis in neutral electrolyte. Advanced Materials, 2018, 30 (18): 1707522.

[80] Zhao G, Li P, Cheng N, et al. An Ir/$Ni(OH)_2$ heterostructured electrocatalyst for the oxygen evolution reaction: Breaking the scaling relation, stabilizing iridium (v), and beyond. Advanced Materials, 2020, 32 (24): 2000872.

[81] Xing Y, Ku J, Fu W, et al. Inductive effect between atomically dispersed iridium and transition-metal hydroxide nanosheets enables highly efficient oxygen evolution reaction. Chemical Engineering Journal, 2020, 395: 125149.

[82] Sun Y, Huang B, Li Y, et al. Trifunctional fishbone-like PtCo/Ir enables high-performance zinc-air batteries to drive the water-splitting catalysis. Chemistry of Materials, 2019, 31

(19): 8136-8144.

[83] Aizaz Ud Din M, Irfan S, Dar S U, et al. Synthesis of 3d IrRuMn sphere as a superior oxygen evolution electrocatalyst in acidic environment. Chemistry-A European Journal, 2020, 26 (25): 5662-5666.

[84] Zhang T, Liao S A, Dai L X, et al. Ir-Pd nanoalloys with enhanced surface-microstructure-sensitive catalytic activity for oxygen evolution reaction in acidic and alkaline media. Science China Materials, 2018, 61 (7): 926-938.

[85] Wang Y, Zhang L, Yin K, et al. Nanoporous iridium-based alloy nanowires as highly efficient electrocatalysts toward acidic oxygen evolution reaction. ACS Applied Materials & Interfaces, 2019, 11 (43): 39728-39736.

[86] Feng J, Lv F, Zhang W, et al. Iridium-based multimetallic porous hollow nanocrystals for efficient overall-water-splitting catalysis. Advanced Materials, 2017, 29 (47): 1703798.

[87] Strickler A L, Flores R A, King L A, et al. Systematic investigation of iridium-based bimetallic thin film catalysts for the oxygen evolution reaction in acidic media. ACS Applied Materials & Interfaces, 2019, 11 (37): 34059-34066.

[88] Zhao Y. Chemically induced cell fate reprogramming and the acquisition of plasticity in somatic cells. Current Opinion in Chemical Biology, 2019, 51: 146-153.

[89] Wang C, Sui Y, Xu M, et al. Synthesis of Ni-Ir nanocages with improved electrocatalytic performance for the oxygen evolution reaction. ACS Sustainable Chemistry & Engineering, 2017, 5 (11): 9787-9792.

[90] Alia S M, Shulda S, Ngo C, et al. Iridium-based nanowires as highly active, oxygen evolution reaction electrocatalysts. ACS Catalysis, 2018, 8 (3): 2111-2120.

[91] Pi Y, Guo J, Shao Q, et al. Highly efficient acidic oxygen evolution electrocatalysis enabled by porous Ir-Cu nanocrystals with three-dimensional electrocatalytic surfaces. Chemistry of Materials, 2018, 30 (23): 8571-8578.

[92] Lim J, Yang S, Kim C, et al. Shaped Ir-Ni bimetallic nanoparticles for minimizing Ir utilization in oxygen evolution reaction. Chemical Communications, 2016, 52 (32): 5641-5644.

[93] Park J, Sa Y J, Baik H, et al. Iridium-based multimetallic nanoframe@nanoframe structure: An efficient and robust electrocatalyst toward oxygen evolution reaction. ACS Nano, 2017, 11 (6): 5500-5509.

[94] Jin H, Hong Y, Yoon J, et al. Lanthanide metal-assisted synthesis of rhombic dodecahedral MNi (M=Ir and Pt) nanoframes toward efficient oxygen evolution catalysis. Nano Energy, 2017, 42: 17-25.

[95] Povia M, Abbott D F, Herranz J, et al. Operando x-ray characterization of high surface area iridium oxides to decouple their activity losses for the oxygen evolution reaction. Energy & Environmental Science, 2019, 12 (10): 3038-3052.

[96] Geiger S, Kasian O, Shrestha B R, et al. Activity and stability of electrochemically and thermally treated iridium for the oxygen evolution reaction. Journal of The Electrochemical Society, 2016, 163 (11): F3132-F3138.

[97] Shan J, Guo C, Zhu Y, et al. Charge-redistribution-enhanced nanocrystalline Ru@IrO_x electrocatalysts for oxygen evolution in acidic media. Chem, 2019, 5 (2): 445-459.

[98] Faustini M, Giraud M, Jones D, et al. Hierarchically structured ultraporous iridium-based materials: A novel catalyst architecture for proton exchange membrane water electrolyzers. Advanced Energy Materials, 2019, 9 (4): 1802136.

[99] Audichon T, Napporn T W, Canaff C, et al. IrO_2 coated on RuO_2 as efficient and stable elee troactive nanocatalysts for electrochemical water splitting. The Journal of Physical Chemistry C, 2016, 120 (5): 2562-2573.

[100] Hu W, Zhong H, Liang W, et al. Ir-surface enriched porous Ir-Co oxide hierarchical architecture for high performance water oxidation in acidic media. ACS Applied Materials & Interfaces, 2014, 6 (15): 12729-12736.

[101] Zaman W Q, Sun W, Zhou Z H, et al. Anchoring of IrO_2 on one-dimensional Co_3O_4 nanorods for robust electrocatalytic water splitting in an acidic environment. ACS Applied Energy Materials, 2018, 1 (11): 6374-6380.

[102] Tariq M, Zaman W Q, Sun W, et al. Unraveling the beneficial electrochemistry of IrO_2/MoO_3 hybrid as a highly stable and efficient oxygen evolution reaction catalyst. ACS Sustainable Chemistry & Engineering, 2018, 6 (4): 4854-4862.

[103] Gou W, Zhang M, Zou Y, et al. Iridium-chromium oxide nanowires as highly performed OER catalysts in acidic media. ChemCatChem, 2019, 11 (24): 6008-6014.

[104] Lee J, Kim I, Park S. Boosting stability and activity of oxygen evolution catalyst in acidic medium: Bimetallic Ir-Fe oxides on reduced graphene oxide prepared through ultrasonic spray pyrolysis. ChemCatChem, 2019, 11 (11): 2615-2623.

[105] Ghadge S D, Velikokhatnyi O I, Datta M K, et al. Experimental and theoretical validation of high efficiency and robust electrocatalytic response of one-dimensional (1d) (Mn, Ir) O_2: 10F nanorods for the oxygen evolution reaction in PEM-based water electrolysis. ACS Catalysis, 2019, 9 (3): 2134-2157.

[106] Audichon T, Morisset S, Napporn T W, et al. Effect of adding CeO_2 to RuO_2-IrO_2 mixed nanocatalysts: Activity towards the oxygen evolution reaction and stability in acidic media. ChemElectroChem, 2015, 2 (8): 1128-1137.

[107] Kumari S, Ajayi B P, Kumar B, et al. A low-noble-metal $W_{1-x}Ir_xO_{3-\delta}$ water oxidation electrocatalyst for acidic media via rapid plasma synthesis. Energy & Environmental Science, 2017, 10 (11): 2432-2440.

[108] Hao S, Wang Y, Zheng G, et al. Tuning electronic correlations of ultra-small IrO_2 nanoparticles with la and pt for enhanced oxygen evolution performance and long-durable stability in acidic media. Applied Catalysis B: Environmental, 2020, 266: 118643.

[109] Zaman W Q, Sun W, Tariq M, et al. Iridium substitution in nickel cobaltite renders high mass specific OER activity and durability in acidic media. Applied Catalysis B: Environmental, 2019, 244: 295-302.

[110] Ghadge S D, Patel P P, Datta M K, et al. First report of vertically aligned (Sn, Ir) O_2: F solid solution nanotubes: Highly efficient and robust oxygen evolution electrocatalysts for proton exchange membrane based water electrolysis. Journal of Power Sources, 2018, 392: 139-149.

[111] Ghadge S D, Patel P P, Datta M K, et al. Fluorine substituted (Mn, Ir) O_2: F high performance solid solution oxygen evolution reaction electro-catalysts for PEM water electrolysis. RSC Advances, 2017, 7 (28): 17311-17324.

[112] Li G, Li K, Yang L, et al. Boosted performance of Ir species by employing tin as the support toward oxygen evolution reaction. ACS Applied Materials & Interfaces, 2018, 10 (44): 38117-38124.

[113] Abbou S, Chattot R, Martin V, et al. Manipulating the corrosion resistance of SnO_2 aerogels through doping for efficient and durable oxygen evolution reaction electrocatalysis in acidic

media. ACS Catalysis，2020，10 (13)：7283-7294.

[114] Zhang H，Yuan Z Y，Li B，et al. Design of mining isolation switch action torque test system; proceedings of the 2019 Chinese Control And Decision Conference，2019.

[115] Chen J，Cui P，Zhao G，et al. Low-coordinate iridium oxide confined on graphitic carbon nitride for highly efficient oxygen evolution. Angewandte Chemie International Edition，2019，58 (36)：12540-12544.

[116] Guan J，Li D，Si R，et al. Synthesis and demonstration of subnanometric iridium oxide as highly efficient and robust water oxidation catalyst. ACS Catalysis，2017，7 (9)：5983-5986.

[117] Pham C V，Bühler M，Knöppel J，et al. IrO_2 coated TiO_2 core-shell microparticles advance performance of low loading proton exchange membrane water electrolyzers. Applied Catalysis B：Environmental，2020，269：118762.

[118] Ledendecker M，Geiger S，Hengge K，et al. Towards maximized utilization of iridium for the acidic oxygen evolution reaction. Nano Research，2019，12 (9)：2275-2280.

[119] Cheng J，Yang J，Kitano S，et al. Impact of Ir-valence control and surface nanostructure on oxygen evolution reaction over a highly efficient Ir-TiO_2 nanorod catalyst. ACS Catalysis，2019，9 (8)：6974-6986.

[120] Böhm D，Beetz M，Schuster M，et al. Efficient OER catalyst with low Ir volume density obtained by homogeneous deposition of iridium oxide nanoparticles on macroporous antimony-doped tin oxide support. Advanced Functional Materials，2020，30 (1)：1906670.

[121] Diaz-Morales O，Raaijman S，Kortlever R，et al. Iridium-based double perovskites for efficient water oxidation in acid media. Nature Communications，2016，7 (1)：12363.

[122] Yang L，Yu G，Ai X，et al. Efficient oxygen evolution electrocatalysis in acid by a perovskite with face-sharing iro6 octahedral dimers. Nature Communications，2018，9 (1)：5236.

[123] Chen Y，Fan Y，Men M，et al. High cystatin c levels predict long-term mortality in patients with St-segment elevation myocardial infarction undergoing late percutaneous coronary intervention：A retrospective study. Clinical Cardiology，2019，42 (5)：572-578.

[124] Yang L，Chen H，Shi L，et al. Enhanced iridium mass activity of 6H-phase，Ir-based perovskite with nonprecious incorporation for acidic oxygen evolution electrocatalysis. ACS Applied Materials & Interfaces，2019，11 (45)：42006-42013.

[125] Lebedev D，Povia M，Waltar K，et al. Highly active and stable iridium pyrochlores for oxygen evolution reaction. Chemistry of Materials，2017，29 (12)：5182-5191.

[126] Shih P C，Kim J，Sun C J，et al. Single-phase pyrochlore $Y_2Ir_2O_7$ electrocatalyst on the activity of oxygen evolution reaction. ACS Applied Energy Materials，2018，1 (8)：3992-3998.

[127] Abbott D F，Pittkowski R K，Macounová K，et al. Design and synthesis of Ir/Ru pyrochlore catalysts for the oxygen evolution reaction based on their bulk thermodynamic properties. ACS Applied Materials & Interfaces，2019，11 (41)：37748-37760.

[128] Shang C，Cao C，Yu D，et al. Oxygen evolution reaction：Electron correlations engineer catalytic activity of pyrochlore iridates for acidic water oxidation. Advanced Materials，2019，31 (6)：1970042.

[129] Anantharaj S，Noda S. Amorphous catalysts and electrochemical water splitting：An untold story of harmony. Small，2020，16 (2)：1905779.

[130] Han H，Choi H，Mhin S，et al. Advantageous crystalline-amorphous phase boundary for enhanced electrochemical water oxidation. Energy & Environmental Science，2019，12 (8)：2443-2454.

[131] Sun W, Ma C, Tian X, et al. An amorphous lanthanum-iridium solid solution with an open structure for efficient water splitting. Journal of Materials Chemistry A, 2020, 8 (25): 12518-12525.
[132] Gao J, Xu C Q, Hung S F, et al. Breaking long-range order in iridium oxide by alkali ion for efficient water oxidation. Journal of the American Chemical Society, 2019, 141 (7): 3014-3023.
[133] Willinger E, Massué C, Schlögl R, et al. Identifying key structural features of IrO_x water splitting catalysts. Journal of the American Chemical Society, 2017, 139 (34): 12093-12101.
[134] Vij V, Sultan S, Harzandi A M, et al. Nickel-based electrocatalysts for energy-related applications: Oxygen reduction, oxygen evolution, and hydrogen evolution reactions. ACS Catalysis, 2017, 7 (10): 7196-7225.
[135] Han L, Guo L, Dong C, et al. Ternary mesoporous cobalt-iron-nickel oxide efficiently catalyzing oxygen/hydrogen evolution reactions and overall water splitting. Nano Research, 2019, 12 (9): 2281-2287.
[136] Lai W H, Zhang L F, Hua W B, et al. General π-electron-assisted strategy for Ir, Pt, Ru, Pd, Fe, Ni single-atom electrocatalysts with bifunctional active sites for highly efficient water splitting. Angewandte Chemie International Edition, 2019, 58 (34): 11868-11873.
[137] Wang H B, Wang J Q, Mintcheva N, et al. Laser synthesis of iridium nanospheres for overall water splitting. Materials, 2019, 12 (18): 3028.
[138] Cheng Z, Huang B, Pi Y, et al. Partially hydroxylated ultrathin iridium nanosheets as efficient electrocatalysts for water splitting. National Science Review, 2020, 7 (8): 1340-1348.
[139] Yang J, Ji Y, Shao Q, et al. A universal strategy to metal wavy nanowires for efficient electrochemical water splitting at pH-universal conditions. Advanced Functional Materials, 2018, 28 (41): 1803722.
[140] Fu L, Yang F, Cheng G, et al. Ultrathin Ir nanowires as high-performance electrocatalysts for efficient water splitting in acidic media. Nanoscale, 2018, 10 (4): 1892-1897.
[141] You H, Wu D, Chen Z N, et al. Highly active and stable water splitting in acidic media using a bifunctional iridium/cucurbit [6] uril catalyst. ACS Energy Letters, 2019, 4 (6): 1301-1307.
[142] Wei S, Cui X, Xu Y, et al. Iridium-triggered phase transition of MoS_2 nanosheets boosts overall water splitting in alkaline media. ACS Energy Letters, 2019, 4 (1): 368-374.
[143] Zhu M, Shao Q, Qian Y, et al. Superior overall water splitting electrocatalysis in acidic conditions enabled by bimetallic Ir-Ag nanotubes. Nano Energy, 2019, 56: 330-337.
[144] Yang A, Su K, Wang S, et al. Self-stabilization of zero-dimensional PdIr nanoalloys at two-dimensional manner for boosting their oer and her performance. Applied Surface Science, 2020, 510: 145408.
[145] Joo J, Jin H, Oh A, et al. An irru alloy nanocactus on $Cu_{2-x}S$@IrS_y as a highly efficient bifunctional electrocatalyst toward overall water splitting in acidic electrolytes. Journal of Materials Chemistry A, 2018, 6 (33): 16130-16138.
[146] Shan J, Ling T, Davey K, et al. Transition-metal-doped RuIr bifunctional nanocrystals for overall water splitting in acidic environments. Advanced Materials, 2019, 31 (17): 1900510.
[147] Pi Y, Shao Q, Wang P, et al. General formation of monodisperse IrM (M=Ni, Co, Fe) bimetallic nanoclusters as bifunctional electrocatalysts for acidic overall water splitting. Advanced Functional Materials, 2017, 27 (27): 1700886.

[148] Zhang S, Zhang X, Shi X, et al. Facile fabrication of ultrafine nickel-iridium alloy nanoparticles/graphene hybrid with enhanced mass activity and stability for overall water splitting. Journal of Energy Chemistry, 2020, 49: 166-173.

[149] Sun X, Liu F, Chen X, et al. Iridium-doped zifs-derived porous carbon-coated IrCo alloy as competent bifunctional catalyst for overall water splitting in acid medium. Electrochimica Acta, 2019, 307: 206-213.

[150] Si Z, Lv Z, Lu L, et al. Nitrogen-doped graphene chainmail wrapped IrCo alloy particles on nitrogen-doped graphene nanosheet for highly active and stable full water splitting. ChemCatChem, 2019, 11 (22): 5457-5465.

[151] Li D, Zong Z, Tang Z, et al. Total water splitting catalyzed by Co@Ir core-shell nanoparticles encapsulated in nitrogen-doped porous carbon derived from metal-organic frameworks. ACS Sustainable Chemistry & Engineering, 2018, 6 (4): 5105-5114.

[152] Fu L, Hu X, Li Y, et al. IrW nanobranches as an advanced electrocatalyst for pH-universal overall water splitting. Nanoscale, 2019, 11 (18): 8898-8905.

[153] Lv F, Zhang W, Yang W, et al. Ir-based alloy nanoflowers with optimized hydrogen binding energy as bifunctional electrocatalysts for overall water splitting. Small Methods, 2020, 4 (6): 1900129.

[154] Kundu M K, Mishra R, Bhowmik T, et al. Three-dimensional hierarchically porous iridium oxide-nitrogen doped carbon hybrid: An efficient bifunctional catalyst for oxygen evolution and hydrogen evolution reaction in acid. International Journal of Hydrogen Energy, 2020, 45 (11): 6036-6046.

[155] You B, Jiang N, Sheng M, et al. High-performance overall water splitting electrocatalysts derived from cobalt-based metal-organic frameworks. Chemistry of Materials, 2015, 27 (22): 7636-7642.

[156] Li J, Zheng G. One-dimensional earth-abundant nanomaterials for water-splitting electrocatalysts. Advanced Science, 2017, 4 (3): 1600380.

[157] Lyu F, Wang Q, Choi S M, et al. Noble-metal-free electrocatalysts for oxygen evolution. Small, 2019, 15 (1): 1804201.

[158] Zheng T, Shang C, He Z, et al. Intercalated iridium diselenide electrocatalysts for efficient ph-universal water splitting. Angewandte Chemie International Edition, 2019, 58 (41): 14764-14769.

[159] Wang D, Li Q, Han C, et al. Atomic and electronic modulation of self-supported nickel-vanadium layered double hydroxide to accelerate water splitting kinetics. Nature Communications, 2019, 10 (1): 3899.

[160] Li S, Xi C, Jin Y Z, et al. Ir-O-V catalytic group in Ir-doped $NiV(OH)_2$ for overall water splitting. ACS Energy Letters, 2019, 4 (8): 1823-1829.

[161] Chen Q Q, Hou C C, Wang C J, et al. Ir^{4+}-doped NiFe LDH to expedite hydrogen evolution kinetics as a Pt-like electrocatalyst for water splitting. Chemical Communications, 2018, 54 (49): 6400-6403.

[162] Fan R, Mu Q, Wei Z, et al. Atomic Ir-doped NiCo layered double hydroxide as a bifunctional electrocatalyst for highly efficient and durable water splitting. Journal of Materials Chemistry A, 2020, 8 (19): 9871-9881.

[163] Li X, Zhang L, Huang M, et al. Cobalt and nickel selenide nanowalls anchored on graphene as bifunctional electrocatalysts for overall water splitting. Journal of Materials Chemistry A, 2016, 4 (38): 14789-14795.

[164] Jiang W J, Tang T, Zhang Y, et al. Synergistic modulation of non-precious-metal electrocatalysts for advanced water splitting. Accounts of Chemical Research, 2020, 53 (6): 1111-1123.
[165] Anantharaj S, Aravindan V. Developments and perspectives in 3d transition-metal-based electrocatalysts for neutral and near-neutral water electrolysis. Advanced Energy Materials, 2020, 10 (1): 1902666.
[166] Chhetri M, Rana M, Loukya B, et al. Mechanochemical synthesis of free-standing platinum nanosheets and their electrocatalytic properties. Advanced Materials, 2015, 27 (30): 4430-4437.
[167] Liu Z, Qi J, Liu M, et al. Aqueous synthesis of ultrathin platinum/non-noble metal alloy nanowires for enhanced hydrogen evolution activity. Angewandte Chemie International Edition, 2018, 57 (36): 11678-11682.
[168] Zhao G, Rui K, Dou S X, et al. Heterostructures for electrochemical hydrogen evolution reaction: A review. Advanced Functional Materials, 2018, 28 (43): 1803291.
[169] Liu J, Ma Q, Huang Z, et al. Recent progress in graphene-based noble-metal nanocomposites for electrocatalytic applications. Advanced Materials, 2019, 31 (9): 1800696.
[170] Seh Z W, Kibsgaard J, Dickens C F, et al. Combining theory and experiment in electrocatalysis: Insights into materials design. Science, 2017, 355 (6321): eaad4998.
[171] Zhang L, Doyle-Davis K, Sun X. Pt-based electrocatalysts with high atom utilization efficiency: From nanostructures to single atoms. Energy & Environmental Science, 2019, 12 (2): 492-517.
[172] Gao Z, Li M, Wang J, et al. Pt nanocrystals grown on three-dimensional architectures made from graphene and mos2 nanosheets: Highly efficient multifunctional electrocatalysts toward hydrogen evolution and methanol oxidation reactions. Carbon, 2018, 139: 369-377.
[173] Kundu M K, Bhowmik T, Mishra R, et al. Platinum nanostructure/nitrogen-doped carbon hybrid: Enhancing its base media HER/HOR activity through bi-functionality of the catalyst. ChemSusChem, 2018, 11 (14): 2388-2401.
[174] Kwon H C, Kim M, Grote J P, et al. Carbon monoxide as a promoter of atomically dispersed platinum catalyst in electrochemical hydrogen evolution reaction. Journal of the American Chemical Society, 2018, 140 (47): 16198-16205.
[175] Yan X, Duan P, Zhang F, et al. Stable single-atom platinum catalyst trapped in carbon onion graphitic shells for improved chemoselective hydrogenation of nitroarenes. Carbon, 2019, 143: 378-384.
[176] Roger I, Shipman M A, Symes M D. Earth-abundant catalysts for electrochemical and photoelectrochemical water splitting. Nature Reviews Chemistry, 2017, 1 (1): 0003.
[177] Zhang B W, Wang Y X, Chou S L, et al. Fabrication of superior single-atom catalysts toward diverse electrochemical reactions. Small Methods, 2019, 3 (9): 1800497.
[178] Esposito D V, Chen J G. Monolayer platinum supported on tungsten carbides as low-cost electrocatalysts: Opportunities and limitations. Energy & Environmental Science, 2011, 4 (10): 3900-3912.
[179] Liu L, Corma A. Metal catalysts for heterogeneous catalysis: From single atoms to nanoclusters and nanoparticles. Chemical Reviews, 2018, 118 (10): 4981-5079.
[180] Li T, Yu B, Liu Z, et al. Homocysteine directly interacts and activates the angiotensin ii type i receptor to aggravate vascular injury. Nature Communications, 2018, 9 (1): 11.
[181] Chao T, Luo X, Chen W, et al. Atomically dispersed copper-platinum dual sites alloyed with

palladium nanorings catalyze the hydrogen evolution reaction. Angewandte Chemie International Edition, 2017, 56 (50): 16047-16051.

[182] Zhu Y, Cao T, Cao C, et al. One-pot pyrolysis to N-doped graphene with high-density Pt single atomic sites as heterogeneous catalyst for alkene hydrosilylation. ACS Catalysis, 2018, 8 (11): 10004-10011.

[183] Liu D, Li X, Chen S, et al. Atomically dispersed platinum supported on curved carbon supports for efficient electrocatalytic hydrogen evolution. Nature Energy, 2019, 4 (6): 512-518.

[184] Yin X P, Wang H J, Tang S F, et al. Engineering the coordination environment of single-atom platinum anchored on graphdiyne for optimizing electrocatalytic hydrogen evolution. Angewandte Chemie International Edition, 2018, 57 (30): 9382-9386.

[185] Wu Y, Ren D, Liu X, et al. High-voltage and high-safety practical lithium batteries with ethylene carbonate-free electrolyte. Advanced Energy Materials, 2021, 11 (47): 2102299.

[186] Jiang Y, Wu X, Yan Y, et al. Coupling PtNi ultrathin nanowires with Mxenes for boosting electrocatalytic hydrogen evolution in both acidic and alkaline solutions. Small, 2019, 15 (12): 1805474.

[187] Chen Q, Cao Z, Du G, et al. Excavated octahedral Pt-Co alloy nanocrystals built with ultrathin nanosheets as superior multifunctional electrocatalysts for energy conversion applications. Nano Energy, 2017, 39: 582-589.

[188] Zhang Z, Liu G, Cui X, et al. Crystal phase and architecture engineering of lotus-thalamus-shaped Pt-Ni anisotropic superstructures for highly efficient electrochemical hydrogen evolution. Advanced Materials, 2018, 30 (30): 1801741.

[189] Lao M, Rui K, Zhao G, et al. Platinum/nickel bicarbonate heterostructures towards accelerated hydrogen evolution under alkaline conditions. Angewandte Chemie International Edition, 2019, 58 (16): 5432-5437.

[190] Chi J Q, Xie J Y, Zhang W W, et al. N-doped sandwich-structured Mo_2C@C@Pt interface with ultralow Pt loading for pH-universal hydrogen evolution reaction. ACS Applied Materials & Interfaces, 2019, 11 (4): 4047-4056.

[191] Germano L D, Marangoni V S, Mogili N V V, et al. Ultrasmall (<2nm) Au@Pt nanostructures: Tuning the surface electronic states for electrocatalysis. ACS Applied Materials & Interfaces, 2019, 11 (6): 5661-5667.

[192] Han X, Wu X, Deng Y, et al. Ultrafine Pt nanoparticle-decorated pyrite-type CoS_2 nanosheet arrays coated on carbon cloth as a bifunctional electrode for overall water splitting. Advanced Energy Materials, 2018, 8 (24): 1800935.

[193] Kim J, Roh C W, Sahoo S K, et al. Highly durable platinum single-atom alloy catalyst for electrochemical reactions. Advanced Energy Materials, 2018, 8 (1): 1701476.

[194] Chen Z J, Cao G X, Gan L Y, et al. Highly dispersed platinum on honeycomb-like NiO@Ni film as a synergistic electrocatalyst for the hydrogen evolution reaction. ACS Catalysis, 2018, 8 (9): 8866-8872.

[195] Tang K, Wang X, Li Q, et al. High edge selectivity of in situ electrochemical Pt deposition on edge-rich layered WS_2 nanosheets. Advanced Materials, 2018, 30 (7): 1704779.

[196] Zhang L, Han L, Liu H, et al. Potential-cycling synthesis of single platinum atoms for efficient hydrogen evolution in neutral media. Angewandte Chemie International Edition, 2017, 56 (44): 13694-13698.

[197] Chen J, Yang Y, Su J, et al. Enhanced activity for hydrogen evolution reaction over CoFe

catalysts by alloying with small amount of pt. ACS Applied Materials & Interfaces, 2017, 9 (4): 3596-3601.

[198] Chen T W, Huang W F, Kang J X, et al. Cycling potential engineering surface configuration of sandwich Au@Ni@PtNiAu for superior catalytic durability. Nano Energy, 2018, 52: 22-28.

[199] Li S, Lee J K, Zhou S, et al. Synthesis of surface grown Pt nanoparticles on edge-enriched MoS_2 porous thin films for enhancing electrochemical performance. Chemistry of Materials, 2019, 31 (2): 387-397.

[200] Wang C, Shi H, Liu H, et al. Quasi-atomic-scale platinum anchored on porous titanium nitride nanorod arrays for highly efficient hydrogen evolution. Electrochimica Acta, 2018, 292: 727-735.

[201] Wang L, Chen M X, Yan Q Q, et al. A sulfur-tethering synthesis strategy toward high-loading atomically dispersed noble metal catalysts. Science Advances, 2019, 5 (10): eaax6322.

[202] Wang Z, Ren X, Luo Y, et al. An ultrafine platinum-cobalt alloy decorated cobalt nanowire array with superb activity toward alkaline hydrogen evolution. Nanoscale, 2018, 10 (26): 12302-12307.

[203] Yang Y, Lun Z, Xia G, et al. Non-precious alloy encapsulated in nitrogen-doped graphene layers derived from MOFs as an active and durable hydrogen evolution reaction catalyst. Energy & Environmental Science, 2015, 8 (12): 3563-3571.

[204] He T, Chen S, Ni B, et al. Zirconium-porphyrin-based metal-organic framework hollow nanotubes for immobilization of noble-metal single atoms. Angewandte Chemie International Edition, 2018, 57 (13): 3493-3498.

[205] Katuri K P, Bettahalli N M S, Wang X, et al. A microfiltration polymer-based hollow-fiber cathode as a promising advanced material for simultaneous recovery of energy and water. Advanced Materials, 2016, 28 (43): 9504-9511.

[206] Nayak P, Jiang Q, Kurra N, et al. Monolithic laser scribed graphene scaffolds with atomic layer deposited platinum for the hydrogen evolution reaction. Journal of Materials Chemistry A, 2017, 5 (38): 20422-20427.

[207] Saha S, Martin B, Leonard B, et al. Probing synergetic effects between platinum nanoparticles deposited via atomic layer deposition and a molybdenum carbide nanotube support through surface characterization and device performance. Journal of Materials Chemistry A, 2016, 4 (23): 9253-9265.

[208] Cheng N, Stambula S, Wang D, et al. Platinum single-atom and cluster catalysis of the hydrogen evolution reaction. Nature Communications, 2016, 7 (1): 13638.

[209] Tan S, Wang L, Saha S, et al. Active site and electronic structure elucidation of Pt nanoparticles supported on phase-pure molybdenum carbide nanotubes. ACS Applied Materials & Interfaces, 2017, 9 (11): 9815-9822.

[210] Yan X, Li H, Sun J, et al. Pt nanoparticles decorated high-defective graphene nanospheres as highly efficient catalysts for the hydrogen evolution reaction. Carbon, 2018, 137: 405-410.

[211] Shang X, Liu Z Z, Lu S S, et al. Pt-C interfaces based on electronegativity-functionalized hollow carbon spheres for highly efficient hydrogen evolution. ACS Applied Materials & Interfaces, 2018, 10 (50): 43561-43569.

[212] Ying J, Jiang G, Paul Cano Z, et al. Nitrogen-doped hollow porous carbon polyhedrons embedded with highly dispersed Pt nanoparticles as a highly efficient and stable hydrogen evolution electrocatalyst. Nano Energy, 2017, 40: 88-94.

[213] Liu G, Qiu Y, Wang Z, et al. Efficiently synergistic hydrogen evolution realized by trace amount of Pt-decorated defect-rich SnS_2 nanosheets. ACS Applied Materials & Interfaces, 2017, 9 (43): 37750-37759.
[214] Wang L, Lin C, Huang D, et al. Optimizing the volmer step by single-layer nickel hydroxide nanosheets in hydrogen evolution reaction of platinum. ACS Catalysis, 2015, 5 (6): 3801-3806.
[215] Yin H, Zhao S, Zhao K, et al. Ultrathin platinum nanowires grown on single-layered nickel hydroxide with high hydrogen evolution activity. Nature Communications, 2015, 6 (1): 6430.
[216] Xing Z, Han C, Wang D, et al. Ultrafine Pt nanoparticle-decorated Co $(OH)_2$ nanosheet arrays with enhanced catalytic activity toward hydrogen evolution. ACS Catalysis, 2017, 7 (10): 7131-7135.
[217] Anantharaj S, Karthick K, Venkatesh M, et al. Enhancing electrocatalytic total water splitting at few layer Pt-NiFe layered double hydroxide interfaces. Nano Energy, 2017, 39: 30-43.
[218] Cui B, Hu B, Liu J, et al. Solution-plasma-assisted bimetallic oxide alloy nanoparticles of Pt and Pd embedded within two-dimensional $Ti_3C_2T_x$ nanosheets as highly active electrocatalysts for overall water splitting. ACS Applied Materials & Interfaces, 2018, 10 (28): 23858-23873.
[219] Li Z, Ge R, Su J, et al. Recent progress in low Pt content electrocatalysts for hydrogen evolution reaction. Advanced Materials Interfaces, 2020, 7 (14): 2000396.
[220] Abbas S A, Kim S H, Iqbal M I, et al. Synergistic effect of nano-Pt and Ni spine for her in alkaline solution: Hydrogen spillover from nano-Pt to Ni spine. Scientific Reports, 2018, 8 (1): 2986.
[221] Anantharaj S, Karthik P E, Subramanian B, et al. Pt nanoparticle anchored molecular self-assemblies of DNA: An extremely stable and efficient her electrocatalyst with ultralow Pt content. ACS Catalysis, 2016, 6 (7): 4660-4672.
[222] Wang S, Gao X, Hang X, et al. Ultrafine Pt nanoclusters confined in a calixarene-based {ni24} coordination cage for high-efficient hydrogen evolution reaction. Journal of the American Chemical Society, 2016, 138 (50): 16236-16239.
[223] Cheng X, Li Y, Zheng L, et al. Highly active, stable oxidized platinum clusters as electrocatalysts for the hydrogen evolution reaction. Energy & Environmental Science, 2017, 10 (11): 2450-2458.
[224] Yan Q, Yan P, Wei T, et al. A highly efficient and durable water splitting system: Platinum sub-nanocluster functionalized nickel-iron layered double hydroxide as the cathode and hierarchical nickel-iron selenide as the anode. Journal of Materials Chemistry A, 2019, 7 (6): 2831-2837.
[225] Tian H, Cui X, Zeng L, et al. Oxygen vacancy-assisted hydrogen evolution reaction of the pt/wo3 electrocatalyst. Journal of Materials Chemistry A, 2019, 7 (11): 6285-6293.
[226] Yang X F, Wang A, Qiao B, et al. Single-atom catalysts: A new frontier in heterogeneous catalysis. Accounts of Chemical Research, 2013, 46 (8): 1740-1748.
[227] Su J, Ge R, Dong Y, et al. Recent progress in single-atom electrocatalysts: Concept, synthesis, and applications in clean energy conversion. Journal of Materials Chemistry A, 2018, 6 (29): 14025-14042.
[228] Zeng X, Shui J, Liu X, et al. Single-atom to single-atom grafting of Pt1 onto FeN_4 center:

Pt1@FeNC multifunctional electrocatalyst with significantly enhanced properties. Advanced Energy Materials, 2018, 8 (1): 1701345.

[229] Zhou M, Bao S, Bard A J. Probing size and substrate effects on the hydrogen evolution reaction by single isolated Pt atoms, atomic clusters, and nanoparticles. Journal of the American Chemical Society, 2019, 141 (18): 7327-7332.

[230] Zhang J, Zhao Y, Guo X, et al. Single platinum atoms immobilized on an Mxene as an efficient catalyst for the hydrogen evolution reaction. Nature Catalysis, 2018, 1 (12): 985-992.

[231] Jiang K, Liu B, Luo M, et al. Single platinum atoms embedded in nanoporous cobalt selenide as electrocatalyst for accelerating hydrogen evolution reaction. Nature Communications, 2019, 10 (1): 1743.

[232] Cao Z, Chen Q, Zhang J, et al. Platinum-nickel alloy excavated nano-multipods with hexagonal close-packed structure and superior activity towards hydrogen evolution reaction. Nature Communications, 2017, 8 (1): 15131.

[233] Yang Y, Lin Z, Gao S, et al. Tuning electronic structures of nonprecious ternary alloys encapsulated in graphene layers for optimizing overall water splitting activity. ACS Catalysis, 2017, 7 (1): 469-479.

[234] Du N, Wang C, Wang X, et al. Trimetallic tristar nanostructures: Tuning electronic and surface structures for enhanced electrocatalytic hydrogen evolution. Advanced Materials, 2016, 28 (10): 2077-2084.

[235] Qin Y, Han X, Gadipelli S, et al. In situ synthesized low-PtCo@porous carbon catalyst for highly efficient hydrogen evolution. Journal of Materials Chemistry A, 2019, 7 (11): 6543-6551.

[236] Bai S, Wang C, Deng M, et al. Surface polarization matters: Enhancing the hydrogen-evolution reaction by shrinking pt shells in Pt-Pd-graphene stack structures. Angewandte Chemie International Edition, 2014, 53 (45): 12120-12124.

[237] Hunt S T, Milina M, Wang Z, et al. Activating earth-abundant electrocatalysts for efficient, low-cost hydrogen evolution/oxidation: Sub-monolayer platinum coatings on titanium tungsten carbide nanoparticles. Energy & Environmental Science, 2016, 9 (10): 3290-3301.

[238] Chen Z, Ye S, Wilson A R, et al. Optically transparent hydrogen evolution catalysts made from networks of copper-platinum core-shell nanowires. Energy & Environmental Science, 2014, 7 (4): 1461-1467.

第六章

非贵金属基纳米催化剂

以 Ir 基和 Ru 基为代表的贵金属催化剂具有优异的催化电解水性能。它们的氧化物（RuO_2 和 IrO_2）不仅有较高的 OER 催化活性还具有较为理想的稳定性，贵金属 Pt 基材料是目前最高效的 HER 电催化剂。但是昂贵的价格和有限的地球储量等缺点严重限制了贵金属催化剂大规模商业化应用。因此，寻找廉价、高效和储量丰富的材料替代贵金属催化剂成了研究的重心。相对于贵金属材料，过渡金属材料在储量、价格、耐腐蚀性和环境友好性等方面都展现出了巨大的优势。因此，过渡金属化合物在近几年受到了广泛的关注，并有望替代贵金属基催化剂。此外，已经有大量的研究证实，以镍、钴和铁为代表的过渡金属化合物具有可以与贵金属催化剂相媲美的催化活性，诸如硼化物、硫化物、磷化物、硒化物以及氧化物/氢氧化物等。本章重点介绍镍基化合物、硼化物和硼酸盐基化合物、硫化物、碳化钼、过渡金属 Fe、Co、Ni 基磷化物在电解水中的应用。

6.1 镍基化合物

6.1.1 镍基硒化物用于 OER

硒化镍是一类重要的硫族镍化物，最近在电解水方面得到了大量的关注。在 HER 催化中，硒化镍在酸性和碱性条件下都是良好的电催化剂。在酸性条件下部分金属溶解，在碱性条件下表面改性为氢氧化镍。在 OER 催化中，硒化镍表现出独特的催化性能，甚至优于简单的氢氧化镍/羟基氧化物。最近具有各种化学计量的硒化镍被用于催化 OER 反应。硒化镍是一类镍的非化学计量氧族化物，可形成各种多晶型物，如 $Ni_{0.85}Se$（通常被简单地称为 NiSe）、$NiSe_2$、Ni_3Se_4 和 Ni_3Se_2。在其简单形式中（如 NiSe），阴离子通常是二价的硒阴离子。在黄铁矿型 $NiSe_2$ 中，阴离子是二价的二硒阴离子。对于更复杂的多晶型，硒阴离子带有非整数电荷[1]。镍和硒的比例不同，它们的电导率也不同。一般而言，较高的镍含量通常具有较好的导电性，较高的硒含量则具有较好的结构柔性，可用于 OER 过程中的表面重构。然而，要想在碱性介质中获得更好的 OER 催化活性，必须在电导率和结构柔性之间进行权衡。与硫化镍不同，硒化镍并不完全将其阴离子全部浸出到电解液中，也不像碲化镍那样保持刚性阻碍表面重构。这些优异的结构特性，使硒化镍在碱性条件下对 OER 的预催化活性优于硫化镍和碲化镍。

此外，硒化镍具有与硫化镍和碲化镍相似的水不溶性，硒化镍是半导体，其导电性比硫化镍好，结构灵活性比碲化镍好，因此成为在碱性条件下电解水较合适的催化剂。

NiSe/$Ni_{0.85}$Se 是简单的非化学计量硒化物，被广泛用作碱性条件下OER电催化剂。Tang等人首次发现了NiSe在1mol/L KOH中具有显著的OER预催化活性[2]。他们以NaHSe为硒源，通过水热法简单地对泡沫镍进行硒化，形成了NiSe纳米线阵列［图6.1(a)］。该材料用于OER催化时，发现表面形成了实时催化剂（real-time catalyst）NiOOH，电流密度为20mA/cm^2时，所需过电位为270mV，Tafel斜率为64mV/dec［图6.1(b)～(d)］。

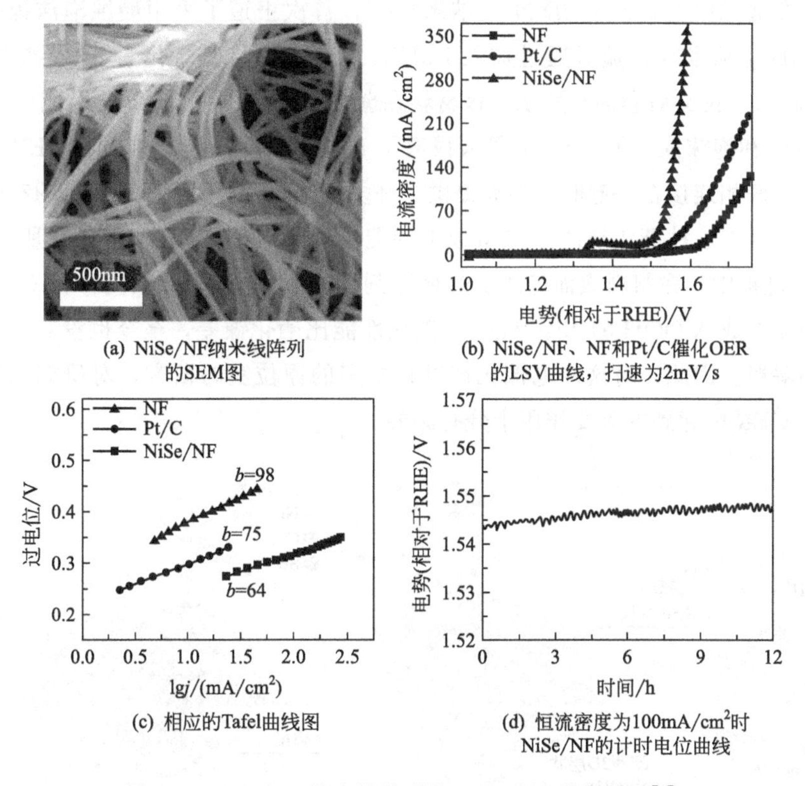

(a) NiSe/NF纳米线阵列的SEM图

(b) NiSe/NF、NF和Pt/C催化OER的LSV曲线，扫速为2mV/s

(c) 相应的Tafel曲线图

(d) 恒流密度为100mA/cm²时NiSe/NF的计时电位曲线

图6.1 NiSe/NF纳米线阵列的SEM和电化学测试[2]

这项工作为后来的许多其他工作奠定了基础。Xu等人以同样的方法在泡沫镍上生长NiSe NW阵列，但反应温度较低[3]，该材料展现出相对较低的Tafel斜率（54mV/dec），但接近起始电位的Tafel斜率区域不是标准的线性。理论上，为了更好地解释动力学，通常在起始过电位至少为 120mV 的线性区域获得其斜率。Han 等人采用类似溶剂热过程，在碳纤维上生长了 NiSe 纳米线阵列[4]，其催化 OER 所需的 Tafel 斜率为 46mV/dec，电流密度为 $50mA/cm^2$ 时所需过电位为 318mV。之后，Li 等人采用两步法在导电石墨烯包覆的镍网基底上生长 NiSe 纳米壁[5]，第一步电沉积 $Ni(OH)_2$ 纳米壁，第二步在惰性气氛中用 Se 高温退火 $Ni(OH)_2$。但相对于其他的 NiSe 催化剂，该材料催化 OER 需要更高的过电位和 Tafel 斜率。与此同时，Gao 等人报道了另一种电沉积法在泡沫镍上制备 NiSe 催化剂[6]，电镀液为 0.035mol/L SeO_2、0.2mol/L LiCl 和 0.065mol/L 乙酸镍，但其 Tafel 斜率高达 112.3mV/dec，可能是由电沉积形成的 NiSe 膜较厚所致。Wu 等人采用一种完全不同的方法在泡沫镍上制备了 2D NiSe 纳米片[7]，首次报道了采用酸蚀泡沫镍生长 $Ni(OH)_2$ 纳米片，随后通过加入 NaHSe 水热处理将其转化为 NiSe 纳米片，如图 6.2（见文后彩插）所示，该材料在碱性条件下的 OER 催化性能优于 $Ni(OH)_2$ 和泡沫镍。另外 Chen 等人报道了 NiSe-$Ni_{0.85}$Se 异质结构对 OER 的影响，当两相连接在一起时，晶界提供了许多缺陷位点作为 OER 催化的有效位点。通常，这种催化剂进行 50 h 以上的催化测试不发生任何降解。众所周知，OER 过程中，在材料表面形成实时催化剂 NiOOH，那为什么同样在 OER 过程中能形成 NiOOH 的 $Ni(OH)_2$ 的催化性能比硒化镍差，至今也没有一个明确的解释。因此，对这些电催化剂进行更多的原位实时研究，对研究硒阴离子在 OER 中起到的重要作用十分有必要。

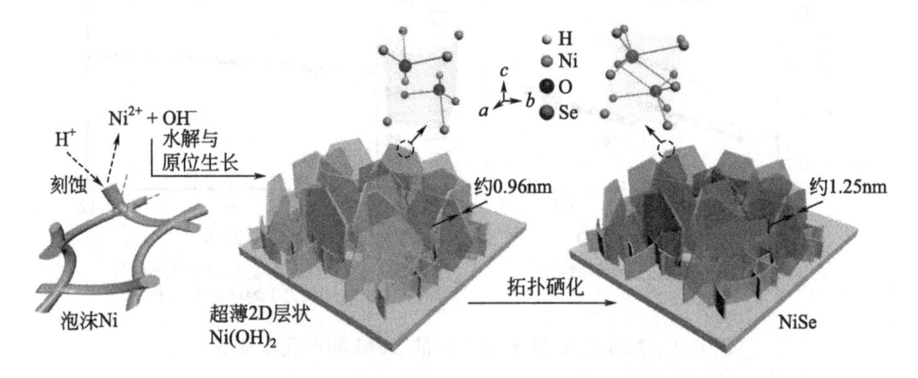

图 6.2 超薄二维层状 $Ni(OH)_2$ 纳米片和超薄 2D 非层状 NiSe 纳米片通过组合酸蚀刻和拓扑硒化的结构演变[7]

黄铁矿型二硒化镍也是一种在碱性条件下催化 OER 的预催化剂。与简单的

$NiSe/Ni_{0.85}Se$ 相比，$NiSe_2$ 具有二价阴离子（Se_2^{2-}）[8]。Pu 等人首次在 Ti 板上电沉积制备了 $NiSe_2$ 催化剂[9]，沉积的电解质与 Gao 等人[6]报道的 Se、Ni、Li 电解质成分相同，但沉积的电位不同，该方法采用饱和甘汞为参比电极，在 0.45V 电位下进行电沉积 1h。该材料在碱性条件下进行 OER 测试时，电流密度为 $20mA/cm^2$ 时需要的过电位为 295mV，Tafel 斜率为 82mV/dec，且具有良好的稳定性。之后，Li 等人[10]发现采用传统水热法直接硒化预氧化的泡沫镍得到的为 $NiSe_2$，而直接硒化泡沫镍则得到 NiSe。测试发现，预氧化泡沫镍形成的 $NiSe_2$ 比由金属泡沫镍形成的 NiSe 具有更低的电荷转移电阻、过电位以及 Tafel 斜率。Yu 等人以 N 掺杂石墨烯包覆泡沫镍为基底生长了 $NiSe_2$，该研究以 N 掺杂石墨烯包覆泡沫镍作为工作电极，在 0.025mol/L $NiCl_2$、0.25mol/L SeO_2 和 0.1mol/L LiCl 的电解液中电化学沉积 $NiSe_2$，得到的催化剂在碱性条件下具有优异的 OER 催化性能。

目前对 Ni_3Se_2 和 Ni_3Se_4 两种催化剂在碱性介质中催化 OER 研究较少。Swesi 等人采用 0.01mol/L 四水醋酸镍、0.01mol/L 二氧化硒和 0.1mol/L 氯化锂为电镀液，电沉积制备了 Ni_3Se_2[11]，研究了不同沉积时间得到的材料对 OER 催化活性的影响，电沉积时间为 300s 的材料显示出的过电位最低。然而，计算的不平整度因子（RF）和 ECSA 与 OER 活性趋势不一致。之后，Sivanantham 和 Shanmugam 采用水热法用硒粉和乙二胺为原料，在泡沫镍上原位生长了 Ni_3Se_2[12]，反应时间为 25h，实现了高质量负载，得到的材料具有良好的 OER 催化活性（$305mV@100mA/cm^2$），Tafel 斜率为 40.2mV/dec。除此之外，Xu 等人采用硫酸镍、亚硒酸钠和水合肼为原料，水热合成 Ni_3Se_2，与 NiSe 和 NiO 相比，Ni_3Se_2 具有较高的 OER 催化活性，该研究还揭示了长时间 OER 催化下催化剂的结构重组[13]。Anantharaj 等人用一种新颖的微波法，在泡沫镍上制备了 Ni_3Se_4 纳米组装体[14]，该方法不需要高温、高压。制备出的 Ni_3Se_4/Ni 电极在不同 pH（13、14、14.5）的碱性溶液和 PBS 中进行 OER 催化测试。碱性条件下，在电流密度为 $50mA/cm^2$ 时需要的过电位分别为 321mV、244mV、232mV；PBS 中电流密度为 $10mA/cm^2$ 时过电位为 480mV。该催化剂展示了硒化镍在碱性条件下催化 OER，具有最低的 Tafel 斜率，也是首次报道了硒化镍在中性 PBS 溶液中的 OER 催化活性。之后该课题组通过改进的水热法制备了 Ni_3Se_4/Ni 电极，所需时间为 2h[15]。制备的催化剂作为非常规酸碱混合电解水槽中的阳极催化剂，Pt 为阴极，利用 pH 驱动的欠电位分解水，电流密度为 $10mA/cm^2$ 时所需的电压为 1.12V，远远低于水分解的理论点位 1.23V。尽管探索了这些化学计量不同的硒化镍在碱性介质中作为 OER 预催化剂的性能，直到 2019 年，仍未清楚地说明这些硒化镍的化学计量对碱性 OER 的影响。最近做了一个系统的尝试，揭示了在碱性条件下相同负载量的镍硒化物的化学计量对 OER 的影响[16]，发现 NiOOH 形成的难易决定了催化剂的活性。催化剂氧化的电位越低，催化 OER 的电流密度越高。

上面所有的研究都揭示了一个共同点：硒化镍在 OER 过程中，确实在表面发生结构变化形成了 NiOOH。这种结构重组的程度主要取决于在 OER 中暴露的时间。但是，如前面所述，没有人知道硒化镍催化行为为什么比镍氧化物/氢氧化物更好，以及硒在增强 OER 活性镍位点中的确切作用。这因此对这些催化剂还有许多研究工作待做。

6.1.2 含有其他金属的镍基硒化物用于 OER

研究发现，当有其他金属（如 Fe、Co 或 V）掺入镍基催化剂时，催化性能更好[17-19]。当 NiOOH 掺入的 Fe 含量范围为 20%～40%时，在碱性条件下表现出优异的 OER 催化活性[20-22]。Tang 等人首次采用常规 NaHSe 水热硒化大孔泡沫 FeNi 合成了 Fe 掺杂的 NiSe[23]，该材料在 1mol/L KOH 中具有优异的催化活性（电流密度 $50mA/cm^2$，过电位 245mV），在 30% KOH 溶液中过电位 263mV 时，电流密度达到了 $1A/cm^2$。之后 Du 等人[24]采用硒和油胺作为硒源，乙酰丙酮镍作为还原剂，合成了海胆状 Fe 掺杂的 $NiSe_2$，其化学计量为 $Ni_{1.12}Fe_{0.49}Se_2$，在催化 OER 过程中具有优异的性能 ($10mA/cm^2$@227mV)。后来，Du 等人[25]提出了另一种制备了 Fe 掺杂的 Ni_3Se_4 的方法，第一步是制备 NiFe LDH 粉末，第二步是采用 NaHSe 对 NiFe LDH 粉末进行常规水热硒化，该材料在电流密度为 $10mA/cm^2$ 时过电位为 225mV，Tafel 斜率为 41mV/dec。与此同时，Gu 等人[26]在合成过程中通过控制 1-十二硫醇与油胺的比例，制备了含 2nm Fe 颗粒的 $NiSe_2$ 超薄纳米线（直径 1.7nm）。在 0.1mol/L KOH 溶液中，该催化剂在电流密度为 $10mA/cm^2$ 时过电位为 268mV。

Xu 等人[27]在 2016 年首次报道了采用两步法合成了 NiFe 二元硒化物

$Ni_{0.8}Fe_{0.2}Se_2$。首先采用水热法以硫酸镍和硫酸亚铁为原料在泡沫镍上生长$Ni_{0.8}Fe_{0.2}(OH)_n$，然后用传统的水热硒化方法（硒粉和肼生成H_2Se）将其硒化。该材料在1mol/L KOH中电流密度为$10mA/cm^2$时过电位为295mV。此后不久，Wang等人[28]通过类似的两步过程在碳纤维布（CFC）上生长了另一种Ni-Fe二元硒化物$Ni_{0.75}Fe_{0.25}Se_2$，但与前面类似的工作相比，该催化剂的催化性能相对较差。Nai等人[29]首先制备NiFe普鲁士蓝类似物，然后用NH_3蚀刻后进行硒化合成了NiFe二硒化物纳米笼，尽管跟同类材料相比同样的电流密度所需的过电位稍高，但其Tafel斜率却最低（24mV/dec），说明该催化剂具有优异的OER动力学。

Zhao等人[30]采用Ni(Ⅱ)和Fe(Ⅱ)为前驱体，SeO_2为硒源，在泡沫镍上合成了3DFe-NiSe/NF，该材料具有良好的催化活性（$10mA/cm^2$@233mV，Tafel斜率48mV/dec），法拉第效率为96.8%，这在其他镍基OER催化剂比较少见。Du等人[31]发现当NiFe的比例为1∶1时，硒化物具有优异的催化OER活性（$10mA/cm^2$@235mV）。Li等人[32]通过对泡沫镍进行硒化后用$FeCl_3$处理，制备成了核壳结构的$Ni_xFe_{1-x}Se$@Ni(Fe)OOH，在1mol/L KOH中具有优异的催化OER活性（$100mA/cm^2$@260mV）。以上所讨论的含铁的硒化镍均与同类组分NiFe LDH的OER活性趋势相似，说明催化OER的实际催化剂为NiFeOOH。但是Ni-Fe硒化物在$10mA/cm^2$时的过电位比NiFe LDH的过电位低，表明硒化物阴离子在提高催化剂OER活性方面确实起了关键作用，但是这些作用还需要进一步的研究。

与Fe一样，将Co掺入镍基催化剂中也能增强其OER催化活性，但不如掺入Fe增强的活性大。Li等人在Ti板上电沉积制备了Co掺杂的NiSe催化剂$Co_{0.13}Ni_{0.87}Se_2$[5]，但$100mA/cm^2$需要320mV过电位，从这个工作看来，Co在KOH中提高Ni基OER催化活性效率不如Fe。之后，Ming等人[33]通过制备Co(Ⅱ)-2-甲基咪唑有机框架化合物，然后将其转化为NiCo LDH，最后于400℃用硒粉硒化，在泡沫镍上生长了用纳米碳片包裹的Co掺杂的$NiSe_2$和Ni_3Se_4混合相，该催化剂催化OER电流密度为$30mA/cm^2$时需要275mV的过电位，Tafel斜率为63mV/dec。Xu等人[34]通过水热硒化，然后再将其浸入含有Co(Ⅱ)离子的溶液中进行阳离子交换，在泡沫镍上原位合成了Co掺杂的NiSe纳米线，该催化剂在众多Co掺杂的硒化镍材料中具有较好的催化活性（$100mA/cm^2$@283mV）。Fang等人[35]考察了不同Ni和Co比例的硒化物催化OER的活性，最终发现Ni/Co

比为 1∶2 时催化性能最佳，通过自由能计算表明，$NiCo_2Se_4$ 的（010）相具有较负的氢氧化物吸附自由能，因此表现出更好的活性，Shinde 等人也做了类似的工作[36]。尽管这两个研究的目的，是要说明镍钴硒化物的成分对 OER 催化活性的影响，但是他们研究的 Ni 和 Co 比例有限，所以这里不能给出一个准确的结论。

后来 Zhu 等人[37]采用两步法制备了 3D $NiCoSe_2$/NF 催化剂，先水热合成相应的双氢氧化物，然后进行热硒化，该材料在碱性条件下获得 10mA/cm^2 的电流密度需要 274mV 的过电位。Xu 等人[38]合成了一种核壳结构的硒化物 NiSe@CoOOH，该催化剂获得 100mA/cm^2 电流密度时仅需 300mV 的过电位。最近，Akbar 等人[39]在泡沫镍上通过电沉积方式制备了 3D $NiCoSe_2$/Ni 催化剂，令人惊叹的是该材料在碱性条件下优异的 OER 催化性能，获得 10mA/cm^2 的电流密度仅需 183mV 的过电位，但是经过确认发现，作者没有扣除催化电极的背景电流密度，该背景电流密度本身就超过了 10mA/cm^2。Ao 等人[40]先通过水热法制备相应的金属有机框架（ZIF-67），然后在 CFP 上制备了蛋黄结构的 Ni-Co-Se 阵列，再通过阳离子交换形成双氢氧化物，最后硒化得到 Ni-Co-Se/CFP 催化剂，但该催化剂在碱性条件下催化 OER 具有较高的过电位（10mA/cm^2@300mV）。以上这些研究是已经报道了的 Co 掺杂的硒化镍作为碱性条件下的 OER 催化剂，但对该类材料的构效关系的研究还需要进一步拓展。

另外，硒化镍通过与其他阴离子结合（比如硫化物和氧化物）形成异质结构用于电催化分解水。Li 等人[41]首次在泡沫镍上制备出了 NiSe-NiS 杂化纳米材料。首先将泡沫镍在 DMF 中硒化，然后在乙醇中硫化，得到的杂化材料在测试 OER 性能之前，先在 KOH 中进行电化学活化，获得 50mA/cm^2 电流密度所需过电位为 200mV，与含 Fe 的镍基 OER 催化剂相近。后来，Gao 等人[42]发现在硒化镍中引入另一种阴离子（O）可以引起其电化学原位活化，在形成 NiSe-NiO_x 核壳结构后，发现提升了其 OER 催化活性。这个结论在 Liu 等人[43]的研究中也有体现，他们通过在 $NiSe_2$ 纳米线上修饰无定形的 NiO_x 纳米颗粒，修饰后的材料催化性能优于 $NiSe_2$、NiO_x 和 IrO_2，原因可能是无定形的 NiO_x 纳米颗粒具有较高的 ECSA，提供了更多的活性位点。相对于纯硒化镍和金属掺杂的硒化镍作为 OER 催化剂的研究而言，采用另一种阴离子制备异质结构的硒化镍的研究相对较少，因此在该领域还可以进行更多研究。

综上，不管是纯的硒化镍、其他金属掺杂的硒化镍还是与其他阴离子杂化的硒化镍，在碱性介质中进行 OER 电催化，这些材料的表面几乎都会发生转化，形成 NiOOH 或 $Ni_xM_{1-x}OOH$，其中 M 可以是任何金属[44]。该类材料结构上的灵活性在提升 OER 催化活性的同时也带来了严重的环境问题。众所周知，硒在一定浓度（$>100\mu g/g$）以上的所有形式释放到环境中，对人类和其他水生生物都有很高的毒性的[45-47]。它是一种致癌物质，在人类和水生物种中引发细胞的异常生长。硒化物阴离子在 OER 过程中发生氧化反应，形成 Se^{4+} 和 Se^{6+} 溶液，其中 Se^{6+} 更易挥发，在电解后处理电解液时有向大气中逸出的风险。因此，要真正以环境友好的方式利用硒化镍在碱性条件下的 OER 催化性能，必须采取严格的措施，避免向大气中释放所有可能的硒。除了环境方面的问题，在电催化性能方面，硒化镍的性能还是不及优异的 NiFeOOH 和贵金属氧化物（IrO_2 和 RuO_2），但这个可以通过与 NiFeOOH 复合进行改性，以及掺杂贵金属来提升其催化性能，从而解决此问题。

6.1.3 NiFe 硫化物

作为非贵金属催化剂，镍基二元过渡金属硫化物（TMCs）对催化电解水具有良好的活性。特别是过渡金属硫化物中的共价键硫化物主体相对于同金属的氢氧化物具有更好的导电性，该共价键还可以防止材料在电解质中的腐蚀，从而使电催化剂具有良好的稳定性。但是要提高镍基硫化物的催化活性，还需要做进一步的研究。从电子结构方面，可以引入其他金属原子来制备二元化合物以提升电催化剂活性。此外，构建具有高比表面积和高活性面的电催化剂也是提升催化性能的重要途径。因此具有良好导电性的镍基双组分金属硫化物纳米材料作为一种良好的电化学催化剂引起了广泛关注。这里重点介绍 NiFe 基硫化物在电解水中的应用。

硫化镍（如 NiS、NiS_2 和 Ni_3S_2），其资源丰富，制备简单，并在 HER 和 OER 中具有潜在的电催化性能[48,49]。遗憾的是硫化镍的本征电导率较低，电催化活性一般，制约了其在电解水中的应用，而通过金属掺杂可以提升硫化镍的电催化性能[50,51]。Fe 是一种很好的掺杂剂，可以通过调节配位价态、增强电子导电性、优化氢/水的吸附能来增强电催化活性[52,53]。比如，Sun 等人[54]采用原位硫化生长在 Ti 网上的镍铁层状双氢氧化物前驱体（NiFe LDH），制备了3D的铁掺杂的二硫化镍纳米阵列（$Fe_{0.1}$-NiS_2 NA/Ti），该材料在0.1mol/L KOH中催化OER，在电流密度为100mA/cm^2时具有较低的过电位231mV，Tafel斜率为43mV/dec，其性能远远优于NiS_2/Ti（100mA/cm^2@420mV，Tafel斜率83mV/dec），同时也具有良好的HER催化活性（20mA/cm^2@243mV）。Liu等人以$NiFeO_x$/$Ni(OH)_2$和硫粉为前驱体，通过温和的煅烧工艺合成了一种新型的Fe-NiS_2纳米片电催化剂[55]。该材料在酸性条件下具有良好的HER催化性能（10mA/cm^2@121mV，Tafel斜率37mV/dec）。DFT计算表明，Fe取代NiS_2中镍的位置，改变了催化活性中心的电子结构，进而引起了不同的HER过程。对于NiS_2，一个质子首先吸附在一个S位点上，然后另一个质子靠近吸附S元素的临近Ni形成过渡态，其激活能（E_a）为0.63eV；对于Fe-NiS_2而言，一个Fe取代一个Ni位，由于没有占据轨道，Fe位可以诱导质子关闭，过渡态E_a为0.41eV，低的E_a可以提升快速的HER过程，进而Fe-NiS_2具有更好的催化性能。

最近，Zhu等人采用一种表面化学气相生长法（SACVT）在三维导电Fe-Ni合金泡沫上原位生长了Fe掺杂的Ni_3S_2纳米颗粒（Fe-Ni_3S_2/AF）[56]。该材料在碱性介质中催化HER和OER均具有良好的活性和稳定性（HER：10mA/cm^2 75mV；OER：10mA/cm^2@267mV），组成全电解水体系时同样具有优异的活性和稳定性。实验结果和理论计算表明，Fe的掺入增大了电化学活性面积、增强了导电性，优化了HER中Ni_3S_2(101)面的氢和H_2O的吸附能，在OER过程中形成了活性双金属Ni-FeOOH，因此提升了材料的催化性能。

众所周知，电导率是决定电催化效率的一个不可或缺的因素，与FeNi基氧化物/氢氧化物相比，金属FeNi基硫化物具有更好的导电性[57]。$FeNi_2S_4$不仅具有良好的导电性，而且具有最佳的Fe/Ni含量比，因此该材料适合作为性能优异的电催化剂。在OER催化过程中，$FeNi_2S_4$的表面重构形成Fe-Ni-O物种，从而促进OER催化。但目前可控相纳米$FeNi_2S_4$催化剂的合成路线较少[58,59]。比如Chen等人以乙酰丙酮镍和乙酰丙酮铁为镍铁源，辛硫醇为硫源，通过控制十六醇和十八烯的比例进行回流，合成了具有可控相和可调形状的金属紫硫镍矿$FeNi_2S_4$纳米材料[60]，该材料在碱性条件下展示了良好的催化OER活性。对于优异的二元金属硫化物电催化剂来说，合适的金属配比是获得独特形貌和催化活性的关键。Dong等人[61]

采用液晶模板（LCT，Triton X-100）辅助一步电沉积方法在泡沫镍上原位制备了不同形貌、不同 Fe/Ni 比的 Fe-Ni 硫化物（FeNi-S/NF），电解液为硝酸铁、硝酸镍、硫脲以及 LCT。他们发现在 Fe/Ni 比为 3/1 时，电催化剂具有最佳的催化活性和稳定性。

Konkena 等人[62]合成了一种无论是在酸性还是在碱性条件下都具有稳定高效催化性能的镍黄铁矿，该材料具有由硫桥连接的 Fe-Ni 中心、高电子电导率、高电流密度、低过电位和长期稳定性等优点。对镍黄铁矿的原位声子研究表明，催化剂表面的硫原子在长期电解过程中发生重排，形成高活性的氢化 Fe-Ni 表面，有助于提高电催化性能[63]。尤其是含有 Fe/Ni 的镍黄铁矿具有类似氢化酶的活性位点，具有高的周转频率（TOF）[64]。

与其他金属硫化物不同，镍黄铁矿（$(Fe_xNi_{1-x})_9S_8$（$x=0\sim1$）的制备方法相对简单，但是发现相对较晚，最早出现是在 1998 年[65]。近年来出现了一些新的合成方法。如 Drebushchak 等人[65]用高纯度的 Fe、Ni 和 S 为原料，在真空中 1100℃进行热处理，得到了两种类型的镍黄铁矿 $Fe_6Ni_3S_8$ 和 $Fe_3Ni_6S_8$。真空环境消除了氧气的影响，防止了样品在制备过程中被氧化而影响材料的纯度。在后面相关的研究中，均采用此方法，只是对煅烧工艺进行了改进，比如通过改变煅烧的温度和时间，Kitakaze 等人制备了 $Fe_{4.5}Ni_{4.5}S_8$[66]，Xia 等人制备了 $Fe_{4.6}Ni_{4.55}S_8$[67]。在 2016 年，Konkena 等人首次研究了天然 $Fe_{4.5}Ni_{4.5}S_8$ 催化 HER 的性能[62]，实验结果表明，该材料在酸性条件下电流密度为 $10mA/cm^2$ 时过电位为 280mV，由于表面硫的消耗，反应 96h 后过电位降至 190mV。这种现象可能是由镍铁活性中心电子结构的变化和高活性氢化铁镍表面的形成所致，在高电流密度下催化 170h 其活性没有损失（图 6.3，见文后彩插）。

一般来讲，与其他富含硫属元素的同类物相比，镍黄铁矿材料具有优异的本征导电性[68]。通过对其进行进一步特殊结构的修饰、掺杂或构建，可以提升其催化活性，因此在电催化水分解方向具有很大的研究潜力。之后出现了一些新的关于使用镍黄铁矿作为 HER 催化剂的报道，研究了其制备过程、元素组成、局部结构和性质之间的关系[69-71]。其中 Puring 等人总结出：在真空加热步骤，升温需要比较缓慢，防止硫黄过热形成高压硫破坏安瓿[69]；另外防止杂质相如单硫化物固溶体的形成也很重要。另外一篇报道研究了 Fe/Ni 的比例对电催化活性的影响[70]，实验结果显示，当 Fe 和 Ni 的比例达到 1∶1 时（$Fe_{4.5}Ni_{4.5}S_8$）具有最优异的催化活性，这可能归

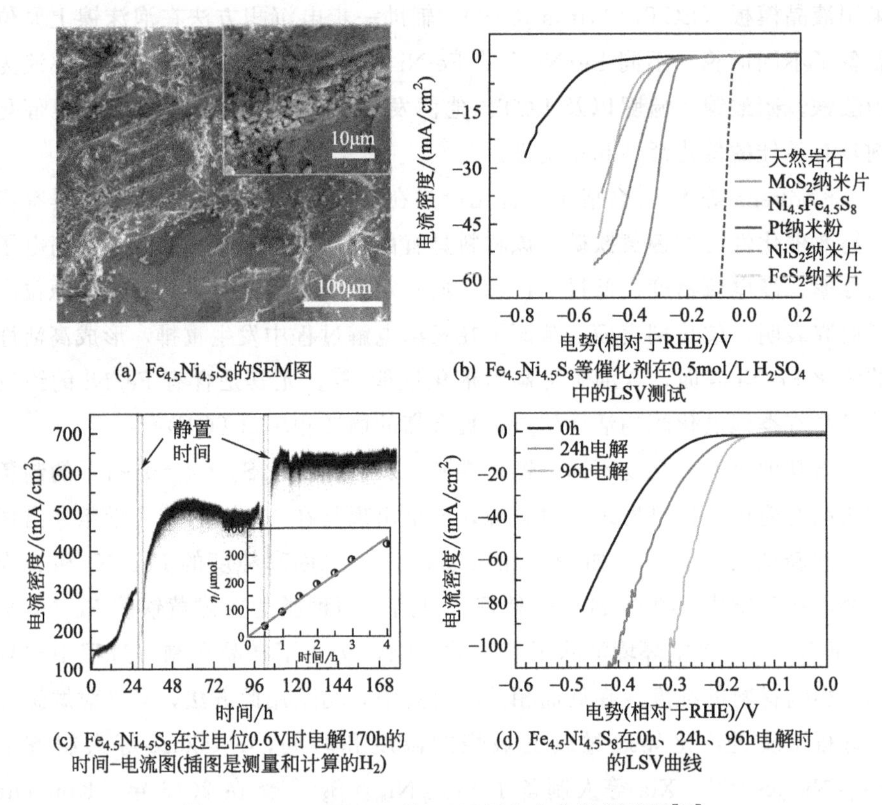

(a) $Fe_{4.5}Ni_{4.5}S_8$的SEM图

(b) $Fe_{4.5}Ni_{4.5}S_8$等催化剂在0.5mol/L H_2SO_4中的LSV测试

(c) $Fe_{4.5}Ni_{4.5}S_8$在过电位0.6V时电解170h的时间-电流图(插图是测量和计算的H_2)

(d) $Fe_{4.5}Ni_{4.5}S_8$在0h、24h、96h电解时的LSV曲线

图 6.3　$Fe_{4.5}Ni_{4.5}S_8$ 的 SEM 图和电化学测试图[62]

因于镍黄铁矿中金属与金属之间的相互作用，该报道发现温度的升高会导致过电位的显著降低和 HER 催化活性的增强。另外，Bentley 等人[71]报道由于材料的“陈化”或改变合成条件，Fe、Ni、S 的表面比的微小变化可能会导致催化剂的显著变化，出现这种现象的原因可以归因于表面缺陷的存在。据推测，由于缺陷边缘硫的释放速度快，因此取代的质子化速度更快导致缺陷位点具有高的催化活性[63]。

以上这些报道对镍黄铁矿的制备以及在电催化析氢性能方面的研究有重要的指导作用。另外除了前面讨论的高温固态法，最近出现了一些其他的方法来合成镍黄铁矿催化剂。比如 Bezverkhy 等人煅烧金属前驱体并用气体硫化来制备 $Fe_{4.5}Ni_{4.5}S_8$[72]，该方法对实验的要求较高，但是制备的催化剂具有良好的微观结构。Ma 等人通过混合固体前驱体，并在 N_2 气氛（以过硫酸铵为硫源）中热处理制备了（Fe，Ni)$_9S_8$[73]。Wu 等人采用化学气相沉积法（CVD）制备了 $Fe_5Ni_4S_8$[74]；Liu 等人以 Fe/Ni 络合物为前驱体，采用先水热后煅烧制备了 $Fe_5Ni_4S_8$[75]；Zhang 等人通过溶剂热共沉淀法制备了 $Fe_5Ni_4S_8$ 纳米花[76]。尽管制备方法开始多样化，但是由于工艺的不稳定，要制备较纯的镍黄铁矿催化剂仍然很困难。实验中微小的偏差都有可能引入杂质，因此，还需进一步探索可控、直接合成具有更优异的催化活性和实用性的镍黄铁矿催化剂的方法[77]。

6.1.4 NiFe 基氧化物/氢氧化物

近年来，涌现了很多对双金属化合物的研究，在这些双金属化合物中，NiFe 基化合物因低成本、含量高引起了越来越多的关注，尤其是 NiFe 基催化剂在电解水催化中表现出优异的 HER 和 OER 催化活性[78-81]。目前有许多工作研究了 NiFe 基化合物的制备和应用，以及 Ni 和 Fe 原子的合理比例对提升催化活性的影响，这对 NiFe 基化合物的研究提供了很多有用的信息。虽然研究者们一致认为 NiFe 共存是提供高催化活性必需的，但是对 NiFe 基化合物的活性位点的识别和催化机理仍然具有争议：水分解反应是通过 Ni 或 Fe 位点进行的，还是两者兼而有之。大家一致强调提高催化活性的关键是 Ni-Fe 之间的协同作用，但是对催化剂原子结构的识别和反应中心原子水平的精确控制，特别是基于 Ni-Fe 的多原子组成活性位点的探索仍是一个巨大的挑战[82]。在这里我们重点讨论 NiFe 基的氧化物、氢氧化物（包括层状双氢氧化物 LDH）在碱性条件下作为电解水催化剂的应用。

6.1.4.1 NiFe 基氧化物

相比于其他氧化物，NiFe 基氧化物（$Ni_{1-x}Fe_xO_y$）具有更优异的 OER 催化活性。尖晶石型 NiFe 氧化物 $NiFe_2O_4$，其中二价阳离子 Ni^{2+} 占据八面体位置，三价阳离子 Fe^{3+} 占据八面体和四面体位置[83]。由于存在多价 Ni^{2+}/Ni^{3+} 和 Fe^{2+}/Fe^{3+} 氧化还原电对，$NiFe_2O_4$ 可作为优异的电催化剂[84]。当向氧化物中引入三价阳离子 M^{3+}（如 Al^{3+}）（$NiFeAlO_4$）时，Fe^{3+} 和 Al^{3+} 可能占据四面体和八面体的位置[83]。这种氧化还原惰性的 Al^{3+} 对最近的 Ni^{2+} 产生了短程的诱导效应，从而干扰了固体氧化物的电子结构延伸。$NiAl_2O_4$ 和 $NiFe_2O_4$ 具有相似的性质，说明催化 OER 的活性与 Fe 基氧化还原活性无关，而 $NiFeAlO_4$ 相对于 $NiAl_2O_4$ 具有更优异的催化 OER 性能，表明 Fe 和 Al 在催化过程中具有协同作用。其他三价阳离子如 Mn^{3+}、Cr^{3+}、V^{3+} 等也可以用来提升 NiFe 基氧化物的催化活性，但是还需进一步的研究来阐明这些三价离子和 Fe 的协同作用。

实际上大多数 NiFe 基氧化物都是由 Fe 掺入 NiO_x 体系得到的，通过原位 X 射线吸收光谱（XAS）和 CV 的研究可以看出，其催化活性依赖于 Ni 和 Fe 的局部环境[85]。例如 Hong 等人合成了 Fe 掺杂的 NiO_x 纳米管作为高效的 OER 催化剂[85]。Fe 掺入 NiO_x 纳米管中引入了 Ni 空位和晶格畸变，从而显著改变了纳米管局部原子和电子结构。扩展 X 射线吸收精细结构表征（EXAFS）表明，Fe 掺杂的 NiO_x 纳米管具有不同的 NiO 局域原子结构，因此 Fe 掺杂 NiO_x 纳米管的电荷传输电阻，因其几何结构和电子结构的变化而明显降低，从而提高了催化活性。此外，在 Kitchin 等人[86]研究的另一个 Fe 掺杂 NiO_x 体系中，EXAFS 中的 Fe K 边谱在 OER 条件下发生了一些变化，表明 Fe 原子发生了进一步配位，并受到反应条件的影响。在 OER 条件下，Fe 的平均配位数也显著增加，这直接说明掺杂在 NiO_x 界面的 Fe 在 OER 催化中积极参与反应。推测铁配位增加的原因包括：①Fe 原子从尖晶石结构内的四面体位置移动到八面体位置，这可以在 $NiFe_2O_4$ 纳米相中找到；②在氧化条件下形成了 Fe_2O_3，但是由于 Fe_2O_3 是较差的电催化剂，与实验中得到的电催化活性结果不一致，因此增加的 Fe 配位不可能是 Fe_2O_3。因此他们认为当 Fe 掺入 NiO_x 时，$NiFe_2O_4$ 中的 Fe 配位是增强 OER 活性的因素[86,87]。但是，Louie 等人采用原位拉曼散射研究 NiFe 氧化物膜时否认了 $NiFe_2O_4$ 的形成[88]。在含铁的电解液中老化 6 天后，即使在足够高的过电位下依然没有观察到 $NiFe_2O_4$ 形成的信号，这个结果表明在 NiFe 氧化膜的表面发生了结构重组。此工作为探索 NiFe 基氧化物表面的局域电子结构开辟了一个全新的方向。但迄今为止，NiFe 基氧化物活性中心的性质以及高催化活性的来源仍然没有一种统一的说法。目前在电解水条件下，关于 Fe 是如何直接影响 NiFe 基氧化物的电化学活性和结构完整性的报道较少，因此了解铁的加入是如何影响 NiFe 基氧化物催化电解水行为十分重要。

众所周知，杂原子（P、B、S 等）的掺杂可以有效地改善催化剂的活性，杂原子的存在可以进一步调控 Ni 和 Fe 周围的电子结构，有利于电荷传输[89]。比如 Wang 等人通过一种简单且经济的一步煅烧法，煅烧 Ni-Fe 基金属有机框架（MOF）和含杂原子的分子，合成了无定形的多孔纳米立方体 NiFe 基材料（Ni-Fe-O-P、Ni-Fe-O-B、Ni-Fe-O-S）[89]。这些杂原子均匀分布在这三种材料的结构中。P、B 或 S 的引入调节了电子结构，加快了电荷转移，在 OER 催化过程中 Ni 的平均氧化价态升高，提升了催化活性。除了掺杂非金属杂原子，也有研究掺杂金属元素于 NiFe 基氧化物中来提升催化活性。比如 Tolbert 等人通过选择性的腐蚀合金制成了纳米多孔结构的 NiFe 基氧化物（$Ni_{60}Fe_{30}Mn_{10}$）作为 OER 催化剂[90]。这种三金属基材料的催化活性高于双金属，他们认为少量的 Mn 可能会提高介孔 NiFeMn 基催化剂在碱性溶液中的耐腐性。一般来讲，3d 过渡金属的亲氧趋势如下：Ni＜Fe＜Mn。因此碱性溶液中钝化残留的 Mn 和 Mn 氧化物可以防止材料在 OER 过程中的进一步腐蚀，从而提高了催化剂的稳定性，但 Mn 起的作用还不完全清楚。Mukerjee 等人[91]研究了 Ni-Fe 基氧化物和 Ni-Fe-Co 基氧化物的 OER 催化性能，发现三金属基氧化物的催化活性高于双金属基氧化物。后来 Zhang 等人[92]采用共晶衍生的自模板法制备了分级多孔 Mo 掺杂的 Ni-Fe 基氧化物纳米线作为电解水的双功能催化剂。在他们的研究中，Mo 的引入没有改变 NiFe 基氧化物催化剂的相组成，但由于 Mo^{4+} 的氧化态较高，他们推测 Mo 的插入导致 Ni—O 和 Ni—Fe 键长缩短，从而促进了两种不同元素原子键的电荷传输，提高了氧化物的导电性，从而提高了催化性能。

6.1.4.2 NiFe 基(氧)氢氧化物

在 1987 年 Corrigan 就发现 Fe 掺入到 $Ni(OH)_2$ 中可以提升其 OER 催化活性[93]。之后出现了大量的工作来解释 Fe 提升 OER 催化的机制及不寻常的动态效应。能确认的是，将 Fe 掺入 Ni（氧)氢氧化物中，可以利用金属-金属间的协同效应来提升 OER 催化活性[94,95]。最直接的证据表明，OER 催化活性随着 β-$Ni(OH)_2$ 晶格中 Fe^{3+} 含量的增加而增加[96]。一般来说在 OER 催化体系中，α-$Ni(OH)_2$ 和 β-$Ni(OH)_2$ 催化剂分别发生相转变，变为 γ-NiOOH 和 β-NiOOH，NiOOH 作为活性相可以提升 OER 催化活性[87]。另外，当 β-NiOOH 过度充电时，也会形成 γ-NiOOH。α-$Ni(OH)_2$ 转化为 β-NiOOH 会经历更多结构变化和多步电子转移[97]。一般的活性趋势为：Fe 掺杂的 β-NiOOH＞β-NiOOH＞Fe 掺杂的 γ-NiOOH＞γ-NiOOH[98]。十多年来，β-NiOOH 一直被认为是碱性电解质中催化 OER 的活性结构[97]。例如，据报道可以通过电化学氧化 β-$Ni(OH)_2$ 形成 β-NiOOH，这种氧化不经过相结构转化，形成的 β-NiOOH 是一种适合催化OER 的氢氧化物催化剂[96]。因此，对 NiFe 基（氧）氢氧化物活性位点的研究更多地集中在 NiOOH 上。如 Trotochaud 等人证明了将 Fe 掺入 $Ni(OH)_2$/NiOOH 薄膜增强了其导电性，并产生了部分电荷转移，显著提高了 OER 催化活性[99]。

另外，六方 $Ni(OH)_2$ 具有层状结构，可以作为特殊价态阳离子的基质[100]。在电化学氧化过程中，Fe^{3+} 占据了 $Ni(OH)_2$ 中 Ni 晶格位点，由于基底稳定、合适，容易形成异常氧化态金属离子（Ni^{4+} 和 Fe^{4+}）[101]。在稳态水氧化过程中，通过现场原位穆斯堡尔光谱，在氢氧化镍催化剂中检测到 Fe^{4+}，而在氧化铁催化剂中未检测到 Fe^{4+}，因此 NiOOH 晶格具有稳定 Fe^{4+} 作用[102]。NiFe（氧）氢氧化物的穆斯堡尔光谱表明，有 21%的铁离子在电催化过程中从 Fe^{3+} 氧化为 Fe^{4+}，即使去除外加电压，Fe^{4+} 仍然存在，因此推测尽管 Fe^{4+} 不具有催化能力，但是对材料的结构和催化行为有影响。

尽管一致认为此类材料要具有高的催化活性，一定同时含有 Ni 和 Fe。但对 NiFe 基（氧）氢氧化物中活性相和活性中心结构的评价仍存在争议，目前只有少数研究报道 Ni-Fe 催化 OER 中间体的能量[79]。因此有必要先了解 Ni 和 Fe 的相互作用，以及这种相互作用对 NiFe 基（氧）氢氧化物高 OER 活性的贡献。目前 NiFe 基（氧）氢氧化物研究的一个中心主题，是确定 Fe 离子在水分解过程中的作用。由于单一成分的 Fe 氧化物催化性能较差，因此一些研究者认为 Ni 中心是 NiFe 基（氧）氢氧化物催化 OER 的活性中心，而 Fe 离子起间接作用[103]。最直接的证据是用 XAFS 和现场原位 XAS 分析水热合成的 NiFe（氧）氢氧化物得到的结果：在催化过程中，铁离子仍处于 Fe^{3+} 状态，而 Ni 的平均氧化态降低，通常被解释为 Fe^{3+} 将 Ni 中心稳定在低氧化态[104]。此外，Görlin 等人发现 Fe 掺杂可能导致部分电子从混合 NiFe（羟基）氢氧化物中心迁移到 Ni，以稳定 Ni^{2+} 原子，促进 Fe^{4+} 的形成[104]。然而，电化学分析和 XANES 对电沉积 NiFe（氧）氢氧化物的分析却出现了相反的结果，随着 Fe 含量从 0 增加到 10%（物质的量分数），镍离子的平均氧化态从 3.2 增加到 3.6[105]。因此，认为 Fe^{3+} 的高路易斯酸促进了 Ni^{2+} 氧化为 Ni^{4+}，从而起到催化作用。另外，通过光谱观察到了被称为吸附的“活性氧”的带负电活性位点，发现了掺入 Fe 形成的 Ni^{4+}，促进了催化氧化还原活性，表明 Ni 是催化 OER 的活性中心[106]。值得注意的是，纯 NiFe（氧）氢氧化物中的 Ni^{4+} 在不引入任何其他元素的情况下很难形成[107]。

当然，Fe 是否是催化活性中心也存在争议。这一观点最初基于 XAFS 研究阴极电沉积 NiFe(氧) 氢氧化物，发现 Fe—O、Ni—O 键长与最近的金属-金属距离有很强的相关性，说明在 α-$Ni(OH)_2$ 和 γ-NiOOH 中 Fe 都可以取代 Ni[79]。此外，在最近的 Ni-Ni 和 Fe-Ni 之间距离的近两倍处，Fe 和 Ni 的 K-edge 的 EXAFS 谱峰相同，这主要是由 Fe-Ni-Ni、Ni-Fe-Ni 和 Ni-Ni-Ni 共线排列的多重散射引起的，这清楚地表明 Fe 没有嵌入六方 [NiO_2] 片层之间，而是取代了晶格中的 Ni。这种改性对 Ni 离子的氧化状态没有影响，但显著提高了 OER 活性。增强催化活性的原因有两点：一是在 γ-NiOOH 晶格中，由于电子环境的变化，取代的 Fe^{3+} 位点变得更加活跃；二是 Fe^{3+} 取代了 γ-NiOOH 晶格并改变了 Ni^{3+} 位点的电子性质，导致 Ni^{3+} 位点的活性增加。这清楚地证实了在 γ-$Ni_{1-x}Fe_xOOH$ 中，OER 的活性中心是 $Ni^{3+/4+}$ 和 Fe^{3+}，而不是单个的 Ni 位点。Diaz-Morales 等人也报道了类似的结论，并在 Ni 位点观察到一个较小的配位效应[108] [图 6.4(a)]，而在掺杂的过渡金属位点尤其是在 Fe 位点上更显著，表明电催化掺杂中的效应不仅取决于金属与晶格氧的相互作用，还取决于拉伸晶格中的主客体金属之间的相互作用 [图 6.4(b)]。通过键长可以直观地观察到这种相互作用，当 Fe 嵌入到 NiOOH 晶格中时，Fe-O 距离与其纯氧化物距离不同。因此，与 NiOOH 相比，Fe 显著提高了 NiFeOOH 催化活性，而不是 Ni。从图 6.4的火山图可以推断：①NiOOH 位于火山的弱侧，因此，其潜在的限

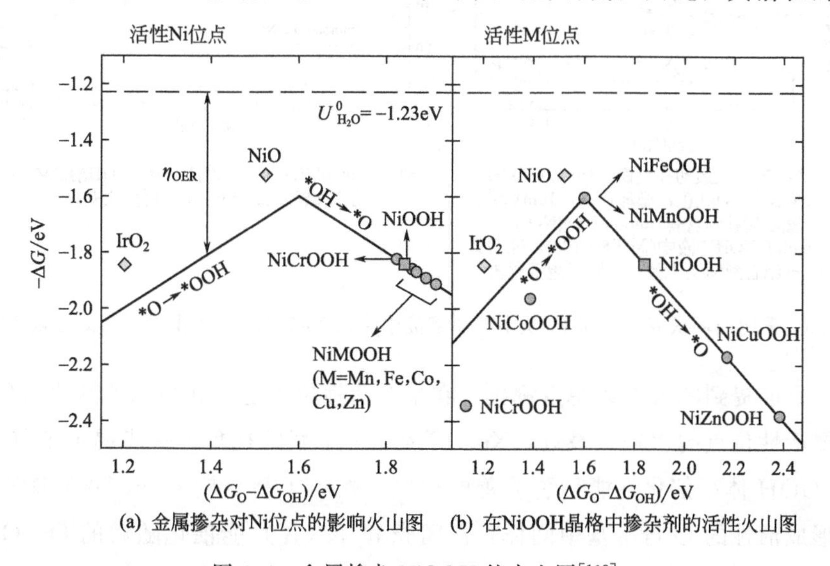

(a) 金属掺杂对Ni位点的影响火山图　(b) 在NiOOH晶格中掺杂剂的活性火山图

图 6.4　金属掺杂 NiOOH 的火山图[108]

制步骤是将*OH 转化为*O，在 OER 条件下 Ni 中心被*OH 覆盖；②Fe 氧化物潜在的限制步骤是将*O 转化为*OOH，因此在 OER 条件下，NiFeOOH 表面的 Fe 位点被*O 覆盖。这种覆盖效应有时很重要，因为侧向吸附-吸附质相互作用可能会减弱或增强吸附能[109]。

下面进一步阐述 Fe 在（Ni,Fe)OOH 体系中的作用。当溶液中 Fe 引入到 NiO_xH_y 中时，Fe-NiO_xH_y 薄膜的电子性质、氧化还原电位、氧化还原峰大小、峰型等与其催化活性无关［图 6.5(a)］[21]。与 Ni 和 Fe 的共沉积不同，Fe 可能首先被引入到 NiO_xH_y 纳米片的边缘/缺陷处，并进一步引入到 NiO_xH_y 纳米片的“块状”中。随着 Fe 的加入，氧化还原峰中每个 Ni 的 e^- 略有增加，但随着 Fe 的大量掺入，即使 e^- 降低，材料仍保持活性［图 6.5(b)］。即使样品变得更加有活性，Fe-NiO_xH_y 在含 Fe 溶液中的电导率也随着循环时间的增加而衰减。这些结果表明“本体”（氧）氢氧化物电子结构不是影响 OER 活性的主要因素。沉积方法和电化学调控与 Fe 的位置和催化剂薄膜的局部环境直接相关，但不会改变催化剂的活性。最近，一项通过脉冲激光烧蚀制备 NiFeOOH 的研究报道中，在末端位置形成了五配位的 Fe^{5+} 物种，这可能是 OER 催化性能增强的原因。

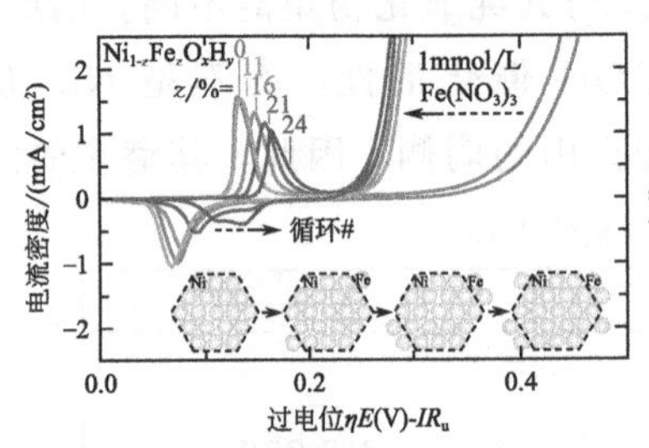

(a) NiO_xH_y最初在无铁的1mol/L KOH水溶液中CV(红色，循环1～4，10mV/s)，然后转移到含有1mmol/L $Fe(NO_3)_3$的1mol/L KOH溶液中(循环5～104)。插图是Fe结合到NiO_xH_y薄片中的可能示意图

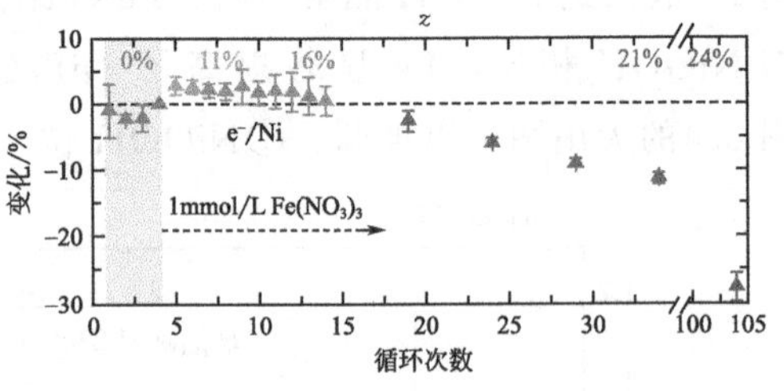

(b) 在1mmol/L $Fe(NO_3)_3$的1mol/L KOH溶液中不同CV循环次数下e^-/Ni的百分比变化

图 6.5 采用循环伏安法证明 Fe 从电解质溶液中掺入 NiO_xH_y 中的作用（见文后彩插）[21]

目前提到的大多数观点都集中在单个活性位点上，并强调了其他组分对主要活性位点的影响。最近，Xiao 等人从一个新的视角进一步揭示了（Ni,Fe)OOH 体系催化性能显著改善的原因，揭示了高自旋 d^4 Fe(Ⅳ）能有效地形成活性的 O 自由基中间体，而闭壳 d^6 Ni(Ⅳ）能催化随后的 O—O 耦合，因此 Fe 和 Ni 之间的协同作用是得到最佳 OER 性能的关键因素[110]。

事实上，在参与反应的同一多组分催化剂中，由两个或两个以上不同活性位点衍生的协同效应，是多步催化反应的关键[111]。因为基元反应发生在不同的活性中心，反应物和中间体被相邻但不同的活性中心吸附和催化，整个反应会消除尺度关系，从而最大化每个基元反应的效率。对中间体具有不同吸附能的两个活性中心可以相互配合，使基元反应顺利进行[103]。

为了弄清 (Ni,Fe)OOH 体系中催化活性位点的特征，需要使用一种具有良好、均匀的原子结构和可调节的复杂环境的催化剂。从基本的角度来看，分散在载体上含有孤立金属原子结构的 (Ni,Fe)OOH 复合材料，有望阐明 Ni 和 Fe 在催化 OER 过程中的相互作用[103]。其中聚合氮化碳（PCN）可作为锚定 Ni 或/和 Fe 金属的理想基质（NiOH@PCN、FeOH@PCN、NiFeOH@PCN)[103]。在假设金属原子为活性中心的基础上，首先计算自由能来研究 Ni 和 Fe 中心的协同效应。对于 NiFeOH@PCN，由于其自由能变化趋势与 FFeOH@PCN 相似，吸附能结果表明 Fe 是活性位点，在高电位下，第三、第四步仍保持上升［图 6.6(a)、(b)］。与 FeOH@PCN 相比，尽管它们的活性中心是相同的 Fe 原子，但 Ni 的引入将 OER 的限速步骤从第四步转移到第三步（$O^* \longrightarrow OOH^*$）。造成这种 OER 反应机制差异的主要原因是 OOH^* 在 NiFeOH@PCN 基底上的吸附强度较弱。此外，发现

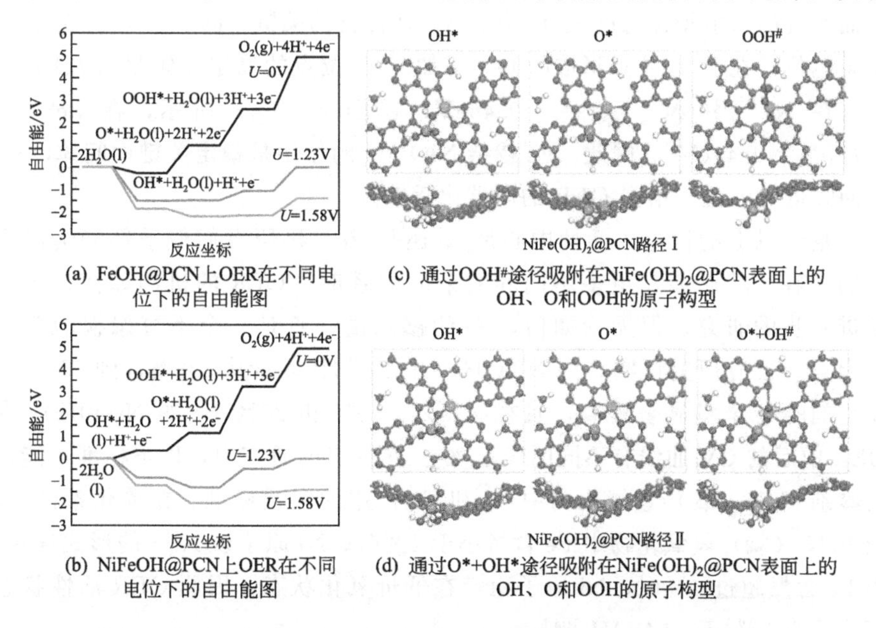

(a) FeOH@PCN上OER在不同电位下的自由能图
(b) NiFeOH@PCN上OER在不同电位下的自由能图
(c) 通过OOH#途径吸附在$NiFe(OH)_2$@PCN表面上的OH、O和OOH的原子构型
(d) 通过O*+OH*途径吸附在$NiFe(OH)_2$@PCN表面上的OH、O和OOH的原子构型

图 6.6　FeOH@PCN 和 NiFeOH@PCN 吸附自由能的 DFT 计算[103]

O^* 和 OH^* 最稳定的吸附位点是 Fe，而 OOH^* 倾向于与 Ni 位点结合［图 6.6(c)］，这说明有两类活性位点激活中间体。因此，OER 的第一步和第二步发生在 Fe 位点，而第三步发生在 Ni 位点，构成反应路径Ⅰ：$H_2O(l) \rightarrow OH^* \rightarrow O^* \rightarrow OOH^{\#} \rightarrow O_2(g)$［图 6.6(c)］。在路径Ⅰ的基础上提出了另一种反应途径（$H_2O(l) \rightarrow OH^* \rightarrow O^* + OH^{\#} \rightarrow O_2(g)$［图 6.6(d)］，即 OH 的另一个中间体（$OH^{\#}$）吸附在 Ni 位点上，与第二步生成的 O^* 反应生成 O_2。因此，具有不同吸附能的 Ni 和 Fe 位点参与了整个 OER 过程。

在 $Ni_{1-x}Fe_xOOH$ 中，也发现 Fe^{4+} 和 Ni^{4+} 都在 OER 中起着重要作用[112]。高旋 d^4 Fe 促进了金属氧（MO）键上具有自由基性质的活性中心，而低旋 d^6Ni 提供了 O—O 键偶联的位点，这就是 $Ni_{1-x}Fe_xOOH$ 比 NiOOH 具有更低的过电位和更好的 OER 催化活性的原因。Shin 等人研究了$Ni_{1-x}Fe_xOOH$和 NiOOH 的 OER 催化机理，发现 Fe 的掺入极大地改变了 NiOOH 在 OER 中的路径和能量［图 6.7(a)，见文后彩插］[113]。在 $Ni_{1-x}Fe_xOOH$中，第一步氧化-去质子化反应所需能量与 NiOOH 接近，而第三步和第四步反应的表观能量相差较大，由于 O· 的未成对电子被高自旋d^4 Fe^{4+}位稳定，在 Fe^{4+} 位点形成关键 O· 中间体时，电位决定步骤(PDS) 所需能量降低到 1.68eV［图 6.7(b)、(c)］。O—O 偶联是通过O·与第三步氧化的另一分子 H_2O 作用，使桥联的 O 在 Ni^{4+} 位点之间发生氢化而实现的。值得注意的是O·作为 OER 催化的关键中间体决定了 OER 的初始电位，在 Ni^{4+} 位点形成的 O·不如 Fe^{4+} 位点的稳定［如图 6.7(b) 所示，在 Ni^{4+} 中 $\Delta G_{4\text{-}4'\text{-}Ni} = -0.46eV$，在图 6.7(c) 所示，在 Fe^{4+} 中 $\Delta G_{4\text{-}4'\text{-}Fe} = 0.41eV$］。因此，Fe 掺杂 NiOOH 的作用是稳定关键中间体O·，从而降低过电位，增强 OER 的电催化活性。

基于以上的讨论，活性中心是 Ni 还是 Fe，科研者们似乎都能给出合理的解释。然而涉及更多不同和复杂的体系时，OER 的机制如何，还需要进一步的研究。但无论如何，Fe 的掺入量一直是一个不容忽视的关键性因素。已有研究证实，当 $Ni(OH)_2$ 中 Fe 的掺入量从 5%增加到 25%时，催化剂的电导率显著提高，且掺杂 5%、15%和 25% Fe 的 $Ni(OH)_2$ 在 OER 前后的 CV 曲线有不同的峰值[98]。在现有的研究中，以下标准可能作为鉴别 NiFe（氧）氢氧化物中 Ni 和 Fe 作用的定量标准。在催化条件下，当 FeNi（氧）氢氧化物中 Fe 含量小于 4%时，Ni 原子以 Ni^{4+} 的形式存在，当 Fe 含量超过 4%时，Ni 原子稳定在低价氧化状态，因此 OER 活性状态可描述为 $Ni^{2+}Fe^{3+}OOH$[104]。

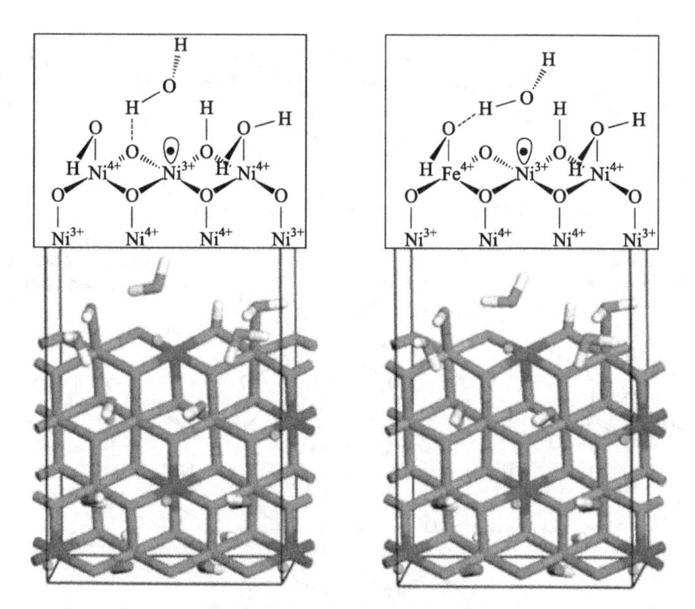

(a) NiOOH和$Ni_{1-x}FeOOH$ OER电催化剂模型，蓝色、紫色、红色、白色、绿色和灰色原子分别表示Ni^{4+}、Ni^{3+}、O、H、K^+和Fe^{4+}

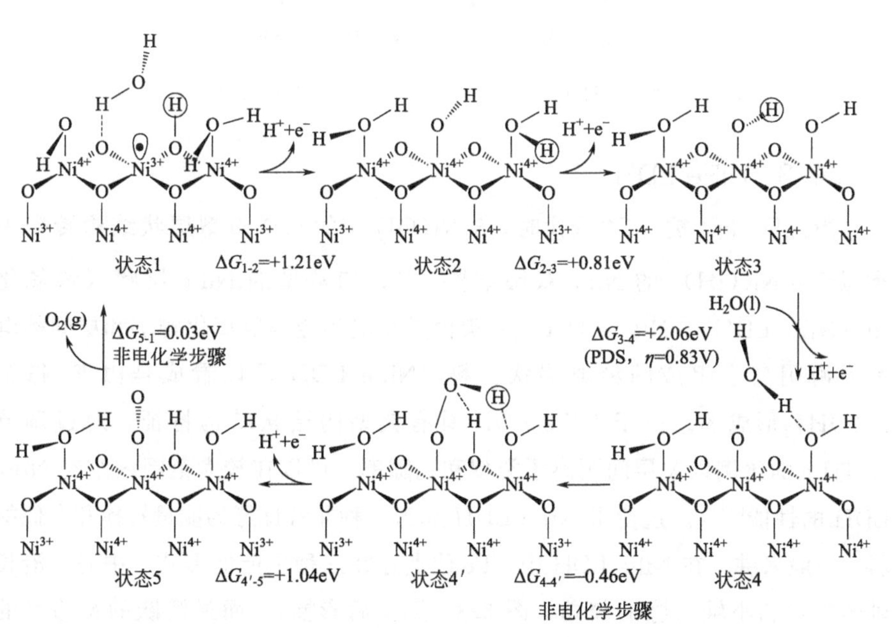

(b) NiOOH催化剂上OER的机理，η=0.83V

图 6.7

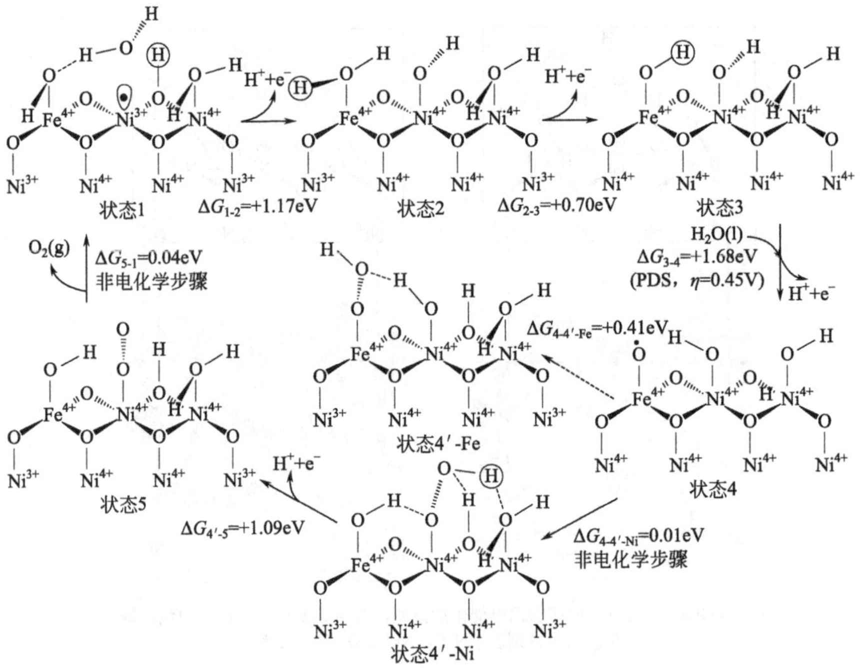

(c) $Ni_{1-x}Fe_xOOH$催化剂上OER的机理，η=0.45V

图 6.7 NiOOH 和 $Ni_{1-x}Fe_xOH$ 模型和催化 OER 机理[113]

6.1.4.3 NiFe LDH

当二价 Ni^{2+} 被 Fe^{3+} 取代时，β-$Ni(OH)_2$ 的水镁石型层状结构转变为类似于 α-$Ni(OH)_2$ 的 NiFe 双层结构[114]，如常见的 NiFe 层状双氢氧化物（NiFe LDH）。NiFe LDH 是一类由带正电的金属氢氧化物层和层间的电荷平衡阴离子组成的经典层状材料。NiFe LDH 可以看成是由 Fe 掺杂 $Ni(OH)_2$形成的。由于 NiFe LDH 具有典型的层状结构特征，通过调节 Ni 与Fe 的比例以及层间阴离子类型和孔隙率，可以在较大范围内改善 NiFe LDH 的性能[115]。这使得 NiFe LDH 成为一种非常良好的能量转换和存储材料。一般来讲，在 NiFe LDH 中，Fe 位点存在三种可能的形式：中心、沿长轴的边缘和小轴边缘的犄角（图 6.8，见文后彩插），而最活跃的是边缘的 Fe 位点[115]。Fe 的引入促使形成了 LDH，从而提高了有效的 ECSA，这种 LDH 晶体结构比焙烧尖晶石相 Ni-Fe 催化剂更具有活性。值得注意的是，大多数报道集中在将 NiFe LDH 与导电碳材料耦合（如石墨烯、碳纳米管），以加速电荷转移动力学和一些所谓的协同效应，从而提高催化活性。遗憾的是，对增强 NiFe LDH 的内在活性和/或增加其活性位点的研究很少。对于大多数过渡金属基 2D 材料，活性中心主要位于边缘，而形成高价态催化相的氧化反应受到紧密堆积的基面的阻碍。因此，这些基面的活化对于进一步提高纳米 NiFe LDH 的催化活性至关重要。

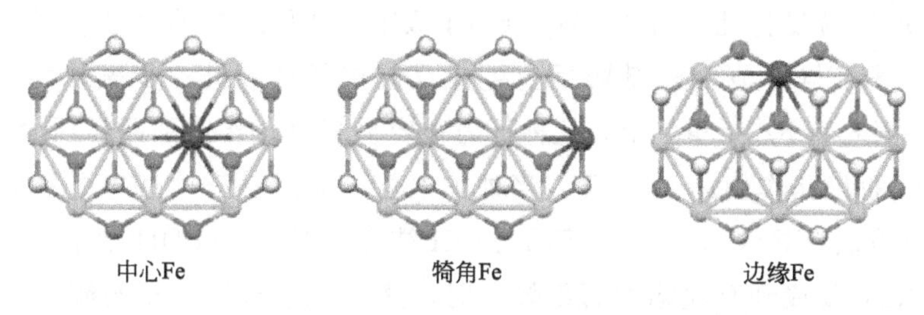

图 6.8 $[Ni_9Fe_1(OH)_{18}]^{3+}$ 簇碎片中的 Fe 位点[115]

颜色：Ni 绿、Fe 栗色、O 红、H 白

目前，NiFe LDH 主要采用一锅法合成，主要包括均相共沉淀法、水热法和电沉积法。但是，这些方法合成的 NiFe LDH 具有紧密堆积的基面，极大地限制了活性位点的暴露。通过阳离子交换法制备的由 Fe 掺杂的 $Ni(OH)_2$ 纳米片 $[Ni_{0.83}Fe_{0.17}(OH)_2]$ 组成的 NiFe LDH，该材料富含缺陷多孔基面，克服了上述的缺点，从而最大限度地提高了电催化性能。多孔纳米片的形貌以及由 Fe^{3+} 的化学蚀刻引起的丰富的 Ni 和 O 空位，有利于暴露更多的离子可进入的位点，从而提供更多的活性位点。其次，$Ni_{0.83}Fe_{0.17}(OH)_2$ 的湿润性较好，保证了电解质的快速渗透，加速了水分解过程中羟基的迁移和氧的释放。

事实上，NiFe LDH 是一种优异的双功能电解水的催化剂，并且这种双功能电催化途径可以用拉曼光谱来解释。在 HER 过程中，水解中间体 H 和 OH 分别吸附在 Ni^{2+} 中心（H_{ad}-NiO）和 Fe^{3+} 中心（HO_{ad}-FeO）上，然后第二个 H_2O 中的 H 与 H_{ad}-NiO 基团中的 H 结合形成 H_2 [图 6.9(a)]。总的来说，NiFe LDH 中形成的物种（即 H_{ad}-NiO 和 OH_{ad}-FeO）与水分解反应中的 Volmer 反应是相容的。实际上，HER 的动力学不仅取决于 H_{ad} 的吸附速率，还取决于电子转移速率和被释放的物种激活到活性位点的能力[116]。在 NiFe LDH 中，Ni-Fe 的协同作用主要是通过形成 FeOOH 来增强 H_{ad}-NiO 和第二氢的键结合来满足过渡能释放要求，从而优化 HER 反应路线。而 OER 反应涉及较多的中间体，其路径比 HER 更为复杂。随着过电位的增加，在 NiFe LDH 表面上 $Ni(OH)_2$ 转变为 NiOOH，从拉曼光谱的结果来看，在 477cm^{-1}到 557cm^{-1}处观察到的谱带与 γ-NiOOH 的光谱特征吻合较好，在较大的过电位下，Ni 的平均氧化态为＋3.3～＋3.7，有助于获得较高的 OER 活性。因此，在较低的电位下，推测 OH_{ad}吸附在 NiFe LDH 电极上形成的 FeOOH 中的 Fe 位点上，而在较高的电位下，随着 NiOOH 拉曼特性的出现，H_{ad}倾向于吸附或迁移到 Ni 位点上，随后吸附的 OH 基团与其他 OH 基团反应生成反应中间体，进一步氧化为 O_2 和 H_2O [图 6.9(b)]。事实上，这种行为很容易解释，作为强路易斯酸，OH 与 Fe^{3+} 的相互作用比 Ni^{2+} 强，当外加电位增加时，Fe^{3+} 的电子效应导致 Ni 价态升高，导致 Ni 中心更容易与 OH 结合。当 NiFe LDH 纳米片（NiFe LDH-NS）集成到有缺陷的石墨烯上时，这种过渡金属原子与碳上缺陷的直接界面接触最大化，这可以通过电子分布来解释。在缺陷中心这种局部电子的积累增强了 HER 催化活性。另外，电子转移导致空穴在 NiFe LDH-NS 上积累，有利于提高 OER 催化活性。因此，NiFe LDH-NS 和 DG 的杂化分别导致 NiFe LDH-NS 上的空穴分离以及 DG 上的电子重新分布，使 HER 和 OER 在有利的活性位点顺利进行。

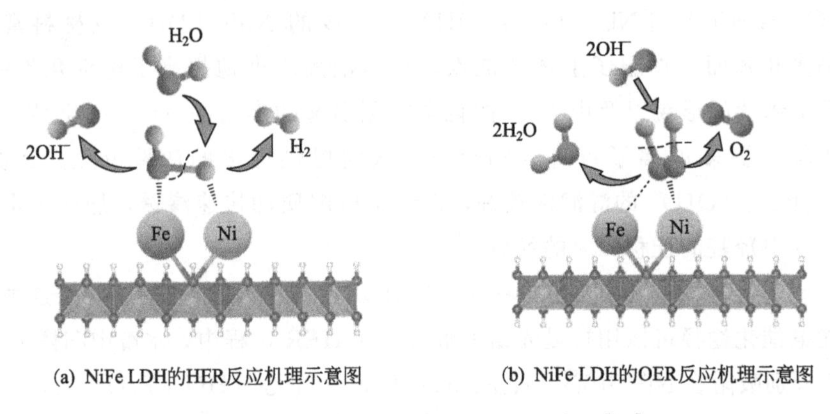

图 6.9　NiFe LDH 催化反应机理示意图[113]

当 NiFe LDH 中部分 Fe^{3+} 被 Al^{3+} 取代时，NiFe LDH 的电导率增加，在碱性条件下，Al^{3+} 的溶解导致 NiFe LDH 表面形成富 Ni 活性中心，形成结构缺陷的低配位 Ni 和 Fe 原子，提升了 OER 的催化活性[117]。尽管 Al^{3+} 可以提高 NiFe LDH 的 OER 催化活性，但它们最终形成的是 LDH[98]。因此，Al^{3+} 提升 OER 催化活性的原因类似于 Fe^{3+} 还是 LDH 的形成呢？如果是前者，即使微量的 Fe，Fe 与 $Ni(OH)_2$ 膜界面的增强作用也是很明显的，但是在微量的 Al^{3+} 水平下，并没有这种增强催化的作用。因此可以看出，在 Al^{3+} 掺杂的 NiFe LDH 催化 OER 性能的提高中，以氢氧化物为核心体系形成的 LDH 相是不可或缺的，即 Al^{3+} 提升 OER 的催化活性并没有像 Fe^{3+} 那样遵循活性结构和反应途径。因此，与 Al^{3+} 或其他三价离子（如 Mn^{3+}、Cr^{3+}、V^{3+} 等）相比，Fe^{3+} 仍然是 LDH 中较好的掺杂剂。

在大多数的研究中，向催化体系中掺入 Fe 的目的是提高催化剂的催化性能，Fe 的价态大部分是＋3 价。尽管有时候合成 NiFe LDH 所用的是 Fe^{2+}，但是最终通过高温水热反应或通过加入一些过氧化物溶液氧化 Fe^{2+} 为 Fe^{3+}。尽管有报道称，形成的 NiFe LDH 中含有 Fe^{2+}，同样具有优异的 OER 性能，但在 NiFe 氧化物或（氧）氢氧化物中，Fe 以杂质或掺杂剂形式的氧化态主要为＋3 价，很少有特殊情况。此外，OER 过程中，在施加的电位下都会发生氧化形成 Fe^{3+}。因此为了获得更好的 OER 催化性能，Fe 的氧化态通常是＋3 价或者更高，在 OER 作用下几乎不可能阻止 Fe^{2+} 氧化为 Fe^{3+}。

6.2 硼化物和硼酸盐催化剂

19 世纪初第一次合成了金属硼化物[118]，而最早利用过渡金属硼化物分解水的报道可追溯到 1974 年，当时 Kuznetsova 等人研究了氢在某些金属硼化物的反应动力学[119]。1981 年，Osaka 等人研究了 Co_xB、Ni_xB、Fe_xB 和 LaB_6 在 6mol/L KOH 中的 OER 行为[120]。他们将金属粉末与 B 粉混合，在氩气气氛下进行不同间隔的压制和烧结得到目标产物。采用同样的烧结方法，合成了 Co-Fe-B、Co-Ni-B、Co-Mn-B 等双金属硼化物作为 OER 催化剂[121]。在这些报道之后，20 世纪 80 年代末，人们对使用商业无定形硼化物催化电解水产生了浓厚的兴趣[122-125]。1986 年，Kreysa 和 Håkansson[126] 测试了一系列含有多种金属（Fe、Ni、Co、Mo、Pd、Cu、Ti）和类金属（B、P、Si）的商用玻璃在 1mol/L KOH 溶液中不同温度下的阴极和阳极反应，其中 $Fe_6Co_{20}Si_{10}B_{10}$ 具有最高的 HER 和 OER 催化活性，而 $Co_{50}Ni_{25}Si_{15}B_{10}$ 具有最高的 OER 催化活性[123]。Schulz 等人使用熔融纺丝技术制备了无定形的 $Ni_{0.65}Al_{0.1}B_{0.25}$ 和 $Ni_{0.7}Mo_{0.2}Si_{0.05}B_{0.05}$ 合金，并在 30%

KOH 溶液中测试了其 HER 催化性能[124]。Thorpe 等人也利用熔融纺丝技术合成了无定形复合硼化物[127]，例如 $Co_{50}Ni_{25}Si_{15}B_{10}$、$Ni_{50}Co_{25}Si_{15}B_{10}$ 和 $Ni_{50}Co_{25}P_{15}B_{10}$，用于 HER 催化。1990 年以后的十年里，关于用于水分解的过渡金属硼化物的报道很少[128]。研究人员开始探索更简单的合成路线来制备硼化物催化剂，而非使用商用玻璃或能源密集型技术（如熔融纺丝）。近十年来的报道中，主要采用化学还原法、化学沉积法、电沉积法、气相沉积/硼化来制备硼化物。

硼化物电催化剂主要有 Ni-B[129]、Co-B[130]、Fe-B[131] 等，B 原子的存在很大程度上促进了缓慢的 OER 过程。尽管几十年来过渡金属硼化物（TMBs）一直被用于电解水，但直到最近，才真正被视为能替代贵金属催化剂的潜在催化剂。事实上，在 2009 年，Daniel Nocera 小组报道了原位形成的类似于磷酸钴[134]的 Co[132]和 Ni[133]硼酸盐催化剂，在近中性条件下电解水。后来，Hu[135]和 Patel[136]小组分别报道了 Mo 硼化物和 Co 硼化物用于催化电解水的研究进展。根据这些报道，后面出现了通过各种技术来合成过渡金属硼化物/硼酸盐[130,137-140]，并广泛用于不同 pH 条件下的催化电解水。一般来说，使用电沉积原位合成的硼基催化剂被称为“硼酸盐”（表示为 M-B_i，M=金属），通过其他技术制备的催化剂被称为“硼化物”（表示为 M-B）。尽管近年来出现了很多硼化物/硼酸盐作为电解水催化剂，但是关于硼化物/硼酸盐还有很多方面尚未完全了解。例如，在无定形 Co 和 Ni 硼化物中发生反向电子转移[136]，但从金属（Co 和 Ni）和硼的电负性来看，并不清楚这一现象背后的化学行为。在一些硼化物中，电催化活性因硼在样品中的比例不同而不同，但这并不是所有硼化物的普遍趋势[139]。在大多数情况下，通过引入第二和第三元素来改善电化学性能的原因也不清楚，还有一些问题涉及纯相材料的合成，特别是钼基硼化物的合成[141,142]。虽然已经有一些较好的报道试图对这类材料进行一个基本的了解[143-145]，但仍存在不足，因此还需要更好的研究方法。最近有相当多的研究集中于开发过渡金属硼化物和硼酸盐基材料（TMBs）作为电解水的电催化剂。

6.2.1 硼化物催化剂

过渡金属硼化物的合成一般采用金属源（金属氧化物或盐）和 B 源（硼烷、卤化硼、金属硼合金、硼氢化物等），通过气固反应或液相反应路线制备[146]。在气固反应过程中，通常需要高温或高压才能形成结晶和超硬硼化物材料[118,147-149]。比如，Zou 等人以镍基底与硼粉为原料，通过气固反应或固体渗硼工艺，在 800℃下制备了硼化镍[150]。另外有研究报道，合成的温度不同，会影响金属硼化物纳米材料的结晶度[151]。对于液相反应途径，通常采用硼氢化钠或硼氢化钾为硼源和还原剂，与金属盐反应得到金属硼化物[152-154]。相比较高温的气固反应，该方法更为高效、简便，得到的金属硼化物大多为无定形的，更利于 OER 催化。另外 Fang 等人采用 Et_2NHBH_3 为硼源，用一种简单的化学镀方法（EP）在不同的基底上沉积了不同的硼化物作为电解水的双功能催化剂[155]。

一般来说，可以通过将催化剂负载在高比表面积的导电载体上来提高其活性。导电载体的使用不仅提供了更好的传质途径，而且使催化剂在工业装置中更易于集成。这些载体包括多孔炭和碳衍生物［石墨烯、氧化石墨烯（GO）］、还原型氧化石墨烯（rGO）、碳纳米管（CNTs）、石墨状氮化碳（g-C_3N_4）等[156-159]。Li 等人利用金属有机骨架（MOF）前驱体合成了具有多孔碳的 Co-B/C[160]，具有很高比表面积（BET＝119 m^2/g）。由于碳的导电性好，Co-B/C 中电荷转移较快，在 1mol/L KOH 溶液中，催化 OER 电流密度为 10mA/cm^2 时，过电位为 320mV(η_{10}＝320mV)。Arivu 等合成了 Ni_3B-rGO 复合材料并将其滴涂在碳纤维纸上作为 OER 催化剂[161]，石墨烯提供了高的比表面积和及时电荷转移途径，有利于提高催化活性。Schuhmann 等人合成了 Co-B 纳米颗粒，然后通过化学气相沉积法将其作为基底生长 N 掺杂的 CNTs[162]。研究发现，大部分的 Co-B 纳米粒子位于碳纳米管的表面，而部分被封装在碳纳米管的内部。该材料在 0.1mol/L KOH 中催化 OER，电流密度为 10mA/cm^2，对应的过电位为 370mV，且具有良好的稳定性。

除了碳衍生物外，金属箔和金属泡沫也是不错的导电基底。对于电解水体系，导电的泡沫金属具有高的表面积及良好的导电性，同时也为产生的气体脱离（通过孔隙）提供了更多的路径。因此催化剂可以原位生长在这种导电泡沫金属上作为电解水催化剂。例如 Liang 等人将 Ni-B 纳米颗粒原位生长在泡沫镍上，得到的催化剂在 1mol/L KOH 溶液中具有良好的催化 HER 和 OER 性能[129]。另外，Co-B 也被原位生长在泡沫镍[163]和镍箔[155]上，其中在泡沫镍上的 Co-B 为几纳米厚度的弯曲纳米片，在镍箔上的 Co-B 为具有大量内部空间的多孔结节状结构，Co-B/NF 具有优异的双功能催化电解水的性能。Wang 等人[164]将 Co-W-B 沉积在泡沫镍上，并在不同温度下煅烧，得到含有 $Co(OH)_2$、CoO 和 Co_3O_4 多晶相的层状形貌。双金属硼化物 Fe-Ni-B 纳米片[165]、Ni-Co-B 纳米棉状材料[166]也原位生长在了泡沫镍上，都具有优异的 OER 催化性能。另外 Ni-B 薄膜也原位生长在 Cu 板[138]和铜箔[167]上，作为优异的 OER 和 HER 双功能催化剂。从图 6.10 可以看出，Ni-B 薄膜由半球状的纳米颗粒与一些纳米孔组成，这些纳米孔是在沉积过程中 H_2 逸出形成的[138]。$NiB_{0.54}$/Cu 板催化剂在 0.5mol/L H_2SO_4、1mol/L PBS 和 1mol/L KOH 中均具有优异的催化 HER 性能，在 pH=7 和 14 条件下，其活性优于 Pt。虽然采用泡沫金属是制备原位自支撑催化剂的良好策略，但是必须要优化催化剂的负载量，以便与其他催化剂进行较为公平的比较。但对于 3D 结构的多孔网络，确定实际活性面积存在一定的挑战，因此难以确定负载在其上的电催化剂的真实活性。

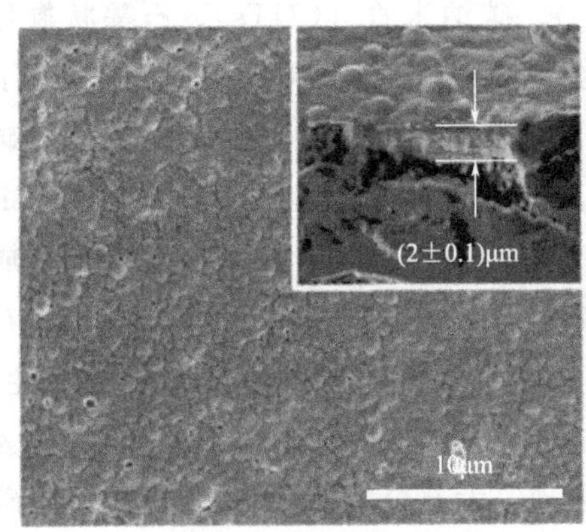

(a) $NiB_{0.54}$薄膜的低倍SEM图（内插图为侧视SEM图）

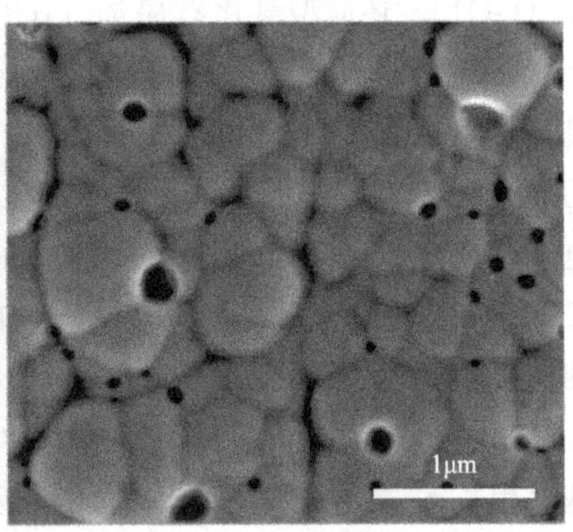

(b) $NiB_{0.54}$薄膜的高倍SEM图

图 6.10 $NiB_{0.54}$ 薄膜的 SEM 图[138]

在电催化中，表面形态和表面积在控制催化剂的电化学性能方面起着重要作用。因此，有必要对传统的合成方法进行改进，以获得具有更多催化活性位点和更高表面积的独特纳米结构的催化剂。具有核壳结构的纳米材料可以提升 OER 催化剂催化活性。Lu 等人合成了 Co-B@CoO 核壳纳米阵列材料[168]，其中被还原的 Co-B 为壳，厚度约为 15～40nm，CoO 为核，这类核壳结构材料具有优异的催化性能。Guo 等人将细小的无定形 Co_2B 纳米片复合在 $CoSe_2$/DETA（二乙烯三胺）纳米片上[169]，该材料在碱性介质中具有良好的双功能催化电解水的性能。他们认为 $CoSe_2$ 纳米片表面丰富的氨基（来自二乙烯三胺）为 Co_2B 生长创造了丰富的成核位点。Yang 等人采用独特的磁场辅助还原法制备了 Ni-Fe-B 纳米链。当 $NaBH_4$ 还原开始时，由于金属的强磁矩和较低的硼含量，纳米颗粒与外加磁场呈线性关系。随着还原的进行，B 的含量增加，磁矩迅速衰减，形成较厚的 Ni-Fe-B 非磁性无定形层[170]（图 6.11）。优化后的催化剂 80Ni-20Fe-B 纳米链（Fe/Ni 原子比＝0.29）在 0.1mol/L KOH 溶液中，0.35V 过电位时具有 64A/g 的 OER 质量活性，是无硼 Ni/Fe 纳米链的 7.6 倍。

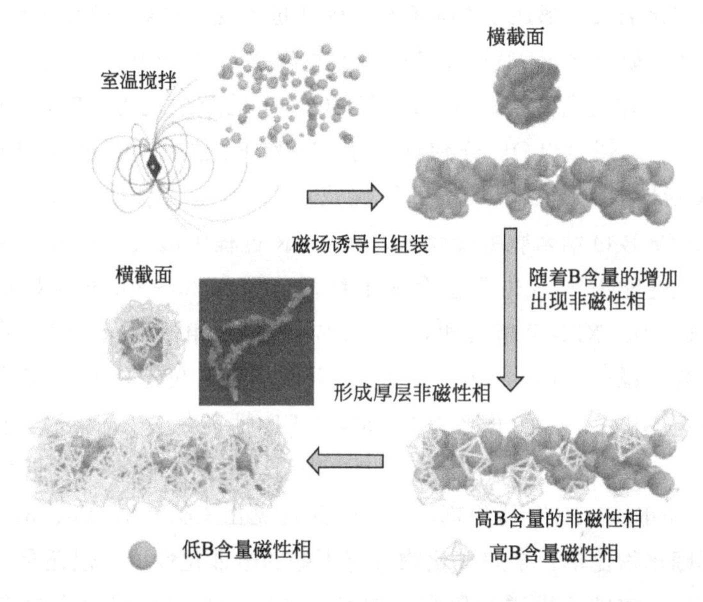

图 6.11　磁场诱导自组装合成 B 掺杂 Fe/Ni 纳米链的示意图[170]

由于单金属组分硼化物在催化电解水上具有优异的性能，促使研究人员研究二元金属硼化物的催化性能。Borodzínski 和 Lasia[171] 开发了一系列用于碱性环境 HER 的金属（Rh、Ru、Cr、Co、Zn、Pt）掺杂的 Ni_2B 催化剂，只需将所需的掺杂剂盐引入反应溶液即可。对于 Rh、Ru、Cr 和 Co，掺杂元素的量限制在 0.5%～10%（质量分数），而对于 Pt，则限制在 0.1%～0.5%（质量分数）。他们发现，在 Ni_2B 中引入 Rh、Ru、Cr、Co 后，HER 活性有所提高，但加入 Pt、Zn 后，性能保持不变甚至更差，电极催化性能提升主要归因于其较大的比表面积。2016 年，Patel 等人合成了 Co-Ni-B，与 Co-B 催化剂相比，没有表现出形态或结构上的差异[172]，但优化后的 Co-Ni-B 具有比 Co-B 更优异的催化性能，且 HER 的催化性能随着溶液 pH

值的增加而增加。为降低过电位，Co-Ni-B 被原位生长在泡沫镍[173]、碳布[174]和 rGo[175]上，比如 Co-Ni-B/CC 催化 HER 具有较低的过电位 $\eta_{10}=80mV$。Patel 等人合成了 Co-Mo-B 作为双功能电解水催化剂[176]。与 Co-B 和 Co-Ni-B 不同，Co-Mo-B 催化剂表现出多晶性质，由 Co-B 和 Co 金属相组成。Mo 的引入［Mo/(Mo+Co)摩尔比=3%］形成了具有高表面积的更小尺寸的纳米粒子，有利于提升催化性能。Co-Mo-B 在碱性（$\eta_{10}=66mV$）和中性（$\eta_{10}=96mV$）条件下展现了优异的 HER 催化性能。由于在碱性介质中，该催化剂表面形成了 CoOOH，因此也表现出良好的 OER 催化性能。

另外还有研究将 Fe 通过简单的还原过程掺入 Co-B 中，制备了 Co-Fe-B 和 Co-Fe-B-O 作为电催化材料，发现当 Co/Fe 的摩尔比为 2 时，OER 催化性能最佳[177]。通过 TOF 分别计算了 Co 和 Fe 位点对催化活性的贡献，推测OER 活性取决于 Co 和 Fe 的活性位点。Liu 等人合成了 20～50nm 的无定形 Co-Fe-B-O 纳米颗粒催化剂，在 OER 过程中形成了超薄纳米片从而提升了催化性能[178]。Liu 等也合成了粒径范围为 30～50nm 颗粒状的无定形 Ni-Fe-B[179]。XPS 分析表明，与 Ni-B 和 Fe-B 相比，Ni 和 Fe 的结合能发生了正移，说明 Ni-Fe-B 中 Ni 和 Fe 电子缺乏，价态较高。实验结果表明，当 Fe/(Ni+Fe) 原子比为 20 时，OER 性能最好。最近，将 rGo 与 NiFeB 纳米片复合，在 Ar 气氛中 400℃烧结，得到了具有优异催化性能的纳米晶型 Ni-Fe-B/rGO 催化剂[180]。由此可见由 Co、Ni、Fe、Mo 组成的二元金属硼化物比单一金属硼化物具有更好的电催化性能。但还是需要对其他地球储量丰富的元素进行研究，如 Cu、Mn、Cr 等，以寻求最合适的元素组合。

从上面的例子可以看出，一元和二元金属硼化物具有优异的催化电解水的性能，且二元硼化物的活性比一元的更优。2018 年，Wang 等人合成了平均尺寸为 30～40nm 的无定形球形纳米颗粒 Fe-Co-Ni-B 形式的三元金属硼化物[181]，该催化剂在 1.0mol/L KOH 中，展示了优异的 OER 催化性能（$\eta_{10}=274mV$）。最近合成了一种空心纳米棱镜的钒掺杂钴镍硼化物（VCNB）三元金属硼化物合金[182]，将催化剂负载在泡沫镍上，在 1mol/L KOH 中具有优异的催化 OER 性能（$\eta_{30}=280mV$），该催化剂综合了金属掺入、纳米结构和多孔泡沫镍的优势。然而，仅仅靠增加金属组分的数量并不是总能够提高催化活性。Yang 等人用 Co-Fe、Ni-Fe、Ba-Sr-Co-Fe 和 La-Sr-Co-Fe 合成了多种二元和四元金属硼化合物[183]，其中二元金属硼化合物表现出比四元金属化合物更高的 OER 质量活性。对于四元金属化合物，活性较小的成分（Ba、La、Sr）导致活性成分高（Ni、Fe）的数量减少，从而导致 OER 活性下降，这表明活性元素的种类和掺入在四元合金的形成过程中非常重要，能让催化活性达到一个新的高度。因此，理解每个元素的作用，对了解这类复杂合金的化学性质具有重要意义。

与金属的掺入类似，P 也被引入形成 Co-P-B 合金。Co-P-B/rGO 催化剂在 pH = 7 条件下 HER 和 OER 过电位分别是 η_{10} = 639mV 和 400mV[184]。Co-B-P 也被沉积在泡沫镍上，优化的 B/P 原子比为 3，该材料在 1mol/L KOH 中 HER 的 η_{10} = 42mV，Tafel 斜率为 42.1mV/dec。Kim 等人在碳纸上电沉积了无定形的 Co-P-B 催化剂[185]，电子从 B 到 Co 和 Co 到 P 的竞争性转移，产生了协同效应，因此 Co-P-B 提供了更好的 HER 催化速率。此材料展示了钴基硼化物在酸性介质中的 HER 活性。众所周知，一般情况下 Co 在酸性介质中是不稳定的，但该作者没有给出在酸性溶液中的稳定性测试。

6.2.2 硼酸盐催化剂

过渡金属硼酸盐（B_i）最早由 Nocera 等人采用电沉积法制备的[132,186]，将电极（ITO、Si 或镀金玻璃）浸入含有过渡金属离子和硼酸盐离子（例如硼酸钾缓冲液）的水溶液中，用恒电位或循环伏安法（CV），在电极上沉积过渡金属硼酸盐。但在这种情况下，电极在极化或 CV 过程中容易被腐蚀，另外金属离子在硼酸盐缓冲液中的溶解度也受到限制，导致制备效率低下。为了解决这个问题，Nocera 等人发展了一种原位电化学衍生或拓扑转化方法，直接使用过渡金属来制备此类材料[187]。基于此，Sun 课题组制备了一系列 3D 过渡金属硼酸盐用于水氧化[188-190]。研究发现，在 OER 过程中或空气中，过渡金属硼化物表面被氧化为金属硼酸盐，形成核（硼化物)-壳（硼酸盐）结构。另外，液相化学还原法也可以从金属硼化物中得到过渡金属硼酸盐作为 OER 催化剂[191,192]。其他方法如固相反应[193]、化学气相沉积[194]、溶胶-凝胶[195]和水热法[196]等，也被应用于制备过渡金属硼酸盐电催化剂。然而，对于所有这些方法，瓶颈在于如何将硼酸盐均匀地分布到过渡金属中，并找到提高过渡金属硼酸盐电荷转移能力的方法。

硼酸是一种高效的质子接受体，它能形成具有持续活性的催化剂[197]。

过渡金属硼酸盐在中性或接近中性的电解液（如硼酸缓冲液）中，具有较高的 OER 催化活性。已经报道了很多 NiB_i 薄膜在硼酸盐中催化 OER 的机理[198]。在硼酸缓冲溶液的 OER 过程中，硼酸盐负离子起着两个重要的作用：一是作为通过质子耦合电子转移（PCET）分子活性的传播者；另一个是作为活性中心吸附物的抑制剂。因此过渡金属硼酸盐在硼酸电解液中催化 OER 的真正活性中心，是金属氧化物和硼酸盐阴离子的复合团簇。硼酸盐的 OER 催化活性可以通过掺入其他外来金属比如 Fe，或者是通过引入碳材料来提高[199,200]。如将 $Co\text{-}B_i$ 纳米片负载在石墨烯上，得到具有分级结构的超薄纳米片结构，与 $Co\text{-}B_i$ 催化剂相比，该材料的 OER 催化性有所提升[201]。碳布、金属等也被广泛用于电沉积硼酸盐催化剂的导电基底[120,121,189,202,203]。Nocera 等人将 $Co\text{-}B_i$ 电沉积在泡沫镍和具有氟氧化锡涂层玻璃（FTO）上[204]，$Co\text{-}B_i/NF$ 在电流密度 $100mA/cm^2$ 时过电位为 360mV，且稳定性良好。借鉴 Nocera 等人的工作，Bai 等人将 $Co\text{-}B_i$ 和 $Co\text{-}P_i$ 催化剂电沉积在 3D 石墨烯泡沫上[205]。与 $Co\text{-}B_i/FTO$ 相比，$Co\text{-}B_i$/石墨烯泡沫复合材料具有更优异的催化性能。Sun 等人利用电沉积技术，在钛网上合成了 $Co\text{-}B_i$ 纳米片阵列[206]。在 0.1mol/L 和 0.5mol/L 的 KB_i 中进行 OER 催化时，其 η_{10} 分别为 469mV 和 400mV。同样，将 $NiSe_2$ 衍生的 $Ni\text{-}B_i$ 纳米片沉积在 Ti 网上也具有良好的催化性能[207]。Xie 等人在 Ti 网上合成了 $Co@Co\text{-}B_i$ 核壳结构的纳米片材料，其中 Co 核以纳米粒子的形式存在，外壳 $Co\text{-}B_i$ 为超薄纳米片，该材料暴露了更多的活性位点，更能提高催化剂的活性[208]。这种核壳结构的材料还有 $Ni_3N@NiB_i$ 纳米片，这种独特的催化剂在 0.1mol/L KB_i 和 0.5mol/L KB_i 溶液中催化 OER 的 η_{10} 分别为 405mV 和 382mV，低于 Ni_3N/Ti 催化剂（0.1mol/L KB_i 溶液中 $\eta_{10}=540mV$）[209]。

2015 年，Boettcher 等人发现在硼酸钾缓冲液中，对电沉积 $Ni\text{-}B_i$ 薄膜进行氧化调节，会促进 OER 催化速率增加[200]。采用 XPS 分析在 0.856V(相对于 SCE）处理后的 $Ni\text{-}B_i$ 薄膜，样品中存在 14%的 Fe，后面发现是使用的硼酸钾中含有约 1μg/g 的 Fe 杂质，导致 Fe 沉积在 $Ni\text{-}B_i$ 薄膜上。因此他们在无铁的硼酸盐中重复相同的 OER 测试，通过测试发现没有 Fe 的存在，但是其催化活性非常低。经过对材料的组成优化后发现，当 $Ni\text{-}B_i$ 薄膜含有 14%的 Fe 时其催化活性最好。与纯 $Ni\text{-}B_i$ 薄膜相比，$Ni\text{-}Fe\text{-}B_i$ 薄膜展示了更高的 TOF 值和更低的塔费尔斜率，具有更优异的催化性能。2017 年，

Sun 等人在碳布上合成了 Ni-Co-B_i 双金属硼酸盐[210]，在近中性条件下进行水氧化。Ni-Co-B_i/CC 在 0.1mol/L 和 0.5mol/L KB_i 中分别显示 $\eta_{10}=388mV$ 和 $\eta_{10}=316mV$，表明这种双金属硼酸盐比一元硼酸盐（Co-B_i[206]、Ni-B_i[206]）具有更好的催化性能。Sun 等人利用多种合成技术合成了 P 掺杂的硼酸盐[211,212]。例如在 KB_i 溶液中通过氧化极化在碳布上原位生长了 Fe-P_i-B_i 纳米片阵列（Fe-P_i-B_i/CC），在中性条件下其催化活性高于 FeP/CC[212]。

尽管硼化物和硼酸盐作为电解水的催化剂已有较多的研究，但关于金属硼化物/硼酸盐基催化剂的许多方面仍然不是很清楚。以无定形金属硼化物为例，最广泛接受的反应机理是金属原子作为活性中心，硼提供电子以丰富其 d 带，防止金属的氧化。这种“反向电子转移”在实验和模拟上得到了证实，但无定形金属硼化物中，这种反向电子转移的具体原因还值得进一步研究。在 OER 催化过程中，B 不作为活性中心，而是辅助形成活性金属氧化物/氢氧化物，但是硼的确切作用以及这种“辅助”的作用机理尚不清楚。硼，作为一种低原子序数元素，采用常用的实验室技术对其检测存在技术上的困难。虽然可以采用诸如 X 射线光电子能谱或电子能量损失谱等方法对其进行研究，但常规实验可能不足以揭示硼在金属硼化物中的真实作用。可以采用基于原位/操作方法深入研究，观察硼在电解水前、中、后的化学行为变化。

从目前的研究来看，与单金属硼化物相比，二元和三元金属硼化物的催化速率有所改善，但对于加入第二或第三金属是如何改变本体金属与硼之间的相互作用还不清楚。在金属硼化物中，加入第二种非金属（如 P）也会完全改变反应机理。尽管现有的研究对这类体系中活性催化中心的变化进行了评述，但却没有直接的证据来证明。因此，要了解外来元素的作用，必须采用原位/操作实验和计算结合，进行更深入的研究。一些报道表明，晶型金属硼化物比它们对应的无定形或部分结晶的催化剂的活性低，但并非在所有的情况下都是这种结果。对于晶型金属硼化物，使用 DFT 很容易识别活性中心。尽管取得了这些成功，但像使用其他催化剂（如 MoS_2）那样，用暴露更多活性平面的方法来调整金属硼化物的催化性能还没有实现。因此需要进行有针对性的研究，以制备具有更多暴露活性位点的纳米结构金属硼化物，这将会给提高晶型金属硼化物催化活性带来巨大的飞跃。在稳定性方面，金属硼化物在酸性介质中不太稳定，这给它们在质子交换膜（PEM）水电解槽中的使用带来了问题，但 Mo 基硼化物是例外，其在酸性介质中表现出良好的稳定性。然而，合成纯相钼硼化物相对来说较为困难，需要更深入地研究来解决。需要采用新的合成策略来合成基于 Mo 的硼化物，包括晶型和无定形的，这对于开发耐酸金属硼化物来说至关重要。

6.3 碳化钼基催化剂

过渡金属碳化物是通过在过渡原子的晶格中插入碳原子而形成的。这些碳化物由于其独特的化学键和晶格结构而变得非常重要。碳化物的性能较母体金属及其氧化物有所改善。基于独特的化学键合和性能，这些碳化物可分为盐类化合物、中间化合物和间隙化合物。在相对较低的温度下，前两者容易被水分解而不稳定。间隙化合物由于碳与过渡金属原子的化学键合，表现出独特的性质[123,213]。这类碳化物具有较高的熔点、硬度、抗拉强度，并且具有与陶瓷和其他贵金属相当的高电导率和热导率。

钼基化合物在 HER 中保持了类似铂的特性，最近几年 Mo_2C 已经成为 HER 催化剂研究的热点[214,215]。Mo_2C 和 MoS_2 由于具有高效的 HER 活性、低成本和高可用性[216,217]，可以代替贵金属 Pt 催化剂。特别是碳化钼，由于在费米能级上具有与 Pt 类似的 d 轨道结构和相同的电子态[215,218,219]，表现出更高的 HER 活性。钼与碳材料的结合在很大程度上增强了 Mo_2C 的导电性。此外，它还通过诱导电荷从钼向碳的转移，降低钼的 d 带中心，从而降低 Mo-H 键能，便于氢的解析[220-222]。但 Mo_2C 的电化学性能取决于其表面积、晶相以及表面组成[135]。在碳化钼中存在三种不同类型的键合：第一种是金属-金属键的重排，称为金属键；第二种是金属-非金属键的形成，称为共价键；第三种是由于电子的转移而形成的离子键[223-225]。由于具有共价化合物、离子晶体和过渡金属的性质，碳化钼在电催化、多相催化和其他一些选择性加氢过程等许多催化反应中，表现出类似铂的行为。Mo_2C 的几何结构和电子结构影响其催化活性和性能[226]。在过渡金属中（如钨和钼）加入碳原子，可能对金属的电负性有效[227]。与其他化合物相比，间隙过渡化合物在湿热和低温条件下表现出更稳定的性质。这些过渡化合物独特而稳定的性质，与其电子和晶体结构有关。过渡化合物晶格中的金属原子有三种不同的形式：面心立方结构、六方密堆积结构和简单六方结构，在这些化合物中，碳原子位于间隙位[227-229]。Mo_2C 存在不同的晶相，如立方 γ-Mo_2C、六方 α-Mo_2C 和正交 β-Mo_2C 相。起初人们认为立方相和六方相是最好的活性催化剂，但最近的发现表明，正交相具有更高的催化活性[230,231]。另外 Mo_2C 的杂化物也引起了许多重要的关注，如 Cui 等人报道了涂覆石墨碳片的 Mo_2C 具有优异的 HER 催化活性，在这种 Mo_2C 与碳的共价键合中，电子得以转移，降低了 HER 过程中的氢键结合能[232]。

6.3.1 Mo_2C 催化剂

为了得到纯的 Mo_2C 催化剂，需要合适的固体碳和钼的前驱体。一个经典的合成方法是采用钼酸铵和多巴胺为原料，步骤如下：将 1.2g 四水合钼酸铵加入 20mL 去离子水中，室温下恒温搅拌约 5min；在上述溶液中加入一定量制备的 SiO_2，超声 30min。由此形成均匀的悬浮液，然后在悬浮液中加入 0.4g 盐酸多巴胺，得到砖红色溶液，向溶液中倒入约 40mL 乙醇，搅拌 5min。在悬浮液中快速注入 0.4mL 的 $NH_3 \cdot H_2O$，调节 pH 在 8.5～9 左右。混合液搅拌约 2h，用去离子水和无水乙醇多次反复洗涤离心后得到橙红色沉淀。进一步将前驱体在温度为 5℃/min 的氩气流量下 800℃加热 3h，然后在 80℃下干燥 10h，如图 6.12 所示。最后将干燥后的样品在 HF (20%) 中浸出 30min，完全除去 SiO_2[233]。Xing 等人报道了用玉米秸秆制备碳化钼催化剂的方法[234]。Hu 等人系统研究了碳化钼用于电催化 HER[135]，他们发现不管是在酸性还是在碱性介质中，β-Mo_2C 都表现出良好的催化活性和稳定性。Leonard 等人通过研究酸性介质中不同晶相的碳化钼对 HER 的催化活性[235]，其结果为：α-MoC_{1-x} < η-MoC < γ-MoC < β-Mo_2C。Chen 等人采用还原固态反应，以生物质大豆为碳源，合成了 γ 型和 β 型混合的碳化钼与生物质碳的复合材料 MoSoy，该材料具有优异的 HER 催化活性[236]。Lou 等人采用 MoO_3 纳米棒为模板，在多巴胺原位聚合的同时加入额外的 Mo 源，通过简单的模板包覆法制备了多孔纳米片构建的新型分级 β-Mo_2C 纳米管，该材料具有高 HER 催化活性且成本低廉。Wang 等人通过两步法（水热法-煅烧）调节葡萄糖和钼酸铵的比例合成了三种钼基化合物（MoO_2、MoC、Mo_2C）[237]，其中 Mo_2C 具有最佳的催化 HER 活性，酸性条件下 $\eta_{10}=35mV$，碱性条件下 $\eta_{10}=96mV$，其优异的催化活性

可能源于其优异的晶体结构、高石墨化度的碳载体、高的电子导电性以及其超小的粒径，提供了较大的比表面积和活性位点。

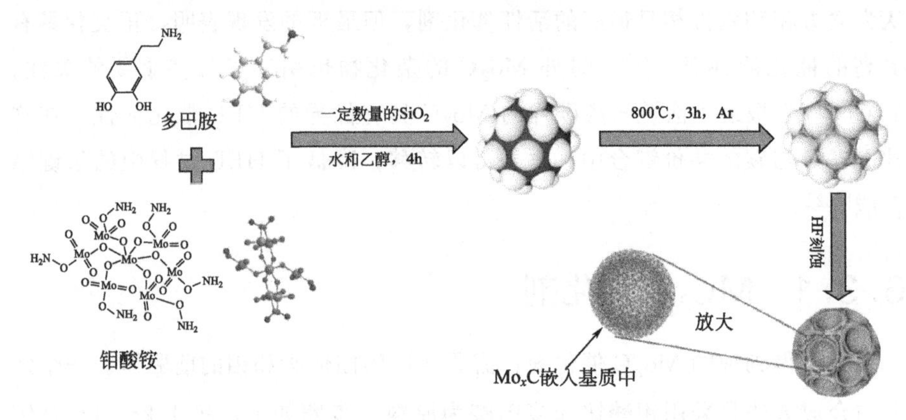

图 6.12 Mo_xC 的合成过程示意图[233]

6.3.2 Mo_2C 杂化催化剂

通常将碳化物和生物质炭、石墨烯或碳纳米管复合，可以提升材料的导电性，增大材料比表面积、防止催化活性材料在催化过程中团聚，从而提升其催化性能。比如 Huang 等人采用方便、可扩展的超声喷雾热解法(USP)，合成了具有少层碳层的碳化钼纳米颗粒（MoC/C），在酸性溶液中对 HER 具有高的催化活性和稳定性[238]，在 0.5mol/L H_2SO_4 中电流密度为 $20mA/cm^2$ 时需要的过电位为 144mV，Tafel 斜率为 63.6mV/dec。孙旭平课题组利用 $(NH_4)_6Mo_7O_{24}\cdot 4H_2O$ 与海藻酸钠在 900℃下 Ar 气氛中，通过固相反应制备了 Mo_2C 纳米颗粒修饰石墨碳片（Mo_2C/GCSs）[232]。作为一种新型的 HER 电催化剂，Mo_2C/GCSs 在酸性溶液中表现出高活性以及良好的稳定性，起始电位为 120mV，塔费尔斜率为 62.6mV/dec，交换电流密度为 $12.5\times10^{-3}mA/cm^2$。Wang 等人采用水和乙醇混合溶液为溶剂，钼酸铵和多巴胺（DA）为 Mo 前驱体和 C/N 源，加入氨水后 DA 自聚合带正电，和带负电的 $[Mo_7O_{24}]^{6-}$ 之间自组装形成含有 Mo 原子的碳纳米片层状球（Mo-HAs)，然后在惰性气体下煅烧得到 3D 结构的 MoC-HAs。该结构中 MoC 纳米晶小于 2nm、具有自由伸展的导电纳米片、高的比表面积和丰富的介孔，这能够暴露更多的活性位点，促进电子/离子的传输，这些都有利于提升催化活性，因此该材料具有优异的催化 HER 活性和稳定性[239]。

许多研究表明，通过构建适当的形貌，如纳米线和纳米颗粒，增加表面活性位点，可以提高 Mo_2C 的 HER 活性[240,241]。另一种是引入其他元素，如氮[242]、磷[243]、硫[244]和硼[139]，这些元素可以调整 Mo 的电子状态，并使活性位点以理想的方式分布。此外，杂原子的引入还可以调节碳骨架的电子排列，从而提高相应的 HER 催化活性[245]。Liu 等人指出，嵌入在氮掺杂碳中的金属碳化物具有很高的 HER 活性，这主要是由于最有利的 H^+ 的吸脱附性能，以及 Mo_2C 与 N 掺杂杂化催化剂之间的协同效应[246,247]。Luo 等人采用了一种环境友好、无模板的方法，合成了纳米带、纳米棒和纳米颗粒 3 种不同形式的 N 掺杂碳化钼[222]。在合成过程中，不使用外加的酸或碱，通过调节含水量和处理时间来控制形貌。该反应主要有两个步骤：步骤 1 是前驱体的链化过程，通过加热四水钼酸铵和乙二醇溶液，将氨离子转化为氨气，剩余的氢离子与 EG 结合形成质子化的 EG(PEG)，再与钼酸根离子形成链状；步骤 2 是替换和聚合过程。随着质子化三聚氰胺（PMA）加入上述溶液中，在 PEG 上发生取代反应，开始出现白色沉淀。其中多孔纳米结构的纳米带，在酸性和碱性电解质中对 HER 均表现出优异的活性和良好的稳定性。在 1.0mol/L KOH 水溶液中，纳米带的起始电位值较小为 −52mV，在较低过电位（110mV）下产生 $10mA/cm^2$ 的电流密度，Tafel 斜率较低（49.7mV/dec）。Chai 等人通过简单的有机-无机热解后还原的方法，合成了具有三明治结构界面的 Mo_2C@NC@Pt 纳米球[248]。该材料在酸性、碱性和中性介质中，表现出优于商业 Pt/C 的 HER 催化活性（η_{10} 分别为 27mV、47mV 和 25mV），并且都具有良好的稳定性。其优异的催化活性主要归因于结构独特的三明治结构界面，Mo_2C 纳米颗粒被碳层包裹，Pt 纳米颗粒在碳载体上分散良好，Mo_2C 纳米颗粒、NC 和 Pt 纳米颗粒之间的协同作用、高的比表面积以及非金属 N 的掺杂等促进了催化性能的提高。

另外，掺杂其他的非贵金属元素诸如铁系元素，也可以提升碳化钼的催化活性。Xiong 等人报道了基于 MOF 制备 Co-NC@Mo_2C 复合物的方法[249]，该材料综合了 Mo_2C 和 Co-NC 用于催化电解水的优点。其中 Mo_2C 作为外涂层结构不仅可以保护 CoNPs 内层免受电解质侵蚀，还可以提供更多的金属和碳的接触位点。He 等人通过高温退火 NiFe 普鲁士蓝类似物和 Mo^{6+} 接枝聚乙烯吡咯烷酮（PVP）杂化前驱体，制备了 NiFe 合金纳米颗粒修饰的 N 掺杂石墨烯包裹的 MoC_2（NG-NiFe@MoC_2）[160]。PVP 包覆 NiFe-PBA NPs 与 Mo^{6+} 接枝形成的独特杂化前驱体，使碳化反应可控且均匀，同时外层 PVP 覆盖物避免了生成的 NiFe 合金 NPs 与 MoC_2 的团聚。MoC_2 和 NiFe NPs 在 HER 和 OER 过程中协同发挥作用，以及其独特的 N 掺杂石墨烯壳层结构，使该材料具有优异的 HER 和 OER 催化活性和稳定性。最近 Dong 等人通过水热反应后再热解，合成了 Ni@MoC_x/NC。壳聚糖和乙二胺分别作为碳载体和有机配体与钼酸铵作用，通过热分解形成氮掺杂碳片（NC）和碳化钼。Mo_2C/MoC 异质结表现出对 H^* 高效吸收和对 H_2 脱附，因此在酸性和碱性条件下具有优异的 HER 催化活性[250]。

尽管已经报道了 Mo_2C 及其杂化物具有独特而优异的表面活性、结构和反应，但对于认识催化 HER 的机制仍存在一些问题和挑战。首先需要研究碳化钼等过渡金属催化 HER 的活性位点中心，主要包括催化剂表面析氢、吸附和扩散。还需要了解 HER 机理与 Mo、C 和其他金属的化学计量之间的关系。另外 Mo_2C 电催化 HER 的活性对于实际应用来说仍然太低。尽管有报道对 Mo_2C 进行改性，在碳化钼结构上合成了具有高性能 HER 活性催化剂，但是其高化学计量比降低了贵金属如 Pt 基催化剂的应用能力。对于大规模应用，需要关注的另一个重要的焦点是增加暴露活性位点以提高 HER 催化活性。由于催化剂活性较低，目前的研究结果仍然难以满足需求。因此，除了依靠晶粒大小、晶相、比例和沉积等因素影响 HER 性能外，还需要将实验研究与模拟结合起来。高活性和潜在协同效应的基础研究可能为开发新型 Mo_2C 电催化剂催化 HER 制氢提供基础。

6.4 Fe、Co、Ni 基磷化物

6.4.1 前驱体的合成

过渡金属磷化物（TMPs）的制备一般包括两步：前驱体的合成和磷化。TMPs 常用的前驱体包括（氢）氧化物、过渡金属以及金属有机骨架化合物（MOFs）等，通常采用水热法、电沉积和其他方法合成。

水热反应是制备纳米材料的一种非常常见的方法，将化学物质充分混合在溶剂中放于高压釜中，在一定的温度下反应一定的时间。高压釜内的高温和高压可加速各种纳米结构的形成[251]。目前，可以通过水热法合成过渡金属（氢）氧化物纳米材料作为 TMPs 前驱体。比如 Yang 等用 $Co(NO_3)_2 \cdot 6H_2O$，$NiCl_2 \cdot 6H_2O$ 和尿素溶解在二次水中，在反应釜中 150℃反应 3h，

得到 CoNi LDH，随后磷化得到磷化钴镍[252]；另外通过 100℃水热反应 12h 得到 Co_3O_4 纳米线，作为合成 CoP 纳米线的前驱体，得到的 CoP 纳米线具有优异的催化活性[253]；水热反应也常用于制备自支撑电催化剂的前驱体，自支撑电极有利于通过电极和电催化剂之间的密切接触，促进 HER 过程中的电荷转移，从而提升 HER 催化性能。Sun 等人通过水热反应在碳毡上原位合成 $NiCo_2O_4$ 纳米线，然后磷化得到 $NiCo_2P_x$ 纳米线，在很宽的 pH 范围内显示了优异的 HER 催化活性[254]。Sun 等人通过水热反应在碳布上原位生长前驱体 Fe_2O_3 纳米阵列，然后磷化得到自支撑的 FeP 纳米棒阵列，该 3D 电极在中性和碱性条件下具有优异的 HER 催化活性[255]；还可以通过水热反应形成分级结构的前驱体来制备具有异质结构的 Fe、Co 和 Ni 基磷化物。比如 Jiang 等人采用两步水热法在碳布上原位生长具有分级结构的 Co_3O_4-FeOOH 前驱体制备了 CoP-FeP 枝状异质结构[256]（图 6.13）。虽然水热反应是最常用的制备前驱体的方法之一，但反应温度较高和处理时间较长（数小时至数十小时），不利于简单的合成磷化物催化剂。

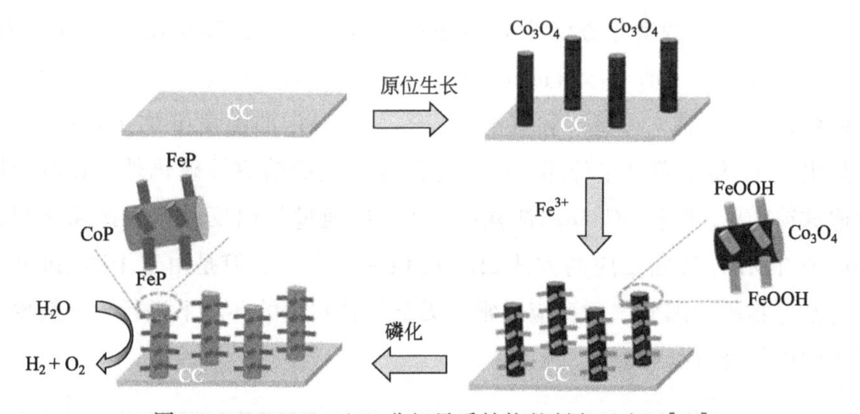

图 6.13　CoP-FeP/CC 分级异质结构的制备示意图[256]

相对于水热法，电沉积是在导电基底上原位生长纳米结构的简单、易行、成本低廉的方法[257-259]。采用电沉积方法可以制备不同组成和形貌的前驱体，然后进一步磷化得到 TMP 基电催化剂。比如 Shao 等人采用电沉积法在泡沫镍上制备了 CoFe-LDH 前驱体，后磷化得到自支撑电极 CoFeP 作为电解水的双功能催化剂[260]。Pu 等人在 Ti 板上电沉积制备了 α-$Co(OH)_2$ 作为制备 CoP 纳米片阵列的前驱体，CoP/Ti 在酸性和中性溶液中均具有较高的电催化活性[261]。Chen 等人采用电沉积的方法在 Ti 片上原位生长了 FeOOH 薄膜作为合成 FeP 的前驱体[262]。

除了水热法和电沉积法，也有一些其他的新颖的方法来合成 Fe、Co 和 Ni 基磷化物的前驱体。比如采用湿化学方法在碳布上制备有序的 ZnO 纳米线作为模板，通过 Fe^{3+} 水解在 ZnO 纳米线上原位生长得到 $Fe(OH)_3$，去除模板后得到 $Fe(OH)_3$ 纳米管作为 FeP 的前驱体[263]。另外，还可以通过沉淀反应得到 MOFs 作为磷化物的前驱体。比如 Zhang 等人通过沉淀法合成了 ZIF-8@ZIF-67 纳米管为前驱体，合成了 CoP/N 掺杂入含有垂直排列 MoS_2 纳米片的多孔碳纳米管复合材料（NCT@CoP@MoS_2），该材料具有优异的双功能催化电解水活性[264]。过渡金属本身也可以作为制备磷化物催化剂的前驱体。比如通过热注射方法在一个胶体相中合成了 ε-Co 纳米粒子作为制备 CoP 纳米粒子的前驱体[265]；Zhang 等人用泡沫镍作为前驱体，采用水诱导磷化合成了 Ni_2P-$Ni_{12}P_5$ 作为 HER 催化剂[266]。

6.4.2 磷化方法

制备 TMPs 的磷化方法主要有三类：气固反应、液相反应和电还原。将预先合成的固态前驱体通过气固反应，可以有效地转化为 TMPs。常用的磷源包括 NaH_2PO_2[267-270] 和红磷[271,272]，在一定的温度下加热这些磷源，释放出 PH_3 和 P 蒸气来使用。除了对预先合成的前驱体磷化外，过渡金属如泡沫镍[271]、镍箔[273]和泡沫铁[274]都可以通过气固反应，直接磷化制备相应的 TMP。气固反应的方式如图 6.14 所示[275]。但是由于 PH_3 和 P 蒸气是剧毒物质，因此气固反应通常在无空气且封闭的环境下进行，一般要对产生的尾气进行处理。

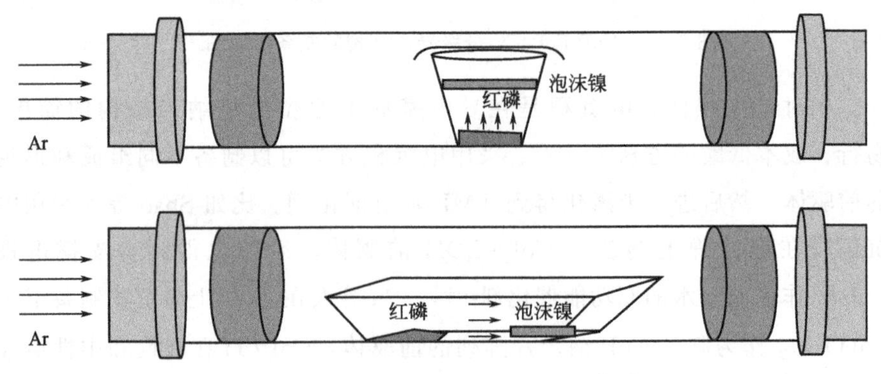

图 6.14 气固反应磷化的两种方式[275]

液相反应是磷化的另一种重要方法，一般采用有机 P 源如三辛基膦（TOP）、三苯基膦（TPP）。以 TOP 为磷源，可以制备不同组成、形貌和结构的 TMPs。比如在加热和无空气的条件下，可以制备 Ni_2P 纳米片[276]和高度分支化的 CoP[277]；用 TPP 为磷源，可以制备 $Ni_{12}P_5$ 纳米粒子[278]和 CoP 纳米棒[279]。这些有机磷源具有相似的磷化工作原理，TMPs 的形态和结构可以通过液相反应和磷化过程进行有效的调整。然而，液相反应通常需要较为复杂的合成技术、多个合成步骤和危险的有机磷化物，限制了这种磷化方法的应用。

电还原合成 TMPs 的电解液中共存有金属和次磷酸盐离子（$H_2PO_2^-$）。$H_2PO_2^-$ 在还原电位下转化为 PH_3，然后与金属离子沉积形成 TMPs[280]。这种磷化方法条件温和、环境友好、省时，是一种绿色简便的制备 TMPs 方法。采用电还原法制备 Fe、Co 和 Ni 基磷化物的研究工作有很多，如 CoP 介孔纳米棒阵列[281]、分级径向 Ni_xP 纳米球[282]、多孔 FeP 立方体[280]、多孔 Co-Fe-P[283] 和微孔无定形 NiFeCoP[284]。

6.4.3 磷化物和 C 的复合材料

一般情况下，非贵金属催化剂在耐酸碱性腐蚀方面不太令人满意，因此导致电催化活性衰减。为了提高 Fe、Co、Ni 基磷化物在电解水体系中的稳定性，可以采用碳纳米材料包覆。另外活性位点的电子电导率和分散性，都是影响电催化 HER 效率的重要问题，较差的导电性会降低材料的催化活性。将 TMPs 和碳复合，不仅可以提高电子导电性，还可以通过碳基底的约束作用，抑制纳米颗粒的聚集，改善分散性。因此，出现了许多 TMPs 和 C 复合的电催化剂，用于高性能的电催化水分解体系。比如 Zhu 等人通过热解 1-羟基亚乙基-1,1-二膦酸和三聚氰胺形成的 Co 配位聚合物，将超细 CoP 纳米晶锚定在高度多孔开放的 N、P 共掺杂骨架上[285]。这种碳骨架提供了更多的活性位点，并加强了质量和电子传递。在 0.5mol/L H_2SO_4 溶液中具有优异的 HER 催化活性，起始电位 33mV，Tafel 斜率为 69mV/dec。Jun 等人采用聚多巴胺为碳源，制备了碳包覆的 FeP 复合材料，该材料在催化电解水体系中展示了良好的稳定性[286]。Wang 等人采用静电纺丝结合可控热解还原的方法，将超细的 Ni_2P 嵌入 N 掺杂多孔碳纳米纤维中，其电阻率降低至单纯的 Ni_2P 的 $1/10^4$ 倍，加快了 HER 过程中的电荷转移速率[287]。Zhang 等人通过电沉积大分子前驱体、浸泡 Co^{2+} 和 H_2 热还原等方法，制备了 Co_2P 量子点嵌入 N、P 双掺杂碳的自支撑电极，Co_2P 量子点与 N、P 双掺杂碳骨架协同作用，提高了电化学活性并降低了电荷转移的电阻[288]。通过这些方法制备的 TMPs 与杂原子掺杂碳的复合材料，普遍具有明确的结构和组成。TMPs 和碳组成的复合材料还有很多，诸如在碳纸上原位生长多孔的 N、P 共掺杂碳包裹 TMP 纳米颗粒的复合材料[17]，在碳布上原位生长 MoP@NPCF/CC[289]，在泡沫镍上原位生长 N 掺杂碳包覆的 Ni_2P 纳米颗粒复合材料（Ni_2P@NC/NF）[290]，嵌入在 N、P 共掺杂碳基质中的核壳 MoP 纳米晶体（MoP@NPCS）[291]，双金属 FeNi 磷化物纳米颗粒复合 N、P 修饰的碳纳米片（FeNiP/NPCS）[292]以及 N 掺杂 C 包裹的 CoP/Ni_2P 纳米颗粒（CoP/Ni_2P@NC）[293]。

除硬碳材料外，石墨化的碳材料比如石墨烯、CNTs 以及石墨也是良好的碳复合材料，它们与 TMP 复合材料产生协同作用，提升催化活性。据报道，将镍基金属-有机框架（MOF-74-Ni）与还原氧化石墨烯（rGO）通过模板限制法结合在一起，在石墨烯上得到了非常小的磷化镍颗粒（约 2.6nm）（Ni_2P/rGO）[294]。与 rGO、Ni_2P 颗粒和从 MOF-74-Ni 衍生来的 Ni_2P/C 相比，Ni_2P/rGO 具有 Ni_2P 和 rGO 的协同作用，rGO 不仅作为 Ni_2P 支撑基底，还作为电子通道，提供更高的电子电导率，因此该材料具有较高的电子电导率和良好的分散性，在 1.0mol/L KOH 中表现出较高的 HER 电催化性能。

多组分异质结构电催化剂的合理设计，需要对电化学催化在分子水平上的认识，但目前对该领域的认识仍然不足，因此对这类催化剂的设计仍受到一定的限制，需要花费更多的研究在理论计算和实验分析上，以便更好地揭示催化的基本反应。比如，将 DFT 计算与实时原位实验相结合，通过对复合电催化剂的组成和结构的研究，可以更好地揭示 HER 机理和中间体的基本信息。

6.4.4 杂原子掺杂的磷化物

一般认为，可以通过掺入外来元素来调控电子结构，从而改变反应中间体的物理化学性质，提高吸附能。金属和非金属掺杂可以调节 Fe、Co 和 Ni 基磷化物的电子结构，加速电荷转移，优化 ΔG_{H^*} 和 ΔG_{H_2O} 对 HER 中间体的吸附和脱附。因此，元素掺杂作为一种有效地改善 TMPs 电子结构和提高电催化性能的方法被广泛研究。

6.4.4.1 非金属掺杂

对于非金属掺杂，一般选用的元素有 N、S 和 Se 等，通过在前驱体磷化的同时加入特定非金属源来实现。例如，前驱体 $Ni(OH)_2$ 纳米片与磷、硒粉的热转化，可以制备 Se 掺杂的 NiP_2[295]。非金属元素通过促进电荷转移以及 HER 中间体的吉布斯自由吸附能，改善 Fe、Co 和 Ni 基磷化物的 HER 催化活性。掺杂剂 S 的 3p 轨道，可以通过增加价带和导带的价态来增强 CoP 的金属特性，从而加速 S 掺杂 CoP 的电荷转移[296]。比如将 S 掺入 MoP 中可以提升 HER 的电催化活性[297]，MoP/S 在 $10mA/cm^2$ 的电流密度下的过电位为 86mV，MoP 则需要 117mV。更深入的研究发现，S 和 P 可以同等地调整电子特性，促使 MoP/S 具有更高的催化活性。Lee 等人采用硫脲-磷酸盐辅助法，在导电基底（碳布或者泡沫镍）上制备了 S 掺杂的 CoP[296]或 Co_2P[298]，在碱性条件下具有优异的双功能催化电解水活性。他们认为 S 掺杂通过增加费米能级附近的态密度来增强金属性质。由于 N 掺杂剂的电负性比 P 更强，因此 d 带中心下移以及 Co 的 d 轨道的电子云密度降低，导致 H_{ads} 的吸附减弱[299]。除了有效地调整 ΔG_{H^*} 值外，氮掺杂还可以优化 ΔG_{H_2O}，从而提升 N 掺杂 CoP 在碱性溶液中的 HER 催化活性[300]。采用高反应性的 N_2/PH_3 等离子气体，可以将 NiCo 氢氧化物原位转化为 NiCo 磷化物，N 被有效地引入泡沫 NiCo 上，原位生长 N 掺杂的 NiCo 磷化物（N-NiCoP/NCF）[301]，该材料在碱性条件下具有优异的双功能电解水活性以及稳定性。通过 DFT 计算分析，有效的 N 掺杂提高了费米能级和异质结 Co_2NiP_4 相暴露的丰富的活性晶面（311），从而提升了催化活性。

另外 NH_3 是常用的掺杂 N 的氮源，比如 Zhang 等人采用 NH_3 为氮源，在碳布上合成了 N 掺杂的 CoP（CC@N-CoP）[302]。在 0.5mol/L H_2SO_4 中，当电流密度为 $10mA/cm^2$ 时，CC@N-CoP 的过电位为 42mV，Tafel 斜率为 41.4mV/dec。与未掺杂的 CC@CoP 相比，催化活性显著提高。由于 N 的强电负性，N 取代 P 导致了 CoP 的 d 带降低，达到了接近热中性状态的 ΔG_{H^*}，反映出良好的 H^* 吸附环境。实验证明，元素掺杂是提高 TMPs 本征活性的有效策略。具有不同电负性的杂原子会改善电子结构，从而导致中间体的有利吸附，提高电导率，从而提升 HER 活性。但是直到现在，尽管结合了理论计算和先进的表征技术，掺杂 TMPs 高活性的根源仍未定论。为了深入理解电子结构与相应的吸附能和 HER 活性之间的关系，需要进行更多的研究。

6.4.4.2 金属掺杂

过渡金属如 Fe、Co、Ni 基磷化物掺杂金属元素相对较为容易，一般通过在前驱体的合成过程中，加入一定计量的掺杂剂来实现，不需任何后处理。通常采用 DFT 计算进行理论研究，并辅以实验论证，来揭示金属掺杂引起的 HER 活性增强。例如，Tang 等人在水热反应中，通过调节 Fe∶Co 的投料比，合成了一系列 $Fe_xCo_{1-x}P$ 纳米线阵列。由于 Fe 的电负性比 Co 的电负性稍低，Fe 的引入可以降低 H_{ads} 的结合强度，导致 ΔG_{H^*} 更接近热中性[303]。在氢氧化钴沉积的过程中，加入 Mn^{2+}，通过磷化得到 Mn 掺杂的 CoP 纳米片阵列，掺杂 Mn 后，材料的 ΔG_{H^*} 更接近热中性[304]。在泡沫镍上生长 CoAl 层状双氢氧化物为前驱体，磷化后得到 Al-CoP/NF 复合催化剂，在酸性、碱性和中性电解液中，展示了比相同条件下制备的 CoP/NF 更优异的催化 HER 活性[305]。其他一些金属元素如 Zn[306]、V[307]、Cu[308]、Cr[309] 和 W[310] 的掺杂，也可以通过改善电子相互作用来提高 HER 性能。此外，通过优化掺杂量来调整组分，对活性位点的电子调控很重要。比如在 N 掺杂碳纳米纤维负载的双金属 $Ni_{2-x}Co_xP$ 中，可以通过调节 Co 的含量，获得最高数量占据 d 轨道并优化 ΔG_{H^*}，从而提升其电催化活性[311]。

具有不同电子填充金属键的掺杂金属元素，也会影响材料的催化活性。Lee 等人制备了不同过渡金属 Mo、Mn、Fe 和 Co 掺杂的磷化镍（NiMP），实验结果表明，掺杂金属元素的 d 轨道电子数［Mo(d6)、Mn(d7)、Fe(d8)、Co(d9) 和 Ni(d10)］和催化活性成反比[312]，由于 Mo 缺乏 d 电子和 4d 伸展轨道，因此 Mo 具有最高程度的 d 电子离域，其催化活性最高。可以通过调节掺杂的金属，来改善电催化剂的电子结构，从而影响电催化活性的局部配位环境。Gao 等人通过电沉积加后续退火处理，制备了 $Ni_{0.1}Co_{0.8}P$ 多孔纳米片[313]，通过 EXAFS 测量确定了其局部配位环境。$Ni_{0.1}Co_{0.8}P$ 的 Co K 边谱振幅低于 CoP，表明 Co 的氧化态增加了。Co 和 P 的键长从 CoP 的 1.78Å（$1Å=10^{-10}m$）降低到 $Ni_{0.1}Co_{0.8}P$ 的 1.63Å，这也说明了局部配位环境随着镍的加入而发生了改变。与 CoP 相比，$Ni_{0.1}Co_{0.8}P$ 的态密度（DOS）中 d 带中心整体下移，表明其电子结构发生了改变。其实验结果和理论结果较为吻合，在中性电解液中，$Ni_{0.1}Co_{0.8}P$ 的过电位为 125mV，Tafel 斜率为 103mV/dec，性能优于 CoP（174mV 和 117mV/dec）。Li 等人报道了第一个由 Ru、Ni_2P 和 Ni 组成的金属-磷化物-金属的独特多相纳米棒（Ni@Ni_2P-Ru），其中 Ru 的加入影响了材料的局部环境，Ni 和 Ru 之间的电子相互作用促进了材料的催化性能[314]。

铁、钴、镍基磷化物具有金属特性（即具有较高的导电性），有利于加速电荷转移。在 TMP 基材料上吸附 HER 中间体的吉布斯自由能适中，表明对 HER 具有较高的本征活性。但还需要更多的研究来开发成分、形貌和电子结构可调的高性能、稳定性好、性价比高的 TMP 基电催化剂，并通过实验研究和理论研究相结合，来预测和证实 TMPs 的电催化活性。

参 考 文 献

[1] Anantharaj S, Kundu S, Noda S. Progress in nickel chalcogenide electrocatalyzed hydrogen evolution reaction. Journal of Materials Chemistry A, 2020, 8 (8): 4174-4192.

[2] Tang C, Cheng N, Pu Z, et al. Nise nanowire film supported on nickel foam: An efficient and stable 3d bifunctional electrode for full water splitting. Angewandte Chemie International Edition, 2015, 54 (32): 9351-9355.

[3] Xu K, Ding H, Jia K, et al. Solution-liquid-solid synthesis of hexagonal nickel selenide nanowire arrays with a nonmetal catalyst. Angewandte Chemie International Edition, 2016, 55 (5): 1710-1713.

[4] Han X, Tong X, Wu G, et al. Carbon fibers supported nise nanowire arrays as efficient and flexible electrocatalysts for the oxygen evolution reaction. Carbon, 2018, 129: 245-251.

[5] Li X, Zhang L, Huang M, et al. Cobalt and nickel selenide nanowalls anchored on graphene as bifunctional electrocatalysts for overall water splitting. Journal of Materials Chemistry A, 2016, 4 (38): 14789-14795.

[6] Gao Z, Qi J, Chen M, et al. An electrodeposited nise for electrocatalytic hydrogen and oxygen evolution reactions in alkaline solution. Electrochimica Acta, 2017, 224: 412-418.

[7] Wu H, Lu X, Zheng G, et al. Topotactic engineering of ultrathin 2d nonlayered nickel selenides for full water electrolysis. Advanced Energy Materials, 2018, 8 (14): 1702704.

[8] Connolly T F. Binary transition metal chalcogenides. 1972: 179-197.

[9] Pu Z, Luo Y, Asiri A M, et al. Efficient electrochemical water splitting catalyzed by electrodeposited nickel diselenide nanoparticles based film. ACS Applied Materials & Interfaces, 2016, 8 (7): 4718-4723.

[10] Li X, Han G Q, Liu Y R, et al. In situ grown pyramid structures of nickel diselenides dependent on oxidized nickel foam as efficient electrocatalyst for oxygen evolution reaction. Electrochimica Acta, 2016, 205: 77-84.

[11] Swesi A T, Masud J, Nath M. Nickel selenide as a high-efficiency catalyst for oxygen evolution

reaction. Energy & Environmental Science, 2016, 9 (5): 1771-1782.

[12] Sivanantham A, Shanmugam S. Nickel selenide supported on nickel foam as an efficient and durable non-precious electrocatalyst for the alkaline water electrolysis. Applied Catalysis B: Environmental, 2017, 203: 485-493.

[13] Xu K, Ding H, Lv H, et al. Understanding structure-dependent catalytic performance of nickel selenides for electrochemical water oxidation. ACS Catalysis, 2017, 7 (1): 310-315.

[14] Anantharaj S, Kennedy J, Kundu S. Microwave-initiated facile formation of Ni_3Se_4 nanoassemblies for enhanced and stable water splitting in neutral and alkaline media. ACS Applied Materials & Interfaces, 2017, 9 (10): 8714-8728.

[15] Anantharaj S, Karthik K, Amarnath T S, et al. Membrane free water electrolysis under 1.23V with Ni_3Se_4/Ni anode in alkali and Pt cathode in acid. Applied Surface Science, 2019, 478: 784-792.

[16] Anantharaj S, Subhashini E, Swaathini K C, et al. Respective influence of stoichiometry and niooh formation in hydrogen and oxygen evolution reactions of nickel selenides. Applied Surface Science, 2019, 487: 1152-1158.

[17] Ren J T, Yuan G G, Weng C C, et al. Ultrafine metal phosphide nanoparticles in situ encapsulated in porous N, P-codoped nanofibrous carbon coated on carbon paper for effective water splitting. Electrochimica Acta, 2018, 261: 454-463.

[18] Zhang Y, Cui B, Zhao C, et al. Co-Ni layered double hydroxides for water oxidation in neutral electrolyte. Physical Chemistry Chemical Physics, 2013, 15 (19): 7363-7369.

[19] Yu X, Zhang M, Yuan W, et al. A high-performance three-dimensional Ni-Fe layered double hydroxide/graphene electrode for water oxidation. Journal of Materials Chemistry A, 2015, 3 (13): 6921-6928.

[20] Zaffran J, Stevens M B, Trang C D M, et al. Influence of electrolyte cations on Ni (Fe) OOH catalyzed oxygen evolution reaction. Chemistry of Materials, 2017, 29 (11): 4761-4767.

[21] Stevens M B, Trang C D M, Enman L J, et al. Reactive Fe-sites in Ni/Fe (oxy) hydroxide are responsible for exceptional oxygen electrocatalysis activity. Journal of the American Chemical Society, 2017, 139 (33): 11361-11364.

[22] Enman L J, Burke M S, Batchellor A S, et al. Effects of intentionally incorporated metal cations on the oxygen evolution electrocatalytic activity of nickel (oxy) hydroxide in alkaline media. ACS Catalysis, 2016, 6 (4): 2416-2423.

[23] Tang C, Asiri A M, Sun X. Highly-active oxygen evolution electrocatalyzed by a Fe-doped NiSe nanoflake array electrode. Chemical Communications, 2016, 52 (24): 4529-4532.

[24] Du Y, Cheng G, Luo W. Colloidal synthesis of urchin-like Fe doped $NiSe_2$ for efficient oxygen evolution. Nanoscale, 2017, 9 (20): 6821-6825.

[25] Du J, Zou Z, Liu C, et al. Hierarchical Fe-doped Ni_3Se_4 ultrathin nanosheets as an efficient electrocatalyst for oxygen evolution reaction. Nanoscale, 2018, 10 (11): 5163-5170.

[26] Gu C, Hu S, Zheng X, et al. Synthesis of sub-2nm iron-doped $NiSe_2$ nanowires and their surface-confined oxidation for oxygen evolution catalysis. Angewandte Chemie International Edition, 2018, 57 (15): 4020-4024.

[27] Xu X, Song F, Hu X. A nickel iron diselenide-derived efficient oxygen-evolution catalyst. Nature Communications, 2016, 7 (1): 12324.

[28] Wang Z, Li J, Tian X, et al. Porous nickel-iron selenide nanosheets as highly efficient electrocatalysts for oxygen evolution reaction. ACS Applied Materials & Interfaces, 2016, 8 (30):

19386-19392.

[29] Nai J，Lu Y，Yu L，et al. Formation of Ni-Fe mixed diselenide nanocages as a superior oxygen evolution electrocatalyst. Advanced Materials，2017，29 (41)：1703870.

[30] Zhao Q，Zhong D，Liu L，et al. Facile fabrication of robust 3d Fe-NiSe nanowires supported on nickel foam as a highly efficient，durable oxygen evolution catalyst. Journal of Materials Chemistry A，2017，5 (28)：14639-14645.

[31] Du Y，Cheng G，Luo W. $NiSe_2/FeSe_2$ Nanodendrites：A highly efficient electrocatalyst for oxygen evolution reaction. Catalysis Science & Technology，2017，7 (20)：4604-4608.

[32] Li Y，Yan D，Zou Y，et al. Rapidly engineering the electronic properties and morphological structure of NiSe nanowires for the oxygen evolution reaction. Journal of Materials Chemistry A，2017，5 (48)：25494-25500.

[33] Ming F，Liang H，Shi H，et al. MOF-derived Co-doped nickel selenide/C electrocatalysts supported on Ni foam for overall water splitting. Journal of Materials Chemistry A，2016，4 (39)：15148-15155.

[34] Xu Y Z，Yuan C Z，Chen X P. Co-doped NiSe nanowires on nickel foam via a cation exchange approach as efficient electrocatalyst for enhanced oxygen evolution reaction. RSC Advances，2016，6 (108)：106832-106836.

[35] Fang Z，Peng L，Lv H，et al. Metallic transition metal selenide holey nanosheets for efficient oxygen evolution electrocatalysis. ACS Nano，2017，11 (9)：9550-9557.

[36] Shinde D V，Trizio L D，Dang Z，et al. Hollow and porous nickel cobalt perselenide nanostructured microparticles for enhanced electrocatalytic oxygen evolution. Chemistry of Materials，2017，29 (16)：7032-7041.

[37] Zhu H，Jiang R，Chen X，et al. 3d nickel-cobalt diselenide nanonetwork for highly efficient oxygen evolution. Science Bulletin，2017，62 (20)：1373-1379.

[38] Xu Y Z，Yuan C Z，Chen X P. A facile strategy for the synthesis of NiSe@CoOOH core-shell nanowires on nickel foam with high surface area as efficient electrocatalyst for oxygen evolution reaction. Applied Surface Science，2017，426：688-693.

[39] Akbar K，Jeon J H，Kim M，et al. Bifunctional electrodeposited 3d $NiCoSe_2$/nickle foam electrocatalysts for its applications in enhanced oxygen evolution reaction and for hydrazine oxidation. ACS Sustainable Chemistry & Engineering，2018，6 (6)：7735-7742.

[40] Ao K，Dong J，Fan C，et al. Formation of yolk-shelled nickel-cobalt selenide dodecahedral nanocages from metal-organic frameworks for efficient hydrogen and oxygen evolution. ACS Sustainable Chemistry & Engineering，2018，6 (8)：10952-10959.

[41] Li X，Yan K L，Rao Y，et al. Electrochemically activated NiSe-Ni_xS_y hybrid nanorods as efficient electrocatalysts for oxygen evolution reaction. Electrochimica Acta，2016，220：536-544.

[42] Gao R，Li G D，Hu J，et al. In situ electrochemical formation of NiSe/NiO_x core/shell nanoelectrocatalysts for superior oxygen evolution activity. Catalysis Science & Technology，2016，6 (23)：8268-8275.

[43] Liu J. Catalysis by supported single metal atoms. ACS Catalysis，2017，7 (1)：34-59.

[44] Are metal chalcogenides，nitrides，and phosphides oxygen evolution catalysts or bifunctional catalysts? ACS Energy Letters，2017，2 (8)：1937-1938.

[45] Brown T A，Shrift A. Selenium：Toxicity and tolerance in higher plants. Biological Reviews，1982，57 (1)：59-84.

[46] Hamilton S J. Review of selenium toxicity in the aquatic food chain. Science of The Total Environment，2004，326 (1)：1-31.

[47] Spallholz J E. On the nature of selenium toxicity and carcinostatic activity. Free Radical Biology and Medicine, 1994, 17 (1): 45-64.
[48] Chen P, Zhou T, Zhang M, et al. 3d nitrogen-anion-decorated nickel sulfides for highly efficient overall water splitting. Advanced Materials, 2017, 29 (30): 1701584.
[49] Yang C, Gao M Y, Zhang Q B, et al. In-situ activation of self-supported 3d hierarchically porous Ni_3S_2 films grown on nanoporous copper as excellent pH-universal electrocatalysts for hydrogen evolution reaction. Nano Energy, 2017, 36: 85-94.
[50] Du H, Kong R, Qu F, et al. Enhanced electrocatalysis for alkaline hydrogen evolution by Mn doping in a Ni_3S_2 nanosheet array. Chemical Communications, 2018, 54 (72): 10100-10103.
[51] Liu Q, Xie L, Liu Z, et al. A Zn-doped Ni_3S_2 nanosheet array as a high-performance electrochemical water oxidation catalyst in alkaline solution. Chemical Communications, 2017, 53 (92): 12446-12449.
[52] Zhang G, Feng YS, Lu WT, et al. Enhanced catalysis of electrochemical overall water splitting in alkaline media by Fe doping in Ni_3S_2 nanosheet arrays. ACS Catalysis, 2018, 8 (6): 5431-5441.
[53] Long X, Li G, Wang Z, et al. Metallic iron-nickel sulfide ultrathin nanosheets as a highly active electrocatalyst for hydrogen evolution reaction in acidic media. Journal of the American Chemical Society, 2015, 137 (37): 11900-11903.
[54] Yang N, Tang C, Wang K, et al. Iron-doped nickel disulfide nanoarray: A highly eff-icient and stable electrocatalyst for water splitting. Nano Research, 2016, 9 (11): 3346-3354.
[55] Yan J, Wu H, Li P, et al. Fe (Ⅲ) doped NiS_2 nanosheet: A highly efficient and low-cost hydrogen evolution catalyst. Journal of Materials Chemistry A, 2017, 5 (21): 10173-10181.
[56] Wang M, Zhang L, Pan J, et al. A highly efficient Fe-doped Ni_3S_2 electrocatalyst for overall water splitting. Nano Research, 2021, 14 (12): 4740-4747.
[57] Gao M R, Xu Y F, Jiang J, et al. Nanostructured metal chalcogenides: Synthesis, modification, and applications in energy conversion and storage devices. Chemical Society Reviews, 2013, 42 (7): 2986-3017.
[58] Jiang J, Lu S, Gao H, et al. Ternary $FeNiS_2$ ultrathin nanosheets as an electrocatalyst for both oxygen evolution and reduction reactions. Nano Energy, 2016, 27: 526-534.
[59] Shen J, Ji J, Dong P, et al. Novel $FeNi_2S_4$/TMD-based ternary composites for supercapacitor applications. Journal of Materials Chemistry A, 2016, 4 (22): 8844-8850.
[60] Jiang J, Zhang Y J, Zhu X J, et al. Nanostructured metallic $FeNi_2S_4$ with reconstruction to generate FeNi-based oxide as a highly-efficient oxygen evolution electrocatalyst. Nano Energy, 2021, 81: 105619.
[61] Shang X, Qin J F, Lin J H, et al. Tuning the morphology and Fe/Ni ratio of a bimetallic Fe-Ni-S film supported on nickel foam for optimized electrolytic water splitting. Journal of Colloid and Interface Science, 2018, 523: 121-132.
[62] Konkena B, Junge Puring K, Sinev I, et al. Pentlandite rocks as sustainable and stable efficient electrocatalysts for hydrogen generation. Nature Communications, 2016, 7 (1): 12269.
[63] Zegkinoglou I, Zendegani A, Sinev I, et al. Operando phonon studies of the protonation mechanism in highly active hydrogen evolution reaction pentlandite catalysts. Journal of the American Chemical Society, 2017, 139 (41): 14360-14363.
[64] Senger M, Laun K, Wittkamp F, et al. Proton-coupled reduction of the catalytic [4Fe-4S] cluster in [FeFe]-hydrogenases. Angewandte Chemie International Edition, 2017, 56 (52): 16503-16506.

[65] Drebushchak V A, Kravchenko T A, Pavlyuchenko V S. Synthesis of pure pentlandite in bulk. Journal of Crystal Growth, 1998, 193 (4): 728-731.

[66] Kitakaze A, Sugaki A. The phase relations between $Fe_{4.5}Ni_{4.5}S_8$ and Co_9S_8 in the system Fe-Ni-Co-S at temperatures from 400 to 1100℃. The Canadian Mineralogist, 2004, 42 (1): 17-42.

[67] Xia F, Pring A, Brugger J. Understanding the mechanism and kinetics of pentlandite oxidation in extractive pyrometallurgy of nickel. Minerals Engineering, 2012, 27: 11-19.

[68] Siegmund D, Blanc N, Smialkowski M, et al. Metal-rich chalcogenides for electrocatalytic hydrogen evolution: Activity of electrodes and bulk materials. ChemElectroChem, 2020, 7 (7): 1514-1527.

[69] Junge Puring K, Piontek S, Smialkowski M, et al. Simple methods for the preparation of non-noble metal bulk-electrodes for electrocatalytic applications. Journal of Visualized Experiments, 2017 (124): 56087.

[70] Piontek S, Andronescu C, Zaichenko A, et al. Influence of the Fe: Ni ratio and reaction temperature on the efficiency of $(Fe_xNi_{1-x})_9S_8$ electrocatalysts applied in the hydrogen evolution reaction. ACS Catalysis, 2018, 8 (2): 987-996.

[71] Bentley C L, Andronescu C, Smialkowski M, et al. Local surface structure and composition control the hydrogen evolution reaction on iron nickel sulfides. Angewandte Chemie International Edition, 2018, 57 (15): 4093-4097.

[72] Bezverkhyy I, Afanasiev P, Danot M. Preparation of highly dispersed pentlandites $(M, M')_9S_8$ (M,M'=Fe,Co,Ni) and their catalytic properties in hydrodesulfurization. The Journal of Physical Chemistry B, 2004, 108 (23): 7709-7715.

[73] Zhu C, Shi Q, Xu B Z, et al. Hierarchically porous M-N-C(M=Co and Fe) single-atom electrocatalysts with robust Mn_x active moieties enable enhanced ORR performance. Advanced Energy Materials, 2018, 8 (29): 1801956.

[74] Wu Y, Li F, Chen W, et al. Coupling interface constructions of $MoS_2/Fe_5Ni_4S_8$ heterostructures for efficient electrochemical water splitting. Advanced Materials, 2018, 30 (38): 1803151.

[75] Liu Y, Yin S, Shen P K. Asymmetric 3d electronic structure for enhanced oxygen evolution catalysis. ACS Applied Materials & Interfaces, 2018, 10 (27): 23131-23139.

[76] Xuan C, Lei W, Wang J, et al. Sea urchin-like Ni-Fe sulfide architectures as efficient electrocatalysts for the oxygen evolution reaction. Journal of Materials Chemistry A, 2019, 7 (19): 12350-12357.

[77] Zhang C, Nan H, Tian H, et al. Recent advances in pentlandites for electrochemical water splitting: A short review. Journal of Alloys and Compounds, 2020, 838: 155685.

[78] Zhang W, Li D, Zhang L, et al. NiFe-based nanostructures on nickel foam as highly efficiently electrocatalysts for oxygen and hydrogen evolution reactions. Journal of Energy Chemistry, 2019, 39: 39-53.

[79] Friebel D, Louie M W, Bajdich M, et al. Identification of highly active Fe sites in (Ni, Fe) OOH for electrocatalytic water splitting. Journal of the American Chemical Society, 2015, 137 (3): 1305-1313.

[80] Wu Z, Wang X, Huang J, et al. A Co-doped Ni-Fe mixed oxide mesoporous nanosheet array with low overpotential and high stability towards overall water splitting. Journal of Materials Chemistry A, 2018, 6 (1): 167-178.

[81] Chen S, Huang D, Zeng G, et al. In-situ synthesis of facet-dependent $BiVO_4/Ag_3PO_4$/PANI photocatalyst with enhanced visible-light-induced photocatalytic degradation performance: Syn-

ergism of interfacial coupling and hole-transfer. Chemical Engineering Journal, 2020, 382: 122840.

[82] Lei L, Huang D, Zhou C, et al. Demystifying the active roles of NiFe-based oxides/(oxy)hydroxides for electrochemical water splitting under alkaline conditions. Coordination Chemistry Reviews, 2020, 408: 213177.

[83] Chen J Y C, Miller J T, Gerken J B, et al. Inverse spinel nifealo4 as a highly active oxygen evolution electrocatalyst: Promotion of activity by a redox-inert metal ion. Energy & Environmental Science, 2014, 7 (4): 1382-1386.

[84] Amiri M, Eskandari K, Salavati-Niasari M. Magnetically retrievable ferrite nanoparticles in thc catalysis application. Advances in Colloid and Interface Science, 2019, 271: 101982.

[85] Wu G, Chen W, Zheng X, et al. Hierarchical Fe-doped NiO_x nanotubes assembled from ultrathin nanosheets containing trivalent nickel for oxygen evolution reaction. Nano Energy, 2017, 38: 167-174.

[86] Landon J, Demeter E, İnoğlu N, et al. Spectroscopic characterization of mixed Fe-Ni oxide electrocatalysts for the oxygen evolution reaction in alkaline electrolytes. ACS Catalysis, 2012, 2 (8): 1793-1801.

[87] Li Y F, Selloni A. Mechanism and activity of water oxidation on selected surfaces of pure and Fe-doped NiO_x. ACS Catalysis, 2014, 4 (4): 1148-1153.

[88] Louie M W, Bell A T. An investigation of thin-film Ni-Fe oxide catalysts for the electrochemical evolution of oxygen. Journal of the American Chemical Society, 2013, 135 (33): 12329-12337.

[89] Xuan C, Wang J, Xia W, et al. Heteroatom (P, B, or S) incorporated NiFe-based nanocubes as efficient electrocatalysts for the oxygen evolution reaction. Journal of Materials Chemistry A, 2018, 6 (16): 7062-7069.

[90] Detsi E, Cook J B, Lesel B K, et al. Mesoporous $Ni_{60}Fe_{30}Mn_{10}$-alloy based metal/metal oxide composite thick films as highly active and robust oxygen evolution catalysts. Energy & Environmental Science, 2016, 9 (2): 540-549.

[91] Bates M K, Jia Q, Doan H, et al. Charge-transfer effects in Ni-Fe and Ni-Fe-Co mixed-metal oxides for the alkaline oxygen evolution reaction. ACS Catalysis, 2016, 6 (1): 155-161.

[92] Chen Y, Dong C, Zhang J, et al. Hierarchically porous Mo-doped Ni-Fe oxide nanowires efficiently catalyzing oxygen/hydrogen evolution reactions. Journal of Materials Chemistry A, 2018, 6 (18): 8430-8440.

[93] Corrigan D A. The catalysis of the oxygen evolution reaction by iron impurities in thin film nickel oxide electrodes. Journal of The Electrochemical Society, 1987, 134 (2): 377-384.

[94] Wang J, Ji L, Chen Z. In situ rapid formation of a nickel-iron-based electrocatalyst for water oxidation. ACS Catalysis, 2016, 6 (10): 6987-6992.

[95] Huang D, Yan X, Yan M, et al. Graphitic carbon nitride-based heterojunction photo-active nanocomposites: Applications and mechanism insight. ACS Applied Materials & Interfaces, 2018, 10 (25): 21035-21055.

[96] Zhu K, Liu H, Li M, et al. Atomic-scale topochemical preparation of crystalline Fe^{3+}-doped β-$Ni(OH)_2$ for an ultrahigh-rate oxygen evolution reaction. Journal of Materials Chemistry A, 2017, 5 (17): 7753-7758.

[97] Gong M, Dai H. A mini review of NiFe-based materials as highly active oxygen evolution reaction electrocatalysts. Nano Research, 2015, 8 (1): 23-39.

[98] Anantharaj S, Karthick K, Kundu S. Evolution of layered double hydroxides (LDH) as high

performance water oxidation electrocatalysts: A review with insights on structure, activity and mechanism. Materials Today Energy, 2017, 6: 1-26.

[99] Trotochaud L, Young S L, Ranney J K, et al. Nickel-iron oxyhydroxide oxygen-evolution electrocatalysts: The role of intentional and incidental iron incorporation. Journal of the American Chemical Society, 2014, 136 (18): 6744-6753.

[100] Axmann P, Glemser O. Nickel hydroxide as a matrix for unusual valencies: The electrochemical behaviour of metal (Ⅲ)-ion-substituted nickel hydroxides of the pyroaurite type. Journal of Alloys and Compounds, 1997, 246 (1): 232-241.

[101] Wang D, Zhou J, Hu Y, et al. In situ x-ray absorption near-edge structure study of advanced NiFe $(OH)_x$ electrocatalyst on carbon paper for water oxidation. The Journal of Physical Chemistry C, 2015, 119 (34): 19573-19583.

[102] Chen J Y C, Dang L, Liang H, et al. Operando analysis of NiFe and Fe oxyhydroxide electrocatalysts for water oxidation: Detection of Fe^{4+} by mössbauer spectroscopy. Journal of the American Chemical Society, 2015, 137 (48): 15090-15093.

[103] Wu C, Zhang X, Xia Z, et al. Insight into the role of Ni-Fe dual sites in the oxygen evolution reaction based on atomically metal-doped polymeric carbon nitride. Journal of Materials Chemistry A, 2019, 7 (23): 14001-14010.

[104] Görlin M, Chernev P, Ferreira De Araújo J, et al. Oxygen evolution reaction dynamics, faradaic charge efficiency, and the active metal redox states of Ni-Fe oxide water splitting electrocatalysts. Journal of the American Chemical Society, 2016, 138 (17): 5603-5614.

[105] Li N, Bediako D K, Hadt R G, et al. Influence of iron doping on tetravalent nickel content in catalytic oxygen evolving films. Proceedings of the National Academy of Sciences, 2017, 114 (7): 1486-1491.

[106] Trześniewski B J, Diaz-Morales O, Vermaas D A, et al. In situ observation of active oxygen species in fe-containing ni-based oxygen evolution catalysts: The effect of pH on electrochemical activity. Journal of the American Chemical Society, 2015, 137 (48): 15112-15121.

[107] Zhong D, Zhang L, Li C, et al. Nanostructured NiFe (oxy) hydroxide with easily oxidized Ni towards efficient oxygen evolution reactions. Journal of Materials Chemistry A, 2018, 6 (35): 16810-16817.

[108] Diaz-Morales O, Ledezma-Yanez I, Koper M T M, et al. Guidelines for the rational design of Ni-based double hydroxide electrocatalysts for the oxygen evolution reaction. ACS Catalysis, 2015, 5 (9): 5380-5387.

[109] Grabow L C, Hvolbæk B, Nørskov J K. Understanding trends in catalytic activity: The effect of adsorbate-adsorbate interactions for co oxidation over transition metals. Topics in Catalysis, 2010, 53 (5): 298-310.

[110] Xiao H, Shin H, Goddard W A. Synergy between Fe and Ni in the optimal performance of (Ni, Fe) OOH catalysts for the oxygen evolution reaction. Proceedings of the National Academy of Sciences, 2018, 115 (23): 5872-5877.

[111] Huang Y B, Liang J, Wang X S, et al. Multifunctional metal-organic framework catalysts: Synergistic catalysis and tandem reactions. Chemical Society Reviews, 2017, 46 (1): 126-157.

[112] Chang F, Guan Y, Chang X, et al. Alkali and alkaline earth hydrides-driven N_2 activation and transformation over Mn nitride catalyst. Journal of the American Chemical Society, 2018, 140 (44): 14799-14806.

[113] Shin H, Xiao H, Goddard W A. In silico discovery of new dopants for Fe-doped Ni oxy-

hydroxide ($Ni_{1-x}Fe_xOOH$) catalysts for oxygen evolution reaction. Journal of the American Chemical Society, 2018, 140 (22): 6745-6748.

[114] Gong M, Li Y, Wang H, et al. An advanced Ni-Fe layered double hydroxide electrocatalyst for water oxidation. Journal of the American Chemical Society, 2013, 135 (23): 8452-8455.

[115] Hunter B M, Hieringer W, Winkler J R, et al. Effect of interlayer anions on [NiFe]-LDH nanosheet water oxidation activity. Energy & Environmental Science, 2016, 9 (5): 1734-1743.

[116] Qiu Z, Tai C W, Niklasson G A, et al. Direct observation of active catalyst surface phases and the effect of dynamic self-optimization in NiFe-layered double hydroxides for alkaline water splitting. Energy & Environmental Science, 2019, 12 (2): 572-581.

[117] Liu H, Wang Y, Lu X, et al. The effects of al substitution and partial dissolution on ultrathin NiFeAl trinary layered double hydroxide nanosheets for oxygen evolution reaction in alkaline solution. Nano Energy, 2017, 35: 350-357.

[118] Carenco S, Portehault D, Boissière C, et al. Nanoscaled metal borides and phosphides: Recent developments and perspectives. Chemical Reviews, 2013, 113 (10): 7981-8065.

[119] Mcmillan D E. Disturbance of serum viscosity in diabetes mellitus. The Journal of Clinical Investigation, 1974, 53 (4): 1071-1079.

[120] Osaka T, Ishibashi H, Endo T, et al. Oxygen evolution reaction on transition metal borides. Electrochimica Acta, 1981, 26 (3): 339-343.

[121] Tetsuya O, Yoshio I, Hiroshi K, et al. Oxygen evolution reaction on composite cobalt borides. Bulletin of the Chemical Society of Japan, 1983, 56 (7): 2106-2111.

[122] Huot J Y, Trudeau M, Brossard L, et al. Electrochemical and electrocatalytic behavior of an iron-base amorphous alloy in alkaline solutions at 70℃. Journal of The Electrochemical Society, 1989, 136 (8): 2224-2230.

[123] Alemu H, Jüttner K. Characterization of the electrocatalytic properties of amorphous metals for oxygen and hydrogen evolution by impedance measurements. Electrochimica Acta, 1988, 33 (8): 1101-1109.

[124] Huor J Y, Trudeau M, Brossard L, et al. Hydrogen evolution on some Ni-base amorphous alloys in alkaline solution. International Journal of Hydrogen Energy, 1989, 14 (5): 319-322.

[125] Kronberger H, Fabjan C, Frithum G. Development of high performance cathodes for hydrogen production from alkaline solutions. International Journal of Hydrogen Energy, 1991, 16 (3): 219-221.

[126] Kreysa G, Håkansson B. Electrocatalysis by amorphous metals of hydrogen and oxygen evolution in alkaline solution. Journal of Electroanalytical Chemistry and Interfacial Electrochemistry, 1986, 201 (1): 61-83.

[127] Lian K, Kirk D W, Thorpe S J. Electrocatalytic behaviour of Ni-base amorphous alloys. Electrochimica Acta, 1991, 36 (3): 537-545.

[128] Ndzebet E, Savadogo O. The hydrogen evolution reaction in 3M KOH on nickel boride electrodeposited with and without $SiW_{12}O_{40}^{4-}$. International Journal of Hydrogen Energy, 1994, 19 (8): 687-691.

[129] Liang Y, Sun X, Asiri A M, et al. Amorphous Ni-B alloy nanoparticle film on Ni foam: Rapid alternately dipping deposition for efficient overall water splitting. Nanotechnology, 2016, 27 (12): 12LT01.

[130] Masa J, Weide P, Peeters D, et al. Amorphous cobalt boride (Co_2B) as a highly efficient

nonprecious catalyst for electrochemical water splitting: Oxygen and hydrogen evolution. Advanced Energy Materials, 2016, 6 (6): 1502313.

[131] Li H, Wen P, Li Q, et al. Earth-abundant iron diboride (FeB_2) nanoparticles as highly active bifunctional electrocatalysts for overall water splitting. Advanced Energy Materials, 2017, 7 (17): 1700513.

[132] Surendranath Y, Dincă M, Nocera D G. Electrolyte-dependent electrosynthesis and activity of cobalt-based water oxidation catalysts. Journal of the American Chemical Society, 2009, 131 (7): 2615-2620.

[133] Dincă M, Surendranath Y, Nocera D G. Nickel-borate oxygen-evolving catalyst that functions under benign conditions. Proceedings of the National Academy of Sciences of the United States of America, 2010, 107 (23): 10337-10341.

[134] Kanan M W, Nocera D G. In situ formation of an oxygen-evolving catalyst in neutral water containing phosphate and Co^{2+}. Science, 2008, 321 (5892): 1072-1075.

[135] Vrubel H, Hu X. Molybdenum boride and carbide catalyze hydrogen evolution in both acidic and basic solutions. Angewandte Chemie International Edition, 2012, 51 (51): 12703-12706.

[136] Gupta S, Patel N, Miotello A, et al. Cobalt-boride: An efficient and robust electrocatalyst for hydrogen evolution reaction. Journal of Power Sources, 2015, 279: 620-625.

[137] Zeng M, Wang H, Zhao C, et al. Nanostructured amorphous nickel boride for high-efficiency electrocatalytic hydrogen evolution over a broad pH range. ChemCatChem, 2016, 8 (4): 708-712.

[138] Zhang P, Wang M, Yang Y, et al. Electroless plated Ni-B_x films as highly active electrocatalysts for hydrogen production from water over a wide pH range. Nano Energy, 2016, 19: 98-107.

[139] Park H, Encinas A, Scheifers J P, et al. Boron-dependency of molybdenum boride electrocatalysts for the hydrogen evolution reaction. Angewandte Chemie International Edition, 2017, 56 (20): 5575-5578.

[140] Chen P, Xu K, Zhou T, et al. Strong-coupled cobalt borate nanosheets/graphene hybrid as electrocatalyst for water oxidation under both alkaline and neutral conditions. Angewandte Chemie International Edition, 2016, 55 (7): 2488-2492.

[141] Park H, Zhang Y, Scheifers J P, et al. Graphene- and phosphorene-like boron layers with contrasting activities in highly active Mo_2B_4 for hydrogen evolution. Journal of the American Chemical Society, 2017, 139 (37): 12915-12918.

[142] Jothi P R, Zhang Y, Scheifers J P, et al. Molybdenum diboride nanoparticles as a highly efficient electrocatalyst for the hydrogen evolution reaction. Sustainable Energy & Fuels, 2017, 1 (9): 1928-1934.

[143] Masa J, Weide P, Peeters D, et al. Amorphous cobalt boride (Co_2B) as a highly efficient nonprecious catalyst for electrochemical water splitting: Oxygen and hydrogen evolution. Advanced Energy Materials, 2016, 6 (12): 1502313.

[144] Farrow C L, Bediako D K, Surendranath Y, et al. Intermediate-range structure of self-assembled cobalt-based oxygen-evolving catalyst. Journal of the American Chemical Society, 2013, 135 (17): 6403-6406.

[145] Yoshida M, Iida T, Mineo T, et al. Electrochromic characteristics of a nickel borate thin film investigated by in situ XAFS and UV/Vis spectroscopy. Electrochemistry, 2014, 82 (5): 355-358.

[146] Ganem B, Osby J O. Synthetically useful reactions with metal boride and aluminide catalysts. Chemical Reviews, 1986, 86 (5): 763-780.

[147] Zhang H, Zhang Q, Tang J, et al. Single-crystalline LaB_6 nanowires. Journal of the American Chemical Society, 2005, 127 (9): 2862-2863.

[148] Amin S S, Li SY, Roth J R, et al. Single crystalline alkaline-earth metal hexaboride one-dimensional (1D) nanostructures: Synthesis and characterization. Chemistry of Materials, 2009, 21 (4): 763-770.

[149] Pol V G, Pol S V, Gedanken A. Dry autoclaving for the nanofabrication of sulfides, selenides, borides, phosphides, nitrides, carbides, and oxides. Advanced Materials, 2011, 23 (10): 1179-1190.

[150] Jiang K, Liu B, Luo M, et al. Single platinum atoms embedded in nanoporous cobalt selenide as electrocatalyst for accelerating hydrogen evolution reaction. Nature Communications, 2019, 10 (1): 1743.

[151] Yuan H, Wang S, Gu X, et al. One-step solid-phase boronation to fabricate self-supported porous FeNiB/FeNi foam for efficient electrocatalytic oxygen evolution and overall water splitting. Journal of Materials Chemistry A, 2019, 7 (33): 19554-19564.

[152] Glavee G N, Klabunde K J, Sorensen C M, et al. Borohydride reductions of metal ions. A new understanding of the chemistry leading to nanoscale particles of metals, borides, and metal borates. Langmuir, 1992, 8 (3): 771-773.

[153] Cao X, Wang X, Cui L, et al. Strongly coupled nickel boride/graphene hybrid as a novel electrode material for supercapacitors. Chemical Engineering Journal, 2017, 327: 1085-1092.

[154] Sun H, Meng J, Jiao L, et al. A review of transition-metal boride/phosphide-based materials for catalytic hydrogen generation from hydrolysis of boron-hydrides. Inorganic Chemistry Frontiers, 2018, 5 (4): 760-772.

[155] Hao W, Wu R, Zhang R, et al. Electroless plating of highly efficient bifunctional boride-based electrodes toward practical overall water splitting. Advanced Energy Materials, 2018, 8 (26): 1801372.

[156] Haag D, Kung H H. Metal free graphene based catalysts: A review. Topics in Catalysis, 2014, 57 (6): 762-773.

[157] Kim J, Hwang J, Chung H. Comparison of near-infrared and Raman spectroscopy for on-line monitoring of etchant solutions directly through a Teflon tube. Analytica Chimica Acta, 2008, 629 (1): 119-127.

[158] Cao S, Low J, Yu J, et al. Polymeric photocatalysts based on graphitic carbon nitride. Advanced Materials, 2015, 27 (13): 2150-2176.

[159] Cao M, Zhang X, Qin J, et al. Enhancement of hydrogen evolution reaction performance of graphitic carbon nitride with incorporated nickel boride. ACS Sustainable Chemistry & Engineering, 2018, 6 (12): 16198-16204.

[160] Li Y, Xu H, Huang H, et al. Synthesis of Co-B in porous carbon using a metal-organic framework (MOF) precursor: A highly efficient catalyst for the oxygen evolution reaction. Electrochemistry Communications, 2018, 86: 140-144.

[161] Arivu M, Masud J, Umapathi S, et al. Facile synthesis of Ni_3B/rGO nanocomposite as an efficient electrocatalyst for the oxygen evolution reaction in alkaline media. Electrochemistry Communications, 2018, 86: 121-125.

[162] Elumeeva K, Masa J, Medina D, et al. Cobalt boride modified with N-doped carbon nanotubes as a high-performance bifunctional oxygen electrocatalyst. Journal of Materials Chemistry

A, 2017, 5 (40): 21122-21129.

[163] Chen Z, Kang Q, Cao G, et al. Study of cobalt boride-derived electrocatalysts for overall water splitting. International Journal of Hydrogen Energy, 2018, 43 (12): 6076-6087.

[164] Cao G X, Xu N, Chen Z J, et al. Cobalt-tungsten-boron as an active electrocatalyst for water electrolysis. ChemistrySelect, 2017, 2 (21): 6187-6193.

[165] Nsanzimana J M V, Reddu V, Peng Y, et al. Ultrathin amorphous iron-nickel boride nanosheets for highly efficient electrocatalytic oxygen production. Chemistry-A European Journal, 2018, 24 (69): 18502-18511.

[166] Wang S, He P, Xie Z, et al. Tunable nanocotton-like amorphous ternary Ni-Co-B: A highly efficient catalyst for enhanced oxygen evolution reaction. Electrochimica Acta, 2019, 296: 644-652.

[167] Jiang J, Wang M, Yan W, et al. Highly active and durable electrocatalytic water oxidation by a $NiB_{0.45}/NiO_x$ core-shell heterostructured nanoparticulate film. Nano Energy, 2017, 38: 175-184.

[168] Lu W, Liu T, Xie L, et al. In situ derived CoB nanoarray: A high-efficiency and durable 3D bifunctional electrocatalyst for overall alkaline water splitting. Small, 2017, 13 (32): 1700805.

[169] Guo Y, Yao Z, Shang C, et al. Amorphous Co_2B grown on $CoSe_2$ nanosheets as a hybrid catalyst for efficient overall water splitting in alkaline medium. ACS Applied Materials & Interfaces, 2017, 9 (45): 39312-39317.

[170] Yang Y, Zhuang L, Lin R, et al. A facile method to synthesize boron-doped Ni/Fe alloy nano-chains as electrocatalyst for water oxidation. Journal of Power Sources, 2017, 349: 68-74.

[171] Borodzínski J J, Lasia A. Electrocatalytic properties of doped nickel boride based electrodes for the hydrogen evolution reaction. Journal of Applied Electrochemistry, 1994, 24 (12): 1267-1275.

[172] Gupta S, Patel N, Fernandes R, et al. Co-Ni-B nanocatalyst for efficient hydrogen evolution reaction in wide pH range. Applied Catalysis B: Environmental, 2016, 192: 126-133.

[173] Xu N, Cao G, Chen Z, et al. Cobalt nickel boride as an active electrocatalyst for water splitting. Journal of Materials Chemistry A, 2017, 5 (24): 12379-12384.

[174] Sheng M, Wu Q, Wang Y, et al. Network-like porous Co-Ni-B grown on carbon cloth as efficient and stable catalytic electrodes for hydrogen evolution. Electrochemistry Communications, 2018, 93: 104-108.

[175] Sun J, Zhang W, Wang S, et al. Ni-Co-B nanosheets coupled with reduced graphene oxide towards enhanced electrochemical oxygen evolution. Journal of Alloys and Compounds, 2019, 776: 511-518.

[176] Gupta S, Patel N, Fernandes R, et al. Co-Mo-B nanoparticles as a non-precious and efficient bifunctional electrocatalyst for hydrogen and oxygen evolution. Electrochimica Acta, 2017, 232: 64-71.

[177] Chen H, Ouyang S, Zhao M, et al. Synergistic activity of Co and Fe in amorphous Co_x-Fe-B catalyst for efficient oxygen evolution reaction. ACS Applied Materials & Interfaces, 2017, 9 (46): 40333-40343.

[178] Liu G, He D, Yao R, et al. Amorphous CoFeBO nanoparticles as highly active electrocatalysts for efficient water oxidation reaction. International Journal of Hydrogen Energy, 2018, 43 (12): 6138-6149.

[179] Liu G, He D, Yao R, et al. Amorphous NiFeB nanoparticles realizing highly active and stable oxygen evolving reaction for water splitting. Nano Research, 2018, 11 (3): 1664-1675.

[180] An L, Sun Y, Zong Y, et al. Nickel iron boride nanosheets on rGO for active electrochemical water oxidation. Journal of Solid State Chemistry, 2018, 265: 135-139.

[181] Nsanzimana J M V, Peng Y, Xu Y Y, et al. An efficient and earth-abundant oxygen-evolving electrocatalyst based on amorphous metal borides. Advanced Energy Materials, 2018, 8 (1): 1701475.

[182] Han H, Hong Y R, Woo J, et al. Electronically double-layered metal boride hollow nanoprism as an excellent and robust water oxidation electrocatalysts. Advanced Energy Materials, 2019, 9 (13): 1803799.

[183] Yang Y, Zhuang L, Rufford T E, et al. Efficient water oxidation with amorphous transition metal boride catalysts synthesized by chemical reduction of metal nitrate salts at room temperature. RSC Advances, 2017, 7 (52): 32923-32930.

[184] Li P, Jin Z, Xiao D. A one-step synthesis of Co-P-B/rGO at room temperature with synergistically enhanced electrocatalytic activity in neutral solution. Journal of Materials Chemistry A, 2014, 2 (43): 18420-18427.

[185] Kim J, Kim H, Kim SK, et al. Electrodeposited amorphous Co-P-B ternary catalyst for hydrogen evolution reaction. Journal of Materials Chemistry A, 2018, 6 (15): 6282-6288.

[186] Dincă M, Surendranath Y, Nocera D G. Nickel-borate oxygen-evolving catalyst that functions under benign conditions. Proceedings of the National Academy of Sciences, 2010, 107 (23): 10337-10341.

[187] Young E R, Nocera D G, Bulović V. Direct formation of a water oxidation catalyst from thin-film cobalt. Energy & Environmental Science, 2010, 3 (11): 1726-1728.

[188] Xie L, Zhang R, Cui L, et al. High-performance electrolytic oxygen evolution in neutral media catalyzed by a cobalt phosphate nanoarray. Angewandte Chemie International Edition, 2017, 56 (4): 1064-1068.

[189] Cui L, Liu D, Hao S, et al. In situ electrochemical surface derivation of cobalt phosphate from a Co $(CO_3)_{0.5}$ (OH) $\cdot 0.11H_2O$ nanoarray for efficient water oxidation in neutral aqueous solution. Nanoscale, 2017, 9 (11): 3752-3756.

[190] Cui L, Qu F, Liu J, et al. Interconnected network of core-shell CoP@CoBiPi for efficient water oxidation electrocatalysis under near neutral conditions. ChemSusChem, 2017, 10 (7): 1370-1374.

[191] Leng X, Wu K H, Su B J, et al. Hydrotalcite-wrapped Co-B alloy with enhanced oxygen evolution activity. Chinese Journal of Catalysis, 2017, 38 (6): 1021-1027.

[192] Liu J, Chen T, Juan P, et al. Hierarchical cobalt borate/Mxenes hybrid with extraordinary electrocatalytic performance in oxygen evolution reaction. ChemSusChem, 2018, 11 (21): 3758-3765.

[193] Gao M R, Xu Y F, Jiang J, et al. Water oxidation electrocatalyzed by an efficient Mn_3O_4/$CoSe_2$ nanocomposite. Journal of the American Chemical Society, 2012, 134 (6): 2930-2933.

[194] Sun Z, Lin L, Nan C, et al. Amorphous boron oxide coated NiCo layered double hydroxide nanoarrays for highly efficient oxygen evolution reaction. ACS Sustainable Chemistry & Engineering, 2018, 6 (11): 14257-14263.

[195] Wang N, Cao Z, Kong X, et al. Activity enhancement via borate incorporation into a NiFe

(oxy) hydroxide catalyst for electrocatalytic oxygen evolution. Journal of Materials Chemistry A, 2018, 6 (35): 16959-16964.

[196] Zhang Z, Zhang T, Lee J Y. Enhancement effect of borate doping on the oxygen evolu-tion activity of α-nickel hydroxide. ACS Applied Nano Materials, 2018, 1 (2): 751-758.

[197] Su L, Du H, Tang C, et al. Borate-ion intercalated NiFe layered double hydroxide to simultaneously boost mass transport and charge transfer for catalysis of water oxidation. Journal of Colloid and Interface Science, 2018, 528: 36-44.

[198] Bediako D K, Lassalle-Kaiser B, Surendranath Y, et al. Structure-activity correlations in a nickel-borate oxygen evolution catalyst. Journal of the American Chemical Society, 2012, 134 (15): 6801-6809.

[199] Fayad R, Dhainy J, Ghandour H, et al. Electrochemical study of the promoting effect of Fe on oxygen evolution at thin 'NiFe-Bi' films and the inhibiting effect of al in borate electrolyte. Catalysis Science & Technology, 2017, 7 (17): 3876-3891.

[200] Smith A M, Trotochaud L, Burke M S, et al. Contributions to activity enhancement via Fe incorporation in Ni-(oxy) hydroxide/borate catalysts for near-neutral pH oxygen evolution. Chemical Communications, 2015, 51 (25): 5261-5263.

[201] Ji X, Hao S, Qu F, et al. Core-shell $CoFe_2O_4$@Co-Fe-Bi nanoarray: A surface-amorphization water oxidation catalyst operating at near-neutral ph. Nanoscale, 2017, 9 (23): 7714-7718.

[202] Jiang W J, Niu S, Tang T, et al. Crystallinity-modulated electrocatalytic activity of a nickel (Ⅱ) borate thin layer on Ni_3B for efficient water oxidation. Angewandte Chemie International Edition, 2017, 56 (23): 6572-6577.

[203] Bediako D K, Surendranath Y, Nocera D G. Mechanistic studies of the oxygen evolution reaction mediated by a nickel-borate thin film electrocatalyst. Journal of the American Chemical Society, 2013, 135 (9): 3662-3674.

[204] Esswein A J, Surendranath Y, Reece S Y, et al. Highly active cobalt phosphate and borate based oxygen evolving catalysts operating in neutral and natural waters. Energy & Environmental Science, 2011, 4 (2): 499-504.

[205] Zeng M, Wang H, Zhao C, et al. 3D graphene foam-supported cobalt phosphate and borate electrocatalysts for high-efficiency water oxidation. Science Bulletin, 2015, 60 (16): 1426-1433.

[206] Yang L, Liu D, Hao S, et al. A cobalt-borate nanosheet array: An efficient and durable non-noble-metal electrocatalyst for water oxidation at near neutral pH. Journal of Materials Chemistry A, 2017, 5 (16): 7305-7308.

[207] Ge R, Ren X, Qu F, et al. Three-dimensional nickel-borate nanosheets array for efficient oxygen evolution at near-neutral pH. Chemistry-A European Journal, 2017, 23 (29): 6959-6963.

[208] Xie C, Wang Y, Yan D, et al. In situ growth of cobalt@cobalt-borate core-shell nanosheets as highly-efficient electrocatalysts for oxygen evolution reaction in alkaline/neutral medium. Nanoscale, 2017, 9 (41): 16059-16065.

[209] Xie L, Qu F, Liu Z, et al. In situ formation of a 3D core/shell structured Ni_3N@Ni-Bi nanosheet array: An efficient non-noble-metal bifunctional electrocatalyst toward full water splitting under near-neutral conditions. Journal of Materials Chemistry A, 2017, 5 (17): 7806-7810.

[210] Ma M, Qu F, Ji X, et al. Bimetallic nickel-substituted cobalt-borate nanowire array: An

earth-abundant water oxidation electrocatalyst with superior activity and durability at near neutral ph. Small, 2017, 13 (25): 1700394.

[211] Ma M, Liu D, Hao S, et al. A nickel-borate-phosphate nanoarray for efficient and durable water oxidation under benign conditions. Inorganic Chemistry Frontiers, 2017, 4 (5): 840-844.

[212] Wang W, Liu D, Hao S, et al. High-efficiency and durable water oxidation under mild pH conditions: An iron phosphate-borate nanosheet array as a non-noble-metal catalyst electrode. Inorganic Chemistry, 2017, 56 (6): 3131-3135.

[213] Rodríguez-Padrón D, Puente-Santiago A R, Balu A M, et al. Environmental catalysis: Present and future. ChemCatChem, 2019, 11 (1): 18-38.

[214] Wang H, Gao L. Recent developments in electrochemical hydrogen evolution reaction. Current Opinion in Electrochemistry, 2018, 7: 7-14.

[215] Kitchin J R, Nørskov J K, Barteau M A, et al. Trends in the chemical properties of early transition metal carbide surfaces: A density functional study. Catalysis Today, 2005, 105 (1): 66-73.

[216] Merki D, Fierro S, Vrubel H, et al. Amorphous molybdenum sulfide films as catalysts for electrochemical hydrogen production in water. Chemical Science, 2011, 2 (7): 1262-1267.

[217] Xing Z, Liu Q, Asiri A M, et al. Closely interconnected network of molybdenum phosphide nanoparticles: A highly efficient electrocatalyst for generating hydrogen from water. Advanced Materials, 2014, 26 (32): 5702-5707.

[218] Kou Z, Zhang L, Ma Y, et al. 2d carbide nanomeshes and their assembling into 3d microflowers for efficient water splitting. Applied Catalysis B: Environmental, 2019, 243: 678-685.

[219] Kou Z, Guo B, Zhao Y, et al. Molybdenum carbide-derived chlorine-doped ordered mesoporous carbon with few-layered graphene walls for energy storage applications. ACS Applied Materials & Interfaces, 2017, 9 (4): 3702-3712.

[220] Jeon J, Park Y, Choi S, et al. Epitaxial synthesis of molybdenum carbide and formation of a Mo_2C/MoS_2 hybrid structure via chemical conversion of molybdenum disulfide. ACS Nano, 2018, 12 (1): 338-346.

[221] Mir R A, Pandey O P. Influence of graphitic/amorphous coated carbon on her activity of low temperature synthesized β-Mo_2C@C nanocomposites. Chemical Engineering Journal, 2018, 348: 1037-1048.

[222] Jing S, Zhang L, Luo L, et al. N-doped porous molybdenum carbide nanobelts as efficient catalysts for hydrogen evolution reaction. Applied Catalysis B: Environmental, 2018, 224: 533-540.

[223] Ma Y, Guan G, Hao X, et al. Molybdenum carbide as alternative catalyst for hydrogen production-a review. Renewable and Sustainable Energy Reviews, 2017, 75: 1101-1129.

[224] Calais J L. Band structure of transition metal compounds. Advances in Physics, 1977, 26 (6): 847-885.

[225] Neckel A. Recent investigations on the electronic structure of the fourth and fifth group transition metal monocarbides, mononitrides, and monoxides. International Journal of Quantum Chemistry, 1983, 23 (4): 1317-1353.

[226] Chen X, Chen X, Qi J, et al. Self-assembly synthesis of lamellar molybdenum carbides with controllable phases for hydrodeoxygenation of diphenyl ether. Molecular Catalysis, 2020, 492: 110972.

[227] He Y, Xu J, Wang F, et al. In-situ carbonization approach for the binder-free Ir-dispersed ordered mesoporous carbon hydrogen evolution electrode. Journal of Energy Chemistry, 2017, 26 (6): 1140-1146.

[228] Hwu H H, Chen J G. Surface chemistry of transition metal carbides. Chemical Reviews, 2005, 105 (1): 185-212.

[229] Liu P, Rodriguez J A. Catalytic properties of molybdenum carbide, nitride and phosphide: A theoretical study. Catalysis Letters, 2003, 91 (3): 247-252.

[230] Chasvin N, Diez A, Pronsato E, et al. Theoretical and experimental study of ethanol adsorption and dissociation on β-Mo_2C surfaces. Molecular Catalysis, 2017, 439: 163-170.

[231] López-Martín A, Caballero A, Colón G. Structural and surface considerations on Mo/ZSM-5 systems for methane dehydroaromatization reaction. Molecular Catalysis, 2020, 486: 110787.

[232] Cui W, Cheng N, Liu Q, et al. Mo_2C nanoparticles decorated graphitic carbon sheets: Biopolymer-derived solid-state synthesis and application as an efficient electrocatalyst for hydrogen generation. ACS Catalysis, 2014, 4 (8): 2658-2661.

[233] Zhang X, Wang J, Guo T, et al. Structure and phase regulation in Mo_xC(α-MoC_{1-x}/β-Mo_2C) to enhance hydrogen evolution. Applied Catalysis B: Environmental, 2019, 247: 78-85.

[234] Xing J, Li Y, Guo S, et al. Molybdenum carbide in-situ embedded into carbon nano-sheets as efficient bifunctional electrocatalysts for overall water splitting. Electro-chimica Acta, 2019, 298: 305-312.

[235] Wan C, Regmi Y N, Leonard B M. Multiple phases of molybdenum carbide as electrocatalysts for the hydrogen evolution reaction. Angewandte Chemie International Edition, 2014, 53 (25): 6407-6410.

[236] Chen W F, Iyer S, Iyer S, et al. Biomass-derived electrocatalytic composites for hydro-gen evolution. Energy & Environmental Science, 2013, 6 (6): 1818-1826.

[237] Guo J, Wang J, Wu Z, et al. Controllable synthesis of molybdenum-based electrocatalysts for a hydrogen evolution reaction. Journal of Materials Chemistry A, 2017, 5 (10): 4879-4885.

[238] Lv C, Huang Z, Yang Q, et al. Ultrafast synthesis of molybdenum carbide nanoparticles for efficient hydrogen generation. Journal of Materials Chemistry A, 2017, 5 (43): 22805-22812.

[239] Wang H, Xu X, Ni B, et al. 3d self-assembly of ultrafine molybdenum carbide confined in N-doped carbon nanosheets for efficient hydrogen production. Nanoscale, 2017, 9 (41): 15895-15900.

[240] Wu H B, Xia B Y, Yu L, et al. Porous molybdenum carbide nano-octahedrons synthesized via confined carburization in metal-organic frameworks for efficient hydrogen production. Nature Communications, 2015, 6 (1): 6512.

[241] Zhao Y, Kamiya K, Hashimoto K, et al. In situ CO_2-emission assisted synthesis of molybdenum carbonitride nanomaterial as hydrogen evolution electrocatalyst. Journal of the American Chemical Society, 2015, 137 (1): 110-113.

[242] Long G F, Wan K, Liu M Y, et al. Active sites and mechanism on nitrogen-doped carbon catalyst for hydrogen evolution reaction. Journal of Catalysis, 2017, 348: 151-159.

[243] Cabán-Acevedo M, Stone M L, Schmidt J R, et al. Efficient hydrogen evolution catalysis using ternary pyrite-type cobalt phosphosulphide. Nature Materials, 2015, 14 (12):

1245-1251.

[244] Reddy S, Du R, Kang L, et al. Three dimensional CNTs aerogel/MoS_x as an electrocatalyst for hydrogen evolution reaction. Applied Catalysis B: Environmental, 2016, 194: 16-21.

[245] Ang H, Wang H, Li B, et al. 3D hierarchical porous $Mo_2C$for efficient hydrogen evolution. Small, 2016, 12 (21): 2859-2865.

[246] Liu Y, Yu G, Li G D, et al. Coupling Mo_2C with nitrogen-rich nanocarbon leads to efficient hydrogen-evolution electrocatalytic sites. Angewandte Chemie International Edition, 2015, 54 (37): 10752-10757.

[247] Liu Y, Li G D, Yuan L, et al. Carbon-protected bimetallic carbide nanoparticles for a highly efficient alkaline hydrogen evolution reaction. Nanoscale, 2015, 7 (7): 3130-3136.

[248] Chi J Q, Xie J Y, Zhang W W, et al. N-doped sandwich-structured Mo_2C@C@Pt interface with ultralow Pt loading for pH-universal hydrogen evolution reaction. ACS Applied Materials & Interfaces, 2019, 11 (4): 4047-4056.

[249] Liang Q, Jin H, Wang Z, et al. Metal-organic frameworks derived reverse-encapsulation Co-NC@Mo_2C complex for efficient overall water splitting. Nano Energy, 2019, 57: 746-752.

[250] Liu C, Sun L, Luo L, et al. Integration of Ni doping and a Mo_2C/MoC heterojunction for hydrogen evolution in acidic and alkaline conditions. ACS Applied Materials & Interfaces, 2021, 13 (19): 22646-22654.

[251] Byrappa K, Adschiri T. Hydrothermal technology for nanotechnology. Progress in Crystal Growth and Characterization of Materials, 2007, 53: 117-166.

[252] Zhang L, Zhuang L, Liu H, et al. Beyond platinum: Defects abundant CoP_3/Ni_2P heterostructure for hydrogen evolution electrocatalysis. Small Science, 2021, 1 (4): 2000027.

[253] Jiang P, Liu Q, Ge C, et al. Cop nanostructures with different morphologies: Synthesis, characterization and a study of their electrocatalytic performance toward the hydrogen evolution reaction. Journal of Materials Chemistry A, 2014, 2 (35): 14634-14640.

[254] Zhang R, Wang X, Yu S, et al. Ternary $NiCo_2P_x$ nanowires as pH-universal electrocatalysts for highly efficient hydrogen evolution reaction. Advanced Materials, 2017, 29 (9): 1605502.

[255] Sun X. Self-supported FeP nanorod arrays: A cost-effective 3D hydrogen evolution cathode with high catalytic activity. 2014, 4 (11), 4065-4069.

[256] Niu Z, Qiu C, Jiang J, et al. Hierarchical CoP-FeP branched heterostructures for highly efficient electrocatalytic water splitting. ACS Sustainable Chemistry & Engineering, 2018, 7: 2335-2342.

[257] Sakita A M P, Noce R D, Vallés E, et al. Pulse electrodeposition of CoFe thin films covered with layered double hydroxides as a fast route to prepare enhanced catalysts for oxygen evolution reaction. Applied Surface Science, 2018, 434: 1153-1160.

[258] Zhang W, Wu Y, Qi J, et al. A thin NiFe hydroxide film formed by stepwise electrodeposition strategy with significantly improved catalytic water oxidation efficiency. Advanced Energy Materials, 2017, 7 (9): 1602547.

[259] Pei Y, Ge Y, Chu H, et al. Controlled synthesis of 3D porous structured cobalt-iron based nanosheets by electrodeposition as asymmetric electrodes for ultra-efficient water splitting. Applied Catalysis B: Environmental, 2019, 244: 583-593.

[260] Zhou L, Shao M, Li J, et al. Two-dimensional ultrathin arrays of CoP: Electronic modulation toward high performance overall water splitting. Nano Energy, 2017, 41: 583-590.

[261] Pu Z, Liu Q, Jiang P, et al. Cop nanosheet arrays supported on a Ti plate: An efficient cath-

ode for electrochemical hydrogen evolution. Chemistry of Materials, 2014, 26: 4326-4329.

[262] Zhao X, Zhang Z, Cao X, et al. Elucidating the sources of activity and stability of FeP electrocatalyst for hydrogen evolution reactions in acidic and alkaline media. Applied Catalysis B: Environmental, 2020, 260: 118156.

[263] Yan Y, Xia B Y, Ge X, et al. A flexible electrode based on iron phosphide nanotubes for overall water splitting. Chemistry-A European Journal, 2015, 21 (50): 18062-18067.

[264] Zhang C L, Xie Y, Liu J T, et al. 1d core-shell MOFs derived CoP nanoparticles-embedded N-doped porous carbon nanotubes anchored with MoS_2 nanosheets as efficient bifunctional electrocatalysts. Chemical Engineering Journal, 2021, 419: 129977.

[265] Popczun E J, Read C G, Roske C W, et al. Highly active electrocatalysis of the hydrogen evolution reaction by cobalt phosphide nanoparticles. Angewandte Chemie International Edition, 2014, 53 (21): 5427-5430.

[266] Wang Z, Wang S, Ma L, et al. Water-induced formation of Ni_2P-$Ni_{12}P_5$ interfaces with superior electrocatalytic activity toward hydrogen evolution reaction. Small, 2021, 17 (6): 2006770.

[267] Tian J, Liu Q, Yanhui L, et al. FeP nanoparticles film grown on carbon cloth: An ultrahighly active 3D hydrogen evolution cathode in both acidic and neutral solutions. ACS Applied Materials & Interfaces, 2014, 6 (23): 20579-20584.

[268] Tian J, Liu Q, Asiri A M, et al. Self-supported nanoporous cobalt phosphide nanowire arrays: An efficient 3d hydrogen-evolving cathode over the wide range of pH 0-14. Journal of the American Chemical Society, 2014, 136 (21): 7587-7590.

[269] Yang X, Lu A Y, Zhu Y, et al. Cop nanosheet assembly grown on carbon cloth: A highly efficient electrocatalyst for hydrogen generation. Nano Energy, 2015, 15: 634-641.

[270] Ma B, Yang Z, Chen Y, et al. Nickel cobalt phosphide with three-dimensional nanostructure as a highly efficient electrocatalyst for hydrogen evolution reaction in both acidic and alkaline electrolytes. Nano Research, 2019, 12 (2): 375-380.

[271] Wang X, Kolen'ko Y V, Bao X Q, et al. One-step synthesis of self-supported nickel phosphide nanosheet array cathodes for efficient electrocatalytic hydrogen generation. Angewandte Chemie International Edition, 2015, 54 (28): 8188-8192.

[272] Han A, Jin S, Chen H, et al. A robust hydrogen evolution catalyst based on crystalline nickel phosphide nanoflakes on three-dimensional graphene/nickel foam: High performance for electrocatalytic hydrogen production from pH 0-14. Journal of Materials Chemistry A, 2015, 3 (5): 1941-1946.

[273] Ledendecker M, Krick Calderón S, Papp C, et al. The synthesis of nanostructured Ni_5P_4 films and their use as a non-noble bifunctional electrocatalyst for full water splitting. Angewandte Chemie International Edition, 2015, 54 (42): 12361-12365.

[274] Wang Y, Ma B, Chen Y. Iron phosphides supported on three-dimensional iron foam as an efficient electrocatalyst for water splitting reactions. Journal of Materials Science, 2019, 54 (24): 14872-14883.

[275] Liu S, Hu C, Lv C, et al. Facile preparation of large-area self-supported porous nickel phosphide nanosheets for efficient electrocatalytic hydrogen evolution. International Journal of Hydrogen Energy, 2019, 44 (33): 17974-17984.

[276] Shi Y, Xu Y, Zhuo S, et al. Ni_2P nanosheets/Ni foam composite electrode for long-lived and pH-tolerable electrochemical hydrogen generation. ACS Applied Materials & Interfaces, 2015, 7 (4), 2376-2384.

[277] Popczun E J, Roske C W, Read C G, et al. Highly branched cobalt phosphide nanostructures for hydrogen-evolution electrocatalysis. Journal of Materials Chemistry A, 2015, 3 (10): 5420-5425.
[278] Huang Z, Chen Z, Chen Z, et al. $Ni_{12}P_5$ nanoparticles as an efficient catalyst for hydrogen generation via electrolysis and photoelectrolysis. ACS Nano, 2014, 8 (8): 8121-8129.
[279] Huang Z, Chen Z, Chen Z, et al. Cobalt phosphide nanorods as an efficient electrocatalyst for the hydrogen evolution reaction. Nano Energy, 2014, 9: 373-382.
[280] Shi J, Qiu F, Yuan W, et al. Novel electrocatalyst of nanoporous FeP cubes prepared by fast electrodeposition coupling with acid-etching for efficient hydrogen evolution. Electrochimica Acta, 2020, 329: 135185.
[281] Zhu Y P, Liu Y P, Ren T Z, et al. Self-supported cobalt phosphide mesoporous nanorod arrays: A flexible and bifunctional electrode for highly active electrocatalytic water reduction and oxidation. Advanced Functional Materials, 2015, 25 (47): 7337-7347.
[282] Cao X, Jia D, Li D, et al. One-step co-electrodeposition of hierarchical radial Ni_xP nanospheres on Ni foam as highly active flexible electrodes for hydrogen evolution reaction and supercapacitor. Chemical Engineering Journal, 2018, 348: 310-318.
[283] Kim H, Oh S, Cho E, et al. 3d porous cobalt-iron-phosphorus bifunctional electrocatalyst for the oxygen and hydrogen evolution reactions. ACS Sustainable Chemistry & Engineering, 2018, 6 (5): 6305-6311.
[284] Zhou M, Sun Q, Shen Y, et al. Fabrication of 3D microporous amorphous metallic phosphides for high-efficiency hydrogen evolution reaction. Electrochimica Acta, 2019, 306: 651-659.
[285] Zhu Y P, Xu X, Su H, et al. Ultrafine metal phosphide nanocrystals in situ decorated on highly porous heteroatom-doped carbons for active electrocatalytic hydrogen evolution. ACS Applied Materials & Interfaces, 2015, 7 (51): 28369-28376.
[286] Chung D Y, Jun S W, Yoon G, et al. Large-scale synthesis of carbon-shell-coated FeP nanoparticles for robust hydrogen evolution reaction electrocatalyst. Journal of the American Chemical Society, 2017, 139 (19): 6669-6674.
[287] Wang M Q, Ye C, Liu H, et al. Nanosized metal phosphides embedded in nitrogen-doped porous carbon nanofibers for enhanced hydrogen evolution at all pH values. Angewandte Chemie International Edition, 2018, 57 (7): 1963-1967.
[288] Zhang C, Pu Z, Amiinu I S, et al. Co2p quantum dot embedded N, P dual-doped carbon self-supported electrodes with flexible and binder-free properties for efficient hydrogen evolution reactions. Nanoscale, 2018, 10 (6): 2902-2907.
[289] Ren J T, Chen L, Weng C C, et al. Self-supported mop nanocrystals embedded in N, P-codoped carbon nanofibers via a polymer-confinement route for electrocatalytic hydrogen production. Materials Chemistry Frontiers, 2019, 3 (9): 1872-1881.
[290] Lv X W, Hu Z P, Chen L, et al. Organic-inorganic metal phosphonate-derived nitrogen-doped core-shell Ni_2P nanoparticles supported on Ni foam for efficient hydrogen evolution reaction at all pH values. ACS Sustainable Chemistry & Engineering, 2019, 7 (15): 12770-12778.
[291] Ren J T, Chen L, Weng CC, et al. Ultrafine molybdenum phosphide nanocrystals on a highly porous N, P-codoped carbon matrix as an efficient catalyst for the hydrogen evolution reaction. Materials Chemistry Frontiers, 2018, 2 (11): 1987-1996.
[292] Ren J T, Wang Y S, Chen L, et al. Binary FeNi phosphides dispersed on N, P-doped carbon nanosheets for highly efficient overall water splitting and rechargeable Zn-air

batteries. Chemical Engineering Journal, 2020, 389: 124408.

[293] Lv X, Tian W, Liu Y, et al. Well-defined CoP/Ni_2P nanohybrids encapsulated in a nitrogen-doped carbon matrix as advanced multifunctional electrocatalysts for efficient overall water splitting and zinc-air batteries. Materials Chemistry Frontiers, 2019, 3 (11): 2428-2436.

[294] Yan L, Jiang H, Xing Y, et al. Nickel metal-organic framework implanted on graphene and incubated to be ultrasmall nickel phosphide nanocrystals acts as a highly efficient water splitting electrocatalyst. Journal of Materials Chemistry A, 2018, 6 (4): 1682-1691.

[295] Zhuo J, Cabán-Acevedo M, Liang H, et al. High-performance electrocatalysis for hydrogen evolution reaction using se-doped pyrite-phase nickel diphosphide nanostructures. ACS Catalysis, 2015, 5 (11): 6355-6361.

[296] Anjum M a R, Okyay M S, Kim M, et al. Bifunctional sulfur-doped cobalt phosphide electrocatalyst outperforms all-noble-metal electrocatalysts in alkaline electrolyzer for overall water splitting. Nano Energy, 2018, 53: 286-295.

[297] Kibsgaard J, Jaramillo T F. Molybdenum phosphosulfide: An active, acid-stable, earth-abundant catalyst for the hydrogen evolution reaction. Angewandte Chemie International Edition, 2014, 53 (52): 14433-14437.

[298] Anjum M a R, Bhatt M D, Lee M H, et al. Sulfur-doped dicobalt phosphide outperforming precious metals as a bifunctional electrocatalyst for alkaline water electrolysis. Chemistry of Materials, 2018, 30 (24): 8861-8870.

[299] Men Y, Li P, Yang F, et al. Nitrogen-doped cop as robust electrocatalyst for high-efficiency pH-universal hydrogen evolution reaction. Applied Catalysis B: Environmental, 2019, 253: 21-27.

[300] Wang L, Wu H, Xi S, et al. Nitrogen-doped cobalt phosphide for enhanced hydrogen evolution activity. ACS applied materials & interfaces, 2019, 11 (19): 17359-17367.

[301] Yi J D, Xu R, Chai G L, et al. Cobalt single-atoms anchored on porphyrinic triazine-based frameworks as bifunctional electrocatalysts for oxygen reduction and hydrogen evolution reactions. Journal of Materials Chemistry A, 2019, 7 (3): 1252-1259.

[302] Zhou Q, Shen Z, Zhu C, et al. Nitrogen-doped cop electrocatalysts for coupled hydrogen evolution and sulfur generation with low energy consumption. Advanced Materials, 2018, 30 (27): 1800140.

[303] Tang C, Gan L, Zhang R, et al. Ternary $Fe_xCo_{1-x}P$ nanowire array as a robust hydrogen evolution reaction electrocatalyst with Pt-like activity: Experimental and theoretical insight. Nano Letters, 2016, 16 (10): 6617-6621.

[304] Sun Y, Hang L, Shen Q, et al. Mo doped Ni_2P nanowire arrays: An efficient electrocatalyst for the hydrogen evolution reaction with enhanced activity at all pH values. Nanoscale, 2017, 9 (43): 16674-16679.

[305] Lv X, Hu Z, Ren J, et al. Self-supported al-doped cobalt phosphide nanosheets grown on three-dimensional Ni foam for highly efficient water reduction and oxidation. Inorganic Chemistry Frontiers, 2019, 6 (1): 74-81.

[306] Liu T, Liu D, Qu F, et al. Enhanced electrocatalysis for energy-efficient hydrogen production over cop catalyst with nonelectroactive Zn as a promoter. Advanced Energy Materials, 2017, 7 (15): 1700020.

[307] Xiao X, Tao L, Li M, et al. Electronic modulation of transition metal phosphide via doping as efficient and pH-universal electrocatalysts for hydrogen evolution reaction. Chemical Science, 2018, 9 (7): 1970-1975.

[308] Song J，Zhu C，Xu B Z，et al. Bimetallic cobalt-based phosphide zeolitic imidazolate framework：CoP_x phase-dependent electrical conductivity and hydrogen atom adsorption energy for efficient overall water splitting. Advanced Energy Materials，2017，7（2）：1601555.

[309] Wu Y，Tao X，Qing Y，et al. Cr-doped FeNi-P nanoparticles encapsulated into n-doped carbon nanotube as a robust bifunctional catalyst for efficient overall water splitting. Advanced Materials，2019，31（15）：1900178.

[310] Lu S S，Zhang L M，Dong Y W，et al. Tungsten-doped Ni-Co phosphides with multiple catalytic sites as efficient electrocatalysts for overall water splitting. Journal of Materials Chemistry A，2019，7（28）：16859-16866.

[311] Mo Q，Zhang W，He L，et al. Bimetallic $Ni_{2-x}Co_xP$/N-doped carbon nanofibers：Solid-solution-alloy engineering toward efficient hydrogen evolution. Applied Catalysis B：Environmental，2019，244：620-627.

[312] Man H W，Tsang CS，Li M M J，et al. Transition metal-doped nickel phosphide nanoparticles as electro- and photocatalysts for hydrogen generation reactions. Applied Catalysis B：Environmental，2019，242：186-193.

[313] Wu R，Xiao B，Gao Q，et al. A janus nickel cobalt phosphide catalyst for high-efficiency neutral-pH water splitting. Angewandte Chemie International Edition，2018，57（47）：15445-15449.

[314] Liu Y，Liu S，Wang Y，et al. Ru modulation effects in the synthesis of unique rod-like Ni@Ni_2P-Ru heterostructures and their remarkable electrocatalytic hydrogen evolution performance. Journal of the American Chemical Society，2018，140（8）：2731-2734.

第七章

单原子催化剂

过去几十年，纳米催化和纳米科学迅速发展，合成尺寸均一的单分散纳米晶，有助于揭示纳米催化中的尺寸效应[1-3]。合适的合成方法能够形成多种类型的纳米材料，有助于建立结构与催化性能的关系，发现纳米催化的潜在规则，如协同效应、配体效应和应变效应。将金属本体或者纳米颗粒缩小为单原子，即单原子催化剂（SACs）是一种非常有效的降低催化剂成本、提高催化活性的方法，可以实现金属原子最大化利用[4-11]。其催化性能的增强是因为以下几个因素：①金属粒子的不饱和配位键增加，这些不饱和配位键容易与反应物相互作用[12-14]；②量子尺寸效应，电子受限导致HOMO-LUMO间隙和离散能级分布的扩大[15-18]；③金属与基底之间的强相互作用，确保了电荷转移[19-21]。然而，原子分散的金属原子比相应的纳米团簇或纳米粒子具有更高的表面能，要制备高金属负载量的纯SACs，相对来说较为困难，因此一般将单原子负载在基底上。由单分散的金属原子负载在各种基底上，形成的SACs可以很好地桥联均相和非均相催化剂[22]。SACs的催化中心不仅包括分散的金属原子，还包括邻位原子或基底的官能团，因此获得了更多的活性位点。同时SACs实现了孤立金属原子的高分散性，原子使用率达到100%[6,23-27]。近年来，各种合成方法制备的SACs被广泛应用于电催化分解水体系，如Pt、Ru、Ir、Fe、Co、Ni单原子催化剂。在这些催化剂上，单原子金属与周围配体一起，对HER或OER具有良好的催化活性[28-32]。一般采用HRTEM、EDS、AC-STEM、STM、XPS、XAS等手段进行表征，再结合DFT计算等方式，揭示SACs对HER和OER的结构-性能关系和催化机理，为合成更稳定的催化剂提供指导。目前合成SACs的主要方法有湿化学法[33-41]、热解法[42-49]、沉积法[13,50-54]、光化学还原法[55,56]等。

7.1 SACs用于HER催化

7.1.1 贵金属基SACs用于HER

与传统方法（如水煤气变换反应）相比，HER是较为理想的生产H_2的方法。而Pt基催化剂是最优异的HER催化剂[57-59]。然而，Pt的珍贵和昂贵限制了其大规模的商业应用[60]。为了降低成本，将Pt纳米粒子和团簇缩小到孤立的单个Pt原子，通过这种方法已经制备了许多Pt-SACs用于

HER 催化[61-63]。Pt-SACs 不仅降低了成本，而且由于孤立 Pt 原子的电子结构可调，有助于调节 HER 活性[64-67]。Pt-SACs 中的单个 Pt 原子，通过孤立的 Pt 中心与邻近的 C 原子之间的强相互作用，以及 Pt 中心与 N 原子的配位，被锚定在碳基支撑体的表面、边缘或缺陷上。Wang 等人通过冰-光化学方法将单个 Pt 原子稳定地固定在介孔碳（MC）边缘和缺陷上[68]。如图 7.1(a) 所示，在紫外灯的照射下，将冷冻的 H_2PtCl_6 溶液与 MC 混合，在介孔碳上制备了单原子级 Pt（Pt_1/MC）。EXAFS 表征得出，在 Pt_1/MC 中存在 Pt-C/O（约 2.1Å）而不是 Pt-Pt（约 2.76Å）[图 7.1(b)]，说明单个 Pt 原子分散在 MC 上，与 Pt 二聚体相比，单个 Pt 原子在 MC 的边缘和缺陷上得到了很好的固定 [图 7.1(c)]。Pt_1/MC 催化剂在酸性介质中表现出优异的 HER 催化活性（η_{100} = 65mV，Tafel 斜率 30mV/dec）[图 7.1(d)]，这归因于 MC 上单原子 Pt 可调的电子结构。Laasonwn 等人通过电沉积方法，将单原子 Pt 固定在单壁碳纳米管（SWNTs）表面，提升

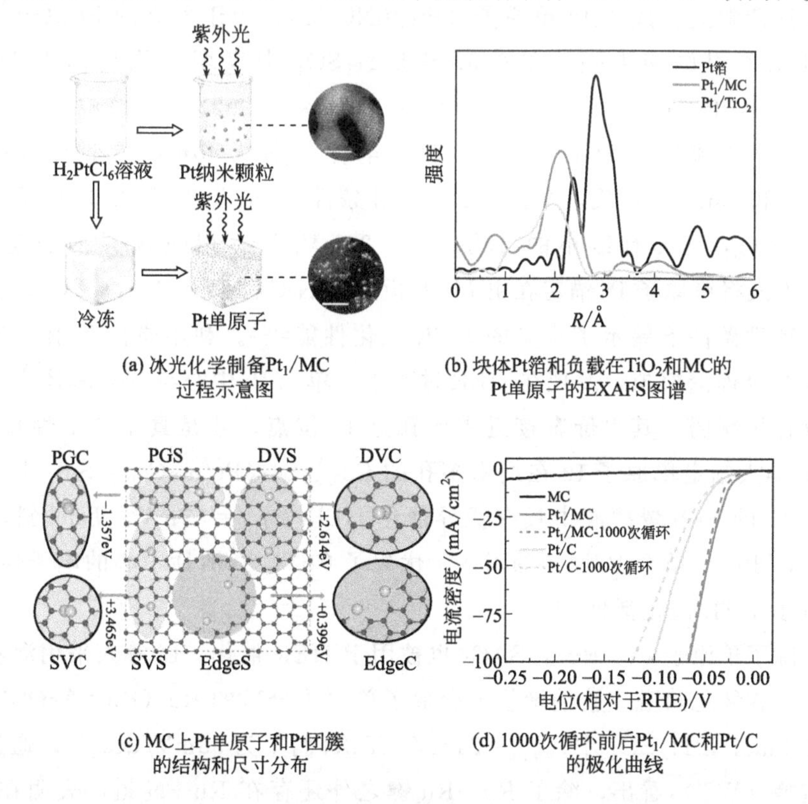

(a) 冰光化学制备Pt_1/MC过程示意图

(b) 块体Pt箔和负载在TiO_2和MC的Pt单原子的EXAFS图谱

(c) MC上Pt单原子和Pt团簇的结构和尺寸分布

(d) 1000次循环前后Pt_1/MC和Pt/C的极化曲线

图 7.1 单原子 Pt_1/MC 的制备过程和相应的表征（见文后彩插）

了 HER 催化性能[69]。

许多研究表明，除了 C 配位的单个 Pt 原子外，锚定在 N 掺杂的碳材料上的 Pt 原子能够提高 HER 动力学，具有优异的催化性能。N 原子很好地调节了单个 Pt 原子的电子结构以促进 HER。Gu 等人采用一种简便的微波还原方法，将单原子 Pt 锚定在苯胺堆积的石墨烯上（Pt SASs/AG），该材料具有优异的 HER 催化性能和稳定性，优于 Pt/C。XAFS 和 DFT 计算表明，单原子 Pt 与苯胺的氮配位优化了 Pt 的电子结构和氢吸附能，促进了 HER 活性[70]。Cheng 等人使用 ALD 方法，在 N 掺杂的石墨烯上合成了原子分散的 Pt，在 0.5mol/L H_2SO_4 中质量活性为 10.1A/mg，表现出优异的 HER 性能[71]。该方法可以通过改变 ALD 循环数来控制 Pt 的尺寸。由于 N 的电负性（3.04）高于 Pt（2.28）和 C（2.55），更多的电子从 Pt 中心转移到 N 原子上，形成更多 Pt 原子的未占据 5d 轨道，导致 H 原子在 Pt 5d 轨道上的化学吸附更强，从而提高 HER 速率。Zhang 等人在N 掺杂碳上合成了 Pt 单原子（Pt_1/hNCNC），由于孤立的 Pt 原子与边缘氮原子配位能力强，在 0.5mol/L H_2SO_4 中表现出优越的 HER 性能[72]。当 Pt 负载量从 0.75%增加到 1.48%、2.92%和 5.68%(质量分数）时，电流密度为 10mA/cm^2 时过电位分别从 40mV 减小到 20mV、15mV 和 13mV。在质量活性方面，酸性条件下 Pt_1/hNCNC-2.92 的活性为 7.60A/mg，且具有最小的 Tafel 斜率 24mV/dec，以及优异的稳定性。Yao 等人将单原子 Pt 锚定在由 MOF 衍生的 N-C 框架上（Pt_1/N-C），在酸性和碱性条件下展示了优异的 HER 催化性能[73]。利用原位 X 射线吸收光谱结合理论模拟，发现在 HER 过程中，单个 Pt 中心与 N-C 基体之间的相互作用减弱，其中价态接近零的孤立 Pt 位点，才是真正的活性位点。Ning 等人锚定单原子 Pt 在纳米多孔的 $Co_{0.85}Se$（Pt/np-$Co_{0.85}Se$）上，在 1mol/L 的 PBS 缓冲液中具有优异催化 HER（η_{10} =55mV，Tafel 斜率为 35mV/dec）。单个 Pt 原子的引入，优化了 $Co_{0.85}Se$ 活性中心的电子性质，提升了 HER 催化活性[74]。

除了单原子 Pt，Ru 基 SACs 也被用于 HER 催化。Li 等人采用湿浸渍方法，在氮化磷酰亚胺纳米管上合成了单原子分散的 Ru（Ru SAs@PN），在 0.5mol/L H_2SO_4 中，η_{10} =24mV，Tafel 斜率为 38mV/dec[75]。通过小波变换（WT）看出，除了 Ru—Ru 键之外还存在 Ru—N 键，表明在 Ru SAs@PN 中没有形成 Ru 纳米颗粒。Ru SAs@PN 中的 P NEXAFS 谱和 ^{31}P

固态 MAS NMR 谱与 PN 的相似，表明 Ru SAs@PN 中 P 骨架非常稳定，Ru 中心与 PN 载体的 N 原子之间相互作用，可以促进 H 原子的强化学吸附以促进 HER。Zhang 等人在室温下通过简单的浸渍方法将单原子 Ru 负载在 MoS_2 上（SA-Ru-MoS_2）作为 HER 催化剂[76]。结果显示，单原子 Ru 掺杂诱导 MoS_2 的相变和 S 空位的产生，有利于提升 HER 的催化性能。Song 等人将单个 Ru 原子结合到 Ni_5P_4 中（Ni_5P_4-Ru），其中 Ni 缺陷和 Ru 阳离子之间的强相互作用有助于降低水还原的能垒[77]，在 1mol/L KOH 中催化 HER，η_{10}＝123mV，Tafel 斜率为 52mV/dec，交换电流密度为 350.9μA/cm^2。

7.1.2 非贵金属基 SACs 用于 HER

非贵金属基的 SACs（诸如 Co、Ni、Fe、Mo 等）也被用于 HER 催化剂[24,26,78-82]。如将这些非贵金属锚定在掺氮的碳材料上，可以得到性能优异的 HER 催化剂。Fei 等人使用水解的方法，将单原子 Co 锚定在 N 掺杂的石墨烯上（Co-NG）[24]，该材料在酸性和碱性条件下都具有优异的催化HER 活性。该工作的表征和电化学测试结合表明，单一的 Co 原子与 N 原子配位对 HER 具有活性，但没有确定 Co 中心的配位特征。有许多报道表明，原子分散的 Co-N_4 位点对 HER 具有高度的活性，可以在一定程度上降低水还原的能垒。Chen 等人通过将 Zn/Co 双金属 MOF 热解，并用酸清洗纯化，得到单原子分散的 Co 基催化剂（Co SACs）[83]，该材料在 0.5mol/L H_2SO_4 溶液中催化 HER，η_{10}＝260mV，Tafel 斜率为 84mV/dec。Cao 等人通过简单的离子热合成，将单原子分散的 Co 均匀地锚定在多孔卟啉三嗪基骨架上（CoSAs/PTF）[84]。通过 XAS 表征，证明有 Co-N_4 活性位点的形成，因此该材料具有优异的催化 HER 性能，η_{10}＝94mV，Tafel 斜率为 50mV/dec。Duan 等人采用微波辅助合成技术，快速地将单原子分散的 Co-N_4 嵌入到石墨烯基质中（Co-NG-MW）[85]，在 0.5mol/L H_2SO_4 溶液中，相比于 Co-G-MW 和 NG-MW，该材料具有更优异的 HER 催化性能（η_{10}＝175mV，Tafel 斜率为 80mV/dec）以及稳定性。

据报道，将 Ni 锚定在 NC 材料上 HER 的活性位点是 Ni 原子与 C 或 N 原子键合。如 Chen 等人将 Ni 原子锚定在纳米多孔石墨烯上（Ni-npG），

其中单个 Ni 原子和少数的 Ni 团簇（1～3nm）嵌入到三维纳米孔中[26]，由于Ni 的 3d 轨道和碳的 sp 轨道之间的电荷转移，该材料在 0.5mol/L H_2SO_4 溶液中，具有优异的 HER 催化活性（η_{10}＝50mV，Tafel 斜率为 45mV/dec，交换电流密度为 0.053mA/cm^2）和良好的稳定性。Fan 等人在石墨碳上制备了 Ni SAs（A-Ni-C），在酸性条件下表现出优异的 HER 性能[86]（η_{10}＝34mV，Tafel 斜率为 41mV/dec，交换电流密度为 1.2mA/cm^2）。Hou 等人通过水热法，制备双氰胺和镍离子的超分子复合物，然后分别进行热解和酸刻蚀合成了一种新型混合电催化剂，它是由锚定在多孔碳上的原子分散的 Ni-N_x 和 Ni 纳米颗粒组成（Ni NP | Ni-N-C）[87]，该材料在碱性条件下具有优异的 HER 催化活性（η_{10}＝147mV，Tafel 斜率为 114mV/dec）。通过 DFT 计算，Ni NP | Ni-N-C 优越的 HER 催化活性是由于 Ni-N-C 与 Ni NP 之间存在较强的相互作用，从而导致构建的界面处电子转移较快。

Huang 等人采用电化学还原方法，将 Fe 和 Ni 原子锚定在石墨二炔上，形成的 Fe/GD 和 Ni/GD 在酸性条下对 HER 表现出较高的活性和稳定性（Fe/GD：η_{10}＝66mV，Tafel 斜率为 37.8mV/dec，交换电流密度 0.29mA/cm^2；Ni/GD：η_{10}＝88mV，Tafel 斜率为 45.8mV/dec，交换电流密度 0.25mA/cm^2）[78]。DFT 计算表明，Fe/GD 和 Ni/GD 优异的 HER 性能，可归因于 Fe（或 Ni）中心与周围 C 原子之间 sp-d 轨道的电荷转移。Li 等人以氮掺杂碳为基底合成了 Mo 和 W SACs 作为 HER 催化剂（$Mo_1N_1C_2$ 和 $W_1N_1C_3$）[79,88]。由于金属中心与周围 N 和 C 原子的强相互作用，使 Mo 和 W SACs 具有优异的 HER 性能。DFT 计算表明，$Mo_1N_1C_2$ 和 $W_1N_1C_3$ 基团独特的结构对 HER 性能起到了至关重要的作用。研究表明，在 MoS_2 载体中引入单分散的金属原子可以提高 MoS_2 对 HER 的催化性能，单个金属原子可以很好地调控电子结构，以提高 HER 活性。

许多单金属原子掺杂 MoS_2 催化剂展示了优异的 HER 催化活性。比如 Wang 等人报道了单原子 Pt 掺杂 MoS_2/NiS_2（Pt@MoS_2/NiS_2），在酸性条件下展现了优异的 HER 催化性能和稳定性（η_{10}＝34mV，Tafel 斜率为 40mV/dec）[89]。Luo 等人合成的单原子 Pd 掺杂的 MoS_2（Pd-MoS_2）在酸性电解液中具有优异的催化 HER 活性（η_{10}＝78mV）[90]。掺杂 Pd 原子可导

致 MoS_2 从 2H 相向 1T 相转变，有利于 HER 过程中界面电子快速转移，从而获得优异的 HER 性能。

此外，非贵金属原子也可以被诱导到 MoS_2 载体上。Gu 等人合成了单原子 Ni 掺杂的 MoS_2（Ni_{SA}-MoS_2/CC），在 1mol/L KOH 和 0.5mol/L H_2SO_4 溶液中具有优异的 HER 催化性能（η_{10} 分别为 98mV 和 110mV，Tafel 斜率分别为 75mV/dec，74mV/dec）[91]。因为单原子 Ni 掺入到 MoS_2 的 S-edge 位点和惰性基面，促进了 HER 反应。Liu 等人通过组装和浸出工艺，合成了 Co 原子掺杂的 MoS_2（SA Co-D 1T MoS_2）[图 7.2(a)]，位于 Mo 原子上方位置的单个 Co 原子与 3 个 S 原子配位，共同形成 Co—S 键，表现出优越的 HER 性能 [图 7.2(b)][92]。在 0.5mol/L H_2SO_4 中，SA Co-D 1T MoS_2 具有较低的起始过电位（42mV）、较小的 Tafel 斜率（32mV/dec）和较好稳定性。DFT 计算表明，除 2H MoS_2 外，单原子 Co 在 1T MoS_2 上优先键合，导致 MoS_2 由 2H 向 1H 相转变。2H 和 1T MoS_2 相的稳定性随着相邻 Co—Co 距离的变化而变化，当相邻 Co 原子之间的键距低于 3.10Å 时，1T MoS_2 相将保持稳定 [图 7.2(c)、(d)]。

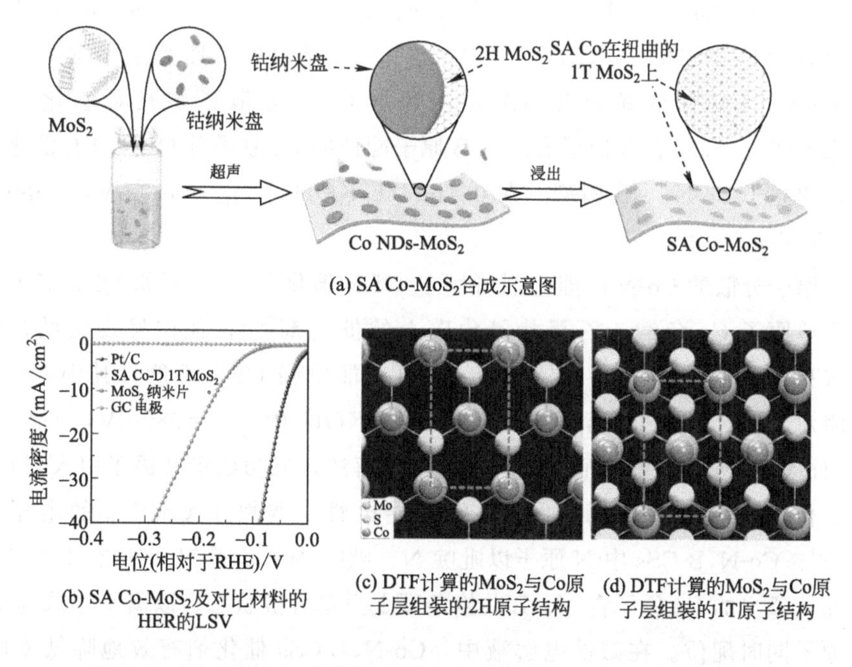

图 7.2　SA Co-MoS_2 的制备过程和相应的表征（见文后彩插）[92]

7.2 SACs用于OER催化

7.2.1 单金属SACs用于OER

OER是电化学技术中的关键反应，作为电解水、电化学还原合成和电池的半反应[93]，其催化路径能垒高，需要较大的过电位。如今，原子分散的非贵金属基（如Fe基、Co基、Ni基、Mn基）催化剂，由于最大化的原子效率而在OER中得到了极大的关注。据报道，与N、C或其他杂原子（如S、B）配位的非贵金属原子，可以极大地提升OER催化活性。并且通过调节配位环境，设计了各种基于非贵金属的SACs。例如，Wu等人制备的原子分散的S、N共掺杂$Fe\text{-}N_x$（S，N-Fe/N/C-CNT），在0.1mol/L KOH中对OER具有良好的催化活性，$\eta_{10}=370mV$，Tafel＝82mV/dec[94]。Hou等人研发了一种Fe-N-C电催化剂，在酸性条件下具有优异的OER催化性能，在质子交换膜电解水槽中具有良好的应用前景[95]。FeN_x/NF/EG中原子分散的$Fe\text{-}N_4$，可以降低OER在0.5mol/L H_2SO_4中的能垒，$\eta_{10}=294mV$，Tafel＝82mV/dec。Zhu等人合成了S掺杂的Fe-N-C催化剂（Fe-NSDC），XRD图谱中Fe-NSDC在26.3°具有强衍射峰，表明Fe-NSDC具有良好的石墨结构，单原子Fe分散在多孔石墨化碳基基底上[96]。与3个N原子和1个S原子配位的Fe原子对OER具有催化活性，建立了$Fe\text{-}N_3|S$构型，在0.1mol/L KOH中$\eta_{10}=410mV$，Tafel＝59mV/dec。

原子分散的Co-N-C催化剂可用于OER催化[97,98]。例如Qiao等人制备了单原子Co负载于石墨状氮化碳上的催化剂[98]，他们发现碳纳米管（CNT）上的$Co\text{-}g\text{-}C_3N_4$（$Co\text{-}C_3N_4$/CNT）能增强OER性能，其中$Co\text{-}N_2$基团对OER具有催化活性，在1mol/L KOH中$\eta_{10}=380mV$，Tafel＝68.4mV/dec。Mu等人通过软模板辅助热解法，成功地将B原子引入到Co-N-C位点中，从而提高了Co-N-C的导电性，调控了Co-N-C的电子结构[99]。Co-N,B-CSs中N原子以吡啶N、吡咯N、石墨N、氧化N的形式存在，孤立的Co原子在N、B共掺杂碳上呈单分散状态，以离子形式与B、N原子同时配位。在碱性电解液中，Co-N,B-CSs催化剂有效地降低OER的能垒，这为杂原子（例如B、S和P原子）掺杂M-N-C催化剂作为高效

OER 催化剂的设计，提供了研究基础。Heumann 等人在碳纳米管上合成了一种功能性聚合离子液体桥接的 Co（CoSSPIL/CNT），其中离子液体（IL）可以调节 Co 配位，并增强 Co 在 CNT 上的分散情况[100]，催化剂的活性与离子液体的含量呈良好的线性相关。

此外，Chen 等人报道了将 Ni SA 负载于 N 掺杂纳米多孔石墨烯上（Ni、N 共掺杂 np-石墨烯），表现出优异的催化活性[101]。从电子显微镜图像来看，单个 Ni 原子和 Ni 团簇分散在 3D 多孔石墨烯基载体上。在碱性条件下，催化剂的 η_{10}=260mV，Tafel=68.4mV/dec，且具有高度的稳定性。Ni、N 共掺杂 np-石墨烯的优异 OER 性能，可归因于多孔结构和被 N 和 C 原子包围的 Ni 原子位点。Wu 等利用双金属离子吸附方法，通过 Ni—O 键合在石墨烯上合成了原子分散的 Ni（Ni-O-G SACs），在 1mol/L KOH 中催化 OER 电流密度为 10mA/cm^2 时过电位为 328mV，Tafel 斜率为 84mV/dec[102]。DFT 计算和原子尺度表征表明，Ni-O-G SACs 的高价态 Ni 位点和 Ni-O 强配位，能够加速电子转移，从而促进 OER 反应。Feng 等人报道与 N 和 S 原子配位的原子分散的 Ni 具有良好的 OER 催化活性[103]。通过热解在电化学剥离石墨烯（EG）上生长三元双氰胺-噻吩-镍配合物，制备了多孔碳（PC）负载的 S | NiN_x 原子。在 1mol/L KOH 溶液中催化 OER，S | NiN_x-PC/EG 催化剂的 η_{10}=280mV，Tafel=45mV/dec。结构表征和 DFT 计算表明，S | NiN_x-PC/EG 优越的 OER 活性，可以归因于 Ni-N_3S 位点的形成。由于 Ni-N_3S 位点的快速电子传递能力，S 的掺杂能增强OER 性能。Wu 等人报道了采用 NaCl 为模板和蔗糖为碳源，合成了 Ni 单原子和碳的 SACs 复合催化剂（Ni-O-G SACs）[图 7.3(a)][104]。多种表征显示，Ni-O-G SACs 由超薄石墨烯片、单个 Ni 原子、Ni 原子与氧配位构筑的三维多孔骨架组成［图 7.3(b)］。因此，该催化剂具有良好的 OER 催化活性和稳定性，η_{10}=224mV，Tafel=42mV/dec，300mV 下氧气产生的 TOF 为 1.44 /s，在大电流密度 115mA/cm^2 下进行 50h 的长时间电解，其催化活性也没有明显的变化［图 7.3(c)、(d)］。

此外，Guan 等人在 N 掺杂石墨烯上，制备了原子分散的 Mn（Mn-NG），其中的 Mn-N_4 位点对水氧化具有很高的活性[105]，在 1mol/L KOH 中，催化 OER 的 η_{10}=337mV，Tafel=55mV/dec。此外，Mn-NG 在 Ce^{IV} 溶液中的 TOF 为 214s^{-1}。

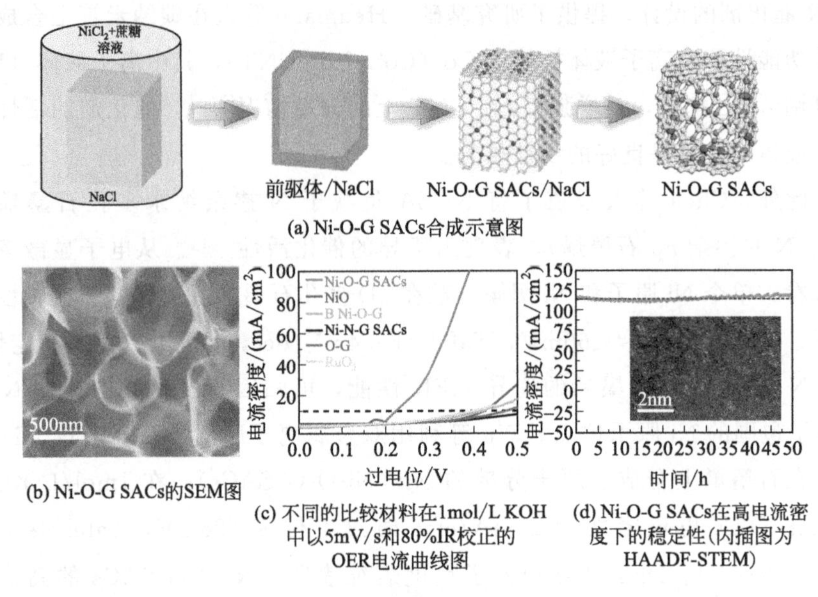

图 7.3 Ni-O-G SACs 的制备过程和相应的表征（见文后彩插）[104]

7.2.2 双金属 SACs 用于 OER

由于双金属原子的协同作用，具有双金属位的原子分散催化剂表现出优越的 OER 性能。Thomas 等人在碳纳米片上采用盐/二氧化硅模板辅助热解合成的原子分散的中/微米 FeCo-N_x-CN 与中/微米 Fe-N_x-CN 相比，具有更高的 OER 催化活性[106]。从元素图谱可以看出，Fe、Co、N 和 C 元素均匀分散在中/微 FeCo-N_x-CN 中。在碱性条件下，催化 OER 的 η_{10} =370mV，Tafel=57mV/dec。Song 等人通过高温聚合的方法在 g-C_3N_4 包覆的 CNT 上合成了原子分散的 Ni 和 Fe（NiFe@g-C_3N_4/CNT）[107]。该双金属催化剂在 1mol/L KOH 中催化 OER 的 η_{10} =326mV，Tafel=67mV/dec 均小于单金属催化剂，NiFe@g-C_3N_4/CNT Fe 和 Ni 位点之间的部分电荷转移促进了 OER 的催化活性。Zhang 等人用不同的 Ni 前驱体 1,10 -菲咯啉镍（Ⅱ）硝酸盐（PNi）和丁二酮肟镍（Ⅱ）（DNi）制备了原子分散的 CoPNi-N/C 和 CoDNi-N/C 催化剂［图 7.4(a)］[108]。从拉曼光谱可以看出，CoPNi-N/C 和 CoDNi-N/C 中存在低石墨化和高缺陷［图 7.4(b)］，且存在原子分散的 Co/Ni 双位点和 Co/Ni 纳米颗粒［图 7.4(c)、(d)］。在 0.1mol/L KOH 中，CoDNi-N/C 在 10mA/cm^2 处的过电位为 360mV，

Tafel 斜率为 72mV/dec。由于 Ni—N 键的引入，与单原子 Co-N/C 催化剂相比，CoDNi-N/C 的 OER 性能更优异［图 7.4(e)］。

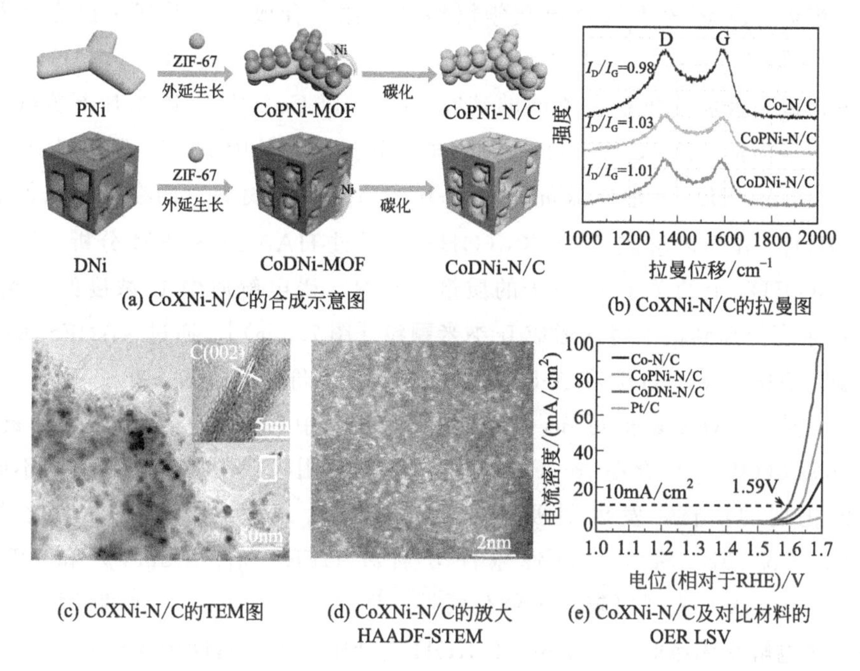

图 7.4 CoXNi-N/C 的制备过程和相应的表征（见文后彩插）[108]

7.3 双功能 SACs 用于电解水

迄今为止，贵金属基 SACs 作为双功能催化剂，对 HER 和 OER 显示出很高的催化活性。Du 等人采用 DFT 计算，考察了石墨二炔负载的各种 SACs 的电催化性能，结果表明石墨二炔负载的原子分散的 Pt，对 HER 和 OER 都具有很高的催化活性[67]。理论研究表明，Pt SACs 在电解水中具有巨大的应用潜力，石墨二炔可以作为锚定各种单一金属原子的载体。Chou 和同事采用了一种 π 电子辅助的方法，在杂化支撑体上负载各种原子金属（Ir、Pt、Ru、Pd、Fe、Ni）。通过这种方法，孤立的金属原子可以稳定在支撑体的两个域上，即四重 N/C 原子（M@NC）和 Co 八面体中心（M@Co）[109]。通过 DFT 计算，发现 Ir_1@Co/NC（下标 1 代表为单原子）对 HER 和 OER 的活性位点不同。Ir@Co(Ir) 和 Ir@NC_3 位点分别对 OER 和

HER 具有高催化活性。在 1mol/L KOH 中，Ir_1@Co/NC 催化电解水需要 1.603V(@10mA/cm^2，相对于 RHE) 的外加电位。

最近，Zeng 等人通过电化学沉积的方法，合成了一系列过渡金属基 SACs [图 7.5(a)][51]，其中 Ru、Rh、Pd、Ag、Ir、Pt、Au 等 SACs，通过阴极或阳极反应沉积在不同的基底上[51]。以 $Co(OH)_2$ 纳米片上负载的 Ir 单原子为例 [$Ir_1/Co(OH)_2$] 阐述电沉积过程，其中 $IrCl^{3+}$ 驱动到阴极，负载在 $Co(OH)_2$ 上形成低价 Ir 的 C-$Ir_1/Co(OH)_2$ 材料，而在碱性条件下 $Ir(OH)_6^{2-}$ 在阳极形成 A-$Ir_1/Co(OH)_2$。通过 HAADF-STEM 分析，沉积在 $Co(OH)_2$ 上的单个 Ir 原子的质量负载量，受电解质中 Ir 浓度的影响。当 Ir 浓度提高到极限时，形成 Ir 纳米颗粒 [图 7.5(a)]。通过 XANES 表征发现，C-$Ir_1/Co(OH)_2$ 和 A-$Ir_1/Co(OH)_2$ 的 Ir 价态分别为＋3 ～＋4 和＋4 以上。从 EXAFS 表征可知，C-$Ir_1/Co(OH)_2$ 中存在 Ir—O 和 Ir—Cl 键，A-$Ir_1/Co(OH)_2$ 中存在 Ir—O 键。通过比较在阴极或阳极沉积制备的不同电催化剂的活性，发现阴极制备的 C-$Ir_1/Co_{0.8}Fe_{0.2}Se_2$ 和阳极制备的 A-Ir_1/ $Co_{0.8}Fe_{0.2}Se_2$，在碱性条件分别对 HER（η_{10} ＝ 8mV） 和 OER（η_{10}＝230mV）表现出较高的催化活性 [图 7.5(b)、(c)]。将两种催化剂组成全电解水体系时，在 1mol/L KOH 溶液中，电流密度为 10mA/cm^2 时需要提供 1.48V（相对于 RHE）的电压。

为了进一步提升电解水的催化活性，在 $Co_{0.8}Fe_{0.2}Se_2$@Ni 上沉积 Ir 原子，制备了 $Ir_1/Co_{0.8}Fe_{0.2}Se_2$@ Ni 催化剂，其在 10mA/$cm^2$ 下的 HER 和 OER 的过电位分别降至 4mV 和 140mV，在 1mol/L KOH 中电解水的电位仅为 1.39V(相对于 RHE)，且具有优异的稳定性。此外，Guan 等人报道了在 N 掺杂的石墨烯上，负载 Ru 单原子（Ru@NG)，对 HER 和 OER 都表现出优越的性能。通过 DFT 计算可知，RuN_4C_x 位点的表面状态，在不同的反应条件下发生不同的变化，在 HER 和 OER 过程中，最稳定的状态分别是 RuN_4C_x 和 $RuN_4(O)C_x$[110]。

此外，过渡金属基 SACs 作为双功能电解水催化剂也表现出优异的性能。比如，由于 Fe 原子和相邻的 N 原子配位，原子分散的 Fe-N-C 展示了优异的 HER 和 OER 催化活性和稳定性。Pan 等人通过热解一种 MOF，合成了原子分散的 Fe-N-C 催化剂（Fe-N_4 SAs/NPC)。该催化剂是一种高效、稳定的 HER 和 OER 双功能催化剂[111]。Fe 原子均匀分散在碳基支撑体上，Fe-N_4 SAs/NPC 中 Fe 的平均价态为＋2～＋3。在 1mol/L KOH 溶液中，

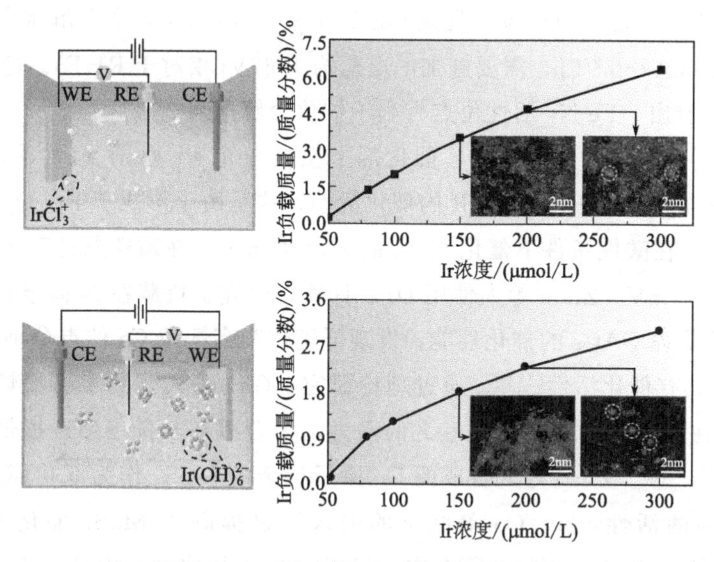

(a) 电沉积Ir基SACs合成示意图及Ir负载量和Ir浓度的关系图

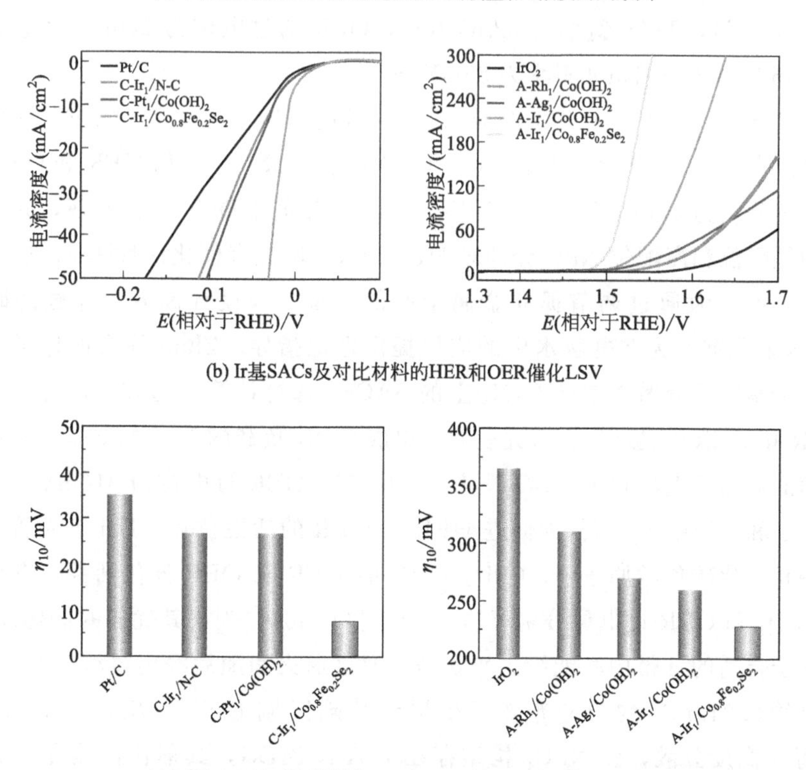

(b) Ir基SACs及对比材料的HER和OER催化LSV

(c) Ir基SACs及对比材料的HER和OER在$10mA/cm^2$的过电位

图 7.5　Ir 基 SACs 双功能催化剂的制备过程和相应的表征（见文后彩插）[51]

催化 HER 的 η_{10} =202mV，催化 OER 的 η_{10} =430mV，全电解水在电流密度达到 10mA/cm^2 时，需要施加的电压为 1.67V(相对于 RHE)。通过 DFT 计算可以看出，Fe-N_4 活性位点有利于催化分解水。

Yu 等人发现，在 Fe-N-C 催化剂上接枝单个 Pt 原子（Pt_1@Fe-N-C）可以明显提高对 HER 和 OER 的催化活性，这是由于形成了 Pt_1-O_2-Fe_1-N_4 基团[112]。在酸性条件下催化 HER 的 η_{10} =60mV，在碱性条件下催化 OER 的 η_{10} =310mV。Zhou 等人使用 DFT 计算，研究了负载在 N 掺杂石墨烯上的过渡金属基 SACs 的催化性能，发现低配位和高配位 Co 位点分别对 HER 和 OER 具有催化活性[113]。通过理论研究，Gao 等人预测了石墨烯负载的 Co 原子在催化电解水上具有巨大的潜力，为合成双功能 SACs 提供了重要的指导[114,115]。Zhao 等人合成的 Co 原子掺杂的 MoS_2 作为双功能催化剂，具有很高的活性[116]。Co 单原子的引入明显提高了 MoS_2 催化 HER 和 OER 性能。在 1mol/L KOH 中，HER 的过电位为 48mV（电流密度 10mA/cm^2），Tafel 斜率为 52mV/dec，OER 的过电位为 260mV（电流密度 10mA/cm^2），Tafel 斜率为 85mV/dec。

Ni SACs 在催化电解水的应用也有报道。Wang 等人通过 DFT 计算预测分散在单层 β_{12} 硼单层上的 Ni 原子（Ni1/β_{12}-BM），催化 HER 和 OER 的过电位分别为 60mV 和 400mV[117]。Gao 等人通过 DFT 计算发现，原子分散的低配位和高配位 Ni-C 分别对 HER 和 OER 具有催化活性[118]。这样的研究，可以为通过调节孤立金属中心的环境，来设计高效、稳健的碳基 SACs 催化剂，为在电解水中的应用提供理论指导。Zhou 等人设计了各种过渡金属原子分散在单层 C_9N_4 上的 SACs。通过计算筛选方法，研究了其 HER 和 OER 性能[119]。研究表明，单层 C_9N_4 负载的 Ni SACs 是一种很有前途的双功能电催化剂，ΔG_{H^*} 为−0.04eV，OER 过电位为 310mV。其中 N 原子和 Ni 原子分别作为促进 HER 和 OER 的活性位点。Dai 等人研究了石墨相氮化碳负载原子 Ni（Ni/g-CN）的 HER 和 OER 催化活性，发现理论 HER 和 OER 过电位分别可达 0.12V 和 0.40V[120]，其优异的催化性能，归因于合适的 d 带中心能级。Yao 等人通过退火和随后浸出过程，将 Ni 原子（约 0.24nm）均匀分布在石墨烯的缺陷区域上[121]。XAS 表征显示，Ni 原子高度分散，以 Ni-C_4 构型存在于 A-Ni@DG，其催化性能优于 DG、Ni@DG 和 Ir/C。

由于 SACs 具有最大化原子利用率、高选择性和高活性等显著特点，在电解水中具有良好的应用前景。然而，原子分散的金属原子具有较高的表面自由能，在制备过程中容易团聚。因此，合成方法在合理设计和理解 SACs 机理方面发挥着重要作用。要合成优异的 SACs 催化剂，载体材料的性能特征也很重要。充分利用载体的特性（孤对、缺陷等），通过单个金属原子与载体之间强相互作用，可稳定载体上金属单原子。然而，目前对 SACs 催化剂的研究仍处于探索阶段[122-124]，大规模的生产仍面临诸多挑战。比如：①增加载体上的单原子金属质量负载。随着单原子质量负载量的增加，活性位点的数量将会增加。由于单原子金属的表面能很高，要实现高负载单原子金属并不容易。一般可以通过探索具有丰富缺陷或配位位点的合适载体，以稳定更多孤立的金属原子，防止其团聚。另外还可以开发高比表面积的 3D 多孔载体，以加速传质，并最大限度地发挥可用的活性位点。可以预见的是，在构建和功能化载体表面产生的重大突破，可以促进 SACs 的大规模合成和应用。因此，迫切需要合成方法来制备高质量负载 SACs 的载体。②为了精确验证活性物质的结构，并表征孤立金属原子的配位环境，需要先进的结构表征技术。目前，很难通过一种技术来表征单个金属原子的配位环境。此外，为了进一步了解催化反应过程中，孤立单原子的氧化态和结合模式，也迫切需要原位表征技术来探索 SACs 的催化机理。③理解单个金属原子的催化机理。与纳米颗粒催化剂相比，由于缺乏金属-金属键，SAC 可以通过不同的途径催化 OER 和/或 HER 反应。应正确评估金属单原子和相邻原子在反应机制中的作用。当然，了解催化机理将有助于我们提高 SACs 的催化活性和选择性。

参 考 文 献

[1] Hvolbæk B, Janssens T V W, Clausen B S, et al. Catalytic activity of au nanoparticles. Nano Today, 2007, 2 (4): 14-18.

[2] Wang X, Peng Q, Li Y. Interface-mediated growth of monodispersed nanostructures. Accounts of Chemical Research, 2007, 40 (8): 635-643.

[3] Kuhn J N, Huang W, Tsung C K, et al. Structure sensitivity of carbon-nitrogen ring opening: Impact of platinum particle size from below 1 to 5nm upon pyrrole hydrogenation product selectivity over monodisperse platinum nanoparticles loaded onto mesoporous silica. Journal of the American Chemical Society, 2008, 130 (43): 14026-14027.

[4] Hunt S T, Milina M, Wang Z, et al. Activating earth-abundant electrocatalysts for efficient, low-cost hydrogen evolution/oxidation: Sub-monolayer platinum coatings on titanium tungsten carbide nanoparticles. Energy & Environmental Science, 2016, 9 (10): 3290-3301.

[5] Yoo E, Okata T, Akita T, et al. Enhanced electrocatalytic activity of Pt subnanoclusters on

graphene nanosheet surface. Nano Letters, 2009, 9 (6): 2255-2259.

[6] Deng J, Li H, Xiao J, et al. Triggering the electrocatalytic hydrogen evolution activity of the inert two-dimensional MoS_2 surface via single-atom metal doping. Energy & Environmental Science, 2015, 8 (5): 1594-1601.

[7] Liu L, Corma A. Metal catalysts for heterogeneous catalysis: From single atoms to nanoclusters and nanoparticles. Chemical Reviews, 2018, 118 (10): 4981-5079.

[8] Wang Y, Mao J, Meng X, et al. Catalysis with two-dimensional materials confining single atoms: Concept, design, and applications. Chemical Reviews, 2019, 119 (3): 1806-1854.

[9] Zhang L, Zhou M, Wang A, et al. Selective hydrogenation over supported metal catalysts: From nanoparticles to single atoms. Chemical Reviews, 2020, 120 (2): 683-733.

[10] Li Z, Ji S, Liu Y, et al. Well-defined materials for heterogeneous catalysis: From nanoparticles to isolated single-atom sites. Chemical Reviews, 2020, 120 (2): 623-682.

[11] Wan G, Zhang G, Lin X M. Toward efficient carbon and water cycles: Emerging opportunities with single-site catalysts made of 3d transition metals. Advanced Materials, 2020, 32 (2): 1905548.

[12] Chen W, Chen S. Oxygen electroreduction catalyzed by gold nanoclusters: Strong core size effects. Angewandte Chemie International Edition, 2009, 48 (24): 4386-4389.

[13] Zhang H, Liu G, Shi L, et al. Single-atom catalysts: Emerging multifunctional mater-ials in heterogeneous catalysis. Advanced Energy Materials, 2018, 8 (1): 1701343.

[14] Lopez N, Janssens T V W, Clausen B S, et al. On the origin of the catalytic acti-vity of gold nanoparticles for low-temperature Co oxidation. Journal of Catalysis, 2004, 223 (1): 232-235.

[15] Liu J. Catalysis by supported single metal atoms. ACS Catalysis, 2017, 7 (1): 34-59.

[16] Valden M, Lai X, Goodman D W. Onset of catalytic activity of gold clusters on titania with the appearance of nonmetallic properties. Science, 1998, 281 (5383): 1647-1650.

[17] Li J, Li X, Zhai HJ, et al. Au_{20}: A tetrahedral cluster. Science, 2003, 299 (5608): 864-867.

[18] Roduner E. Size matters: Why nanomaterials are different. Chemical Society Reviews, 2006, 35 (7): 583-592.

[19] Yoon B, Häkkinen H, Landman U, et al. Charging effects on bonding and catalyzed oxidation of Co on Au_8 clusters on MgO. Science, 2005, 307 (5708): 403-407.

[20] Liu J. Advanced electron microscopy of metal-support interactions in supported metal catalysts. Chem Cat Chem, 2011, 3 (6): 934-948.

[21] Campbell C T. Electronic perturbations. Nature Chemistry, 2012, 4 (8): 597-598.

[22] Liu J C, Wang Y G, Li J. Toward rational design of oxide-supported single-atom catalysts: Atomic dispersion of gold on ceria. Journal of the American Chemical Society, 2017, 139 (17): 6190-6199.

[23] Huang Z, Gu X, Cao Q, et al. Catalytically active single-atom sites fabricated from silver particles. Angewandte Chemie International Edition, 2012, 51 (17): 4198-4203.

[24] Fei H, Dong J, Arellano-Jiménez M J, et al. Atomic cobalt on nitrogen-doped graphene for hydrogen generation. Nature Communications, 2015, 6 (1): 8668.

[25] Hackett S F J, Brydson R M, Gass M H, et al. High-activity, single-site mesoporous Pd/Al_2O_3 catalysts for selective aerobic oxidation of allylic alcohols. Angewandte Chemie International Edition, 2007, 46 (45): 8593-8596.

[26] Qiu H J, Ito Y, Cong W, et al. Nanoporous graphene with single-atom nickel dopants: An efficient and stable catalyst for electrochemical hydrogen production. Angewandte Chemie International Edition, 2015, 54 (47): 14031-14035.

[27] Wang X, Li Z, Qu Y, et al. Review of metal catalysts for oxygen reduction reaction: From

nanoscale engineering to atomic design. Chem, 2019, 5 (6): 1486-1511.

[28] Yang X F, Wang A, Qiao B, et al. Single-atom catalysts: A new frontier in heterogeneous catalysis. Accounts of Chemical Research, 2013, 46 (8): 1740-1748.

[29] Zhao D, Zhuang Z, Cao X, et al. Atomic site electrocatalysts for water splitting, oxygen reduction and selective oxidation. Chemical Society Reviews, 2020, 49 (7): 2215-2264.

[30] Zeng Z, Su Y, Quan X, et al. Single-atom platinum confined by the interlayer nanospace of carbon nitride for efficient photocatalytic hydrogen evolution. Nano Energy, 2020, 69: 104409.

[31] Cheng Y, He S, Veder J P, et al. Atomically dispersed bimetallic FeNi catalysts as highly efficient bifunctional catalysts for reversible oxygen evolution and oxygen reduction reactions. ChemElectroChem, 2019, 6 (13): 3478-3487.

[32] Yan H, Lin Y, Wu H, et al. Bottom-up precise synthesis of stable platinum dimers on graphene. Nature Communications, 2017, 8 (1): 1070.

[33] Ye L, Duan X, Wu S, et al. Self- regeneration of Au/CeO_2 based catalysts with enhanced activity and ultra-stability for acetylene hydrochlorination. Nature Communications, 2019, 10 (1): 914.

[34] Wei H, Liu X, Wang A, et al. FeO_x-supported platinum single-atom and pseudo-single-atom catalysts for chemoselective hydrogenation of functionalized nitroarenes. Nature Communications, 2014, 5 (1): 5634.

[35] Zhang X, Sun Z, Wang B, et al. C—C coupling on single-atom-based heterogeneous catalyst. Journal of the American Chemical Society, 2018, 140 (3): 954-962.

[36] Guo L W, Du P P, Fu X P, et al. Contributions of distinct gold species to catalytic reactivity for carbon monoxide oxidation. Nature Communications, 2016, 7 (1): 13481.

[37] Xin P, Li J, Xiong Y, et al. Revealing the active species for aerobic alcohol oxidation by using uniform supported palladium catalysts. Angewandte Chemie International Edition, 2018, 57 (17): 4642-4646.

[38] Huang W, Zhang S, Tang Y, et al. Low-temperature transformation of methane to methanol on Pd_1O_4 single sites anchored on the internal surface of microporous silicate. Angewandte Chemie International Edition, 2016, 55 (43): 13441-13445.

[39] Matsubu J C, Yang V N, Christopher P. Isolated metal active site concentration and stability control catalytic CO_2 reduction selectivity. Journal of the American Chemical Society, 2015, 137 (8): 3076-3084.

[40] Zhao S, Chen F, Duan S, et al. Remarkable active-site dependent H_2O promoting effect in Co oxidation. Nature Communications, 2019, 10 (1): 3824.

[41] Sun G, Zhao Z J, Mu R, et al. Breaking the scaling relationship via thermally stable Pt/Cu single atom alloys for catalytic dehydrogenation. Nature Communications, 2018, 9 (1): 4454.

[42] Su J, Ge R, Dong Y, et al. Recent progress in single-atom electrocatalysts: Concept, synthesis, and applications in clean energy conversion. Journal of Materials Chemistry A, 2018, 6 (29): 14025-14042.

[43] Zhang C, Sha J, Fei H, et al. Single-atomic ruthenium catalytic sites on nitrogen-doped graphene for oxygen reduction reaction in acidic medium. ACS Nano, 2017, 11 (7): 6930-6941.

[44] Fu X, Zamani P, Choi J Y, et al. In situ polymer graphenization ingrained with nanoporosity in a nitrogenous electrocatalyst boosting the performance of polymer-electrolyte-membrane fuel cells. Advanced Materials, 2017, 29 (7): 1604456.

[45] He X, He Q, Deng Y, et al. A versatile route to fabricate single atom catalysts with high che-

moselectivity and regioselectivity in hydrogenation. Nature Communications, 2019, 10 (1): 3663.

[46] Wang J, Huang Z, Liu W, et al. Design of n-coordinated dual-metal sites: A stable and active Pt-free catalyst for acidic oxygen reduction reaction. Journal of the American Chemical Society, 2017, 139 (48): 17281-17284.

[47] Peng Y, Lu B, Chen S. Carbon-supported single atom catalysts for electrochemical energy conversion and storage. Advanced Materials, 2018, 30 (48): 1801995.

[48] Liang Z, Qu C, Xia D, et al. Atomically dispersed metal sites in MOF-based materials for electrocatalytic and photocatalytic energy conversion. Angewandte Chemie International Edition, 2018, 57 (31): 9604-9633.

[49] Zhu C, Shi Q, Xu B Z, et al. Hierarchically porous M-N-C(M=Co and Fe) single-atom electrocatalysts with robust Mn_x active moieties enable enhanced ORR performance. Advanced Energy Materials, 2018, 8 (29): 1801956.

[50] Piernavieja-Hermida M, Lu Z, White A, et al. Towards ALD thin film stabilized single-atom Pd_1 catalysts. Nanoscale, 2016, 8 (33): 15348-15356.

[51] Zhang Z, Feng C, Liu C, et al. Electrochemical deposition as a universal route for fabricating single-atom catalysts. Nature Communications, 2020, 11 (1): 1215.

[52] Cai Z, Liu B, Zou X, et al. Chemical vapor deposition growth and applications of two-dimensional materials and their heterostructures. Chemical Reviews, 2018, 118 (13): 6091-6133.

[53] Liu J C, Xiao H, Li J. Constructing high-loading single-atom/cluster catalysts via an electrochemical potential window strategy. Journal of the American Chemical Society, 2020, 142 (7): 3375-3383.

[54] Grillo F, Van Bui H, La Zara D, et al. From single atoms to nanoparticles: Autocatalysis and metal aggregation in atomic layer deposition of Pt on TiO_2 nanopowder. Small, 2018, 14 (23): 1800765.

[55] Liu P, Zhao Y, Qin R, et al. Photochemical route for synthesizing atomically dispersed palladium catalysts. Science, 2016, 352 (6287): 797-800.

[56] Wei H, Wu H, Huang K, et al. Ultralow-temperature photochemical synthesis of atomically dispersed Pt catalysts for the hydrogen evolution reaction. Chemical Science, 2019, 10 (9): 2830-2836.

[57] Liu M, Wang L, Zhao K, et al. Atomically dispersed metal catalysts for the oxygen reduction reaction: Synthesis, characterization, reaction mechanisms and electrochemical energy applications. Energy & Environmental Science, 2019, 12 (10): 2890-2923.

[58] Sheng W, Myint M, Chen J G, et al. Correlating the hydrogen evolution reaction activity in alkaline electrolytes with the hydrogen binding energy on monometallic surfaces. Energy & Environmental Science, 2013, 6 (5): 1509-1512.

[59] Zhu Y, Peng W, Li Y, et al. Modulating the electronic structure of single-atom catalysts on 2D nanomaterials for enhanced electrocatalytic performance. Small Methods, 2019, 3 (9): 1800438.

[60] Shi Y, Zhang B. Recent advances in transition metal phosphide nanomaterials: Synthesis and applications in hydrogen evolution reaction. Chemical Society Reviews, 2016, 45 (6): 1529-1541.

[61] Yin X P, Wang H J, Tang S F, et al. Engineering the coordination environment of single-atom platinum anchored on graphdiyne for optimizing electrocatalytic hydrogen evolution. Angewandte Chemie International Edition, 2018, 57 (30): 9382-9386.

[62] Zhang H, An P, Zhou W, et al. Dynamic traction of lattice-confined platinum atoms into me-

soporous carbon matrix for hydrogen evolution reaction. Science Advances, 2018, 4 (1): eaao6657.
[63] Liu D, Li X, Chen S, et al. Atomically dispersed platinum supported on curved carbon supports for efficient electrocatalytic hydrogen evolution. Nature Energy, 2019, 4 (6): 512-518.
[64] Zhang L, Han L, Liu H, et al. Potential-cycling synthesis of single platinum atoms for efficient hydrogen evolution in neutral media. Angewandte Chemie International Edition, 2017, 56 (44): 13694-13698.
[65] Zhang J, Zhao Y, Guo X, et al. Single platinum atoms immobilized on an Mxene as an efficient catalyst for the hydrogen evolution reaction. Nature Catalysis, 2018, 1 (12): 985-992.
[66] Li T, Liu J, Song Y, et al. Photochemical solid-phase synthesis of platinum single atoms on nitrogen-doped carbon with high loading as bifunctional catalysts for hydrogen evolution and oxygen reduction reactions. ACS Catalysis, 2018, 8 (9): 8450-8458.
[67] He T, Matta S K, Will G, et al. Transition-metal single atoms anchored on graphdiyne as high-efficiency electrocatalysts for water splitting and oxygen reduction. Small Methods, 2019, 3 (9): 1800419.
[68] Wei H, Huang K, Wang D, et al. Iced photochemical reduction to synthesize atomically dispersed metals by suppressing nanocrystal growth. Nature Communications, 2017, 8 (1): 1490.
[69] Tavakkoli M, Holmberg N, Kronberg R, et al. Electrochemical activation of single-walled carbon nanotubes with pseudo-atomic-scale platinum for the hydrogen evolution reaction. ACS Catalysis, 2017, 7 (5): 3121-3130.
[70] Ye S, Luo F, Zhang Q, et al. Highly stable single pt atomic sites anchored on aniline-stacked graphene for hydrogen evolution reaction. Energy & Environmental Science, 2019, 12 (3): 1000-1007.
[71] Cheng N, Stambula S, Wang D, et al. Platinum single-atom and cluster catalysis of the hydrogen evolution reaction. Nature Communications, 2016, 7 (1): 13638.
[72] Zhang Z, Chen Y, Zhou L, et al. The simplest construction of single-site catalysts by the synergism of micropore trapping and nitrogen anchoring. Nature Communications, 2019, 10 (1): 1657.
[73] Fang S, Zhu X, Liu X, et al. Uncovering near-free platinum single-atom dynamics during electrochemical hydrogen evolution reaction. Nature Communications, 2020, 11 (1): 1029.
[74] Jiang K, Liu B, Luo M, et al. Single platinum atoms embedded in nanoporous cobalt selenide as electrocatalyst for accelerating hydrogen evolution reaction. Nature Communica-tions, 2019, 10 (1): 1743.
[75] Yang J, Chen B, Liu X, et al. Efficient and robust hydrogen evolution: Phosphorus nitride imide nanotubes as supports for anchoring single ruthenium sites. Angewandte Chemie International Edition, 2018, 57 (30): 9495-9500.
[76] Zhang J, Xu X, Yang L, et al. Single-atom Ru doping induced phase transition of MoS_2 and s vacancy for hydrogen evolution reaction. Small Methods, 2019, 3 (12): 1900653.
[77] He Q, Tian D, Jiang H, et al. Achieving efficient alkaline hydrogen evolution reaction over a Ni_5P_4 catalyst incorporating single-atomic Ru sites. Advanced Materials, 2020, 32 (11): 1906972.
[78] Xue Y, Liu Y, Liao H, et al. Evaluation of electrophysiological mechanisms of post-surgical atrial tachycardias using an automated ultra-high-density mapping system. JACC: Clinical Electrophysiology, 2018, 4 (11): 1460-1470.

[79] Chen W, Pei J, He C T, et al. Single tungsten atoms supported on MOF-derived N-doped carbon for robust electrochemical hydrogen evolution. Advanced Materials, 2018, 30 (30): 1800396.
[80] Sun T, Zhao S, Chen W, et al. Single-atomic cobalt sites embedded in hierarchically ordered porous nitrogen-doped carbon as a superior bifunctional electrocatalyst. Proceedings of the National Academy of Sciences, 2018, 115 (50): 12692-12697.
[81] Song X, Zhang H, Yang Y, et al. Bifunctional nitrogen and cobalt codoped hollow carbon for electrochemical syngas production. Advanced Science, 2018, 5 (7): 1800177.
[82] Wen X, Bai L, Li M, et al. Atomically dispersed cobalt- and nitrogen-codoped graphene toward bifunctional catalysis of oxygen reduction and hydrogen evolution reactions. ACS Sustainable Chemistry & Engineering, 2019, 7 (10): 9249-9256.
[83] Zhao W, Wan G, Peng C, et al. Key single-atom electrocatalysis in metal-organic framework (mof)-derived bifunctional catalysts. ChemSusChem, 2018, 11 (19): 3473-3479.
[84] Yi JD, Xu R, Chai GL, et al. Cobalt single-atoms anchored on porphyrinic triazine-based frameworks as bifunctional electrocatalysts for oxygen reduction and hydrogen evolution reactions. Journal of Materials Chemistry A, 2019, 7 (3): 1252-1259.
[85] Fei H, Dong J, Wan C, et al. Microwave-assisted rapid synthesis of graphene-supported single atomic metals. Advanced Materials, 2018, 30 (35): 1802146.
[86] Fan L, Liu P F, Yan X, et al. Atomically isolated nickel species anchored on graphitized carbon for efficient hydrogen evolution electrocatalysis. Nature Communications, 2016, 7 (1): 10667.
[87] Lei C, Wang Y, Hou Y, et al. Efficient alkaline hydrogen evolution on atomically dispersed Ni-N_x species anchored porous carbon with embedded Ni nanoparticles by accelerating water dissociation kinetics. Energy & Environmental Science, 2019, 12 (1): 149-156.
[88] Chen W, Pei J, He C T, et al. Rational design of single molybdenum atoms anchored on n-doped carbon for effective hydrogen evolution reaction. Angewandte Chemie International Edition, 2017, 56 (50): 16086-16090.
[89] Guan Y, Feng Y, Wan J, et al. Ganoderma-like MoS_2/NiS_2 with single platinum atoms doping as an efficient and stable hydrogen evolution reaction catalyst. Small, 2018, 14 (27): 1800697.
[90] Luo Z, Ouyang Y, Zhng H, et al. Chemically activating MoS_2 via spontaneus atomic palladium interfacial doping towards efficient hydrogen evolution. Nature Communications, 2018, 9 (1): 2120.
[91] Wang Q, Zhao Z L, Dong S, et al. Design of active nickel single-atom decorated MoS_2 as a pH-universal catalyst for hydrogen evolution reaction. Nano Energy, 2018, 53: 458-467.
[92] Qi K, Cui X, Gu L, et al. Single-atom cobalt array bound to distorted 1T MoS_2 with ensemble effect for hydrogen evolution catalysis. Nature Communications, 2019, 10 (1): 5231.
[93] Suen N T, Hung S F, Quan Q, et al. Electrocatalysis for the oxygen evolution reaction: Recent development and future perspectives. Chemical Society Reviews, 2017, 46 (2): 337-365.
[94] Chen P, Zhou T, Xing L, et al. Atomically dispersed iron-nitrogen species as electrocatalysts for bifunctional oxygen evolution and reduction reactions. Angewandte Chemie International Edition, 2017, 56 (2): 610-614.
[95] Lei C, Chen H, Cao J, et al. Fe-n_4 sites embedded into carbon nanofiber integrated with electrochemically exfoliated graphene for oxygen evolution in acidic medium. Advanced Energy Materials, 2018, 8 (26): 1801912.

[96] Zhang J, Zhang M, Zeng Y, et al. Single Fe atom on hierarchically porous S, N-codoped nanocarbon derived from porphyra enable boosted oxygen catalysis for rechargeable Zn-air batteries. Small, 2019, 15 (24): 1900307.

[97] Kent C A, Concepcion J J, Dares C J, et al. Water oxidation and oxygen monitoring by cobalt-modified fluorine-doped tin oxide electrodes. Journal of the American Chemical Society, 2013, 135 (23): 8432-8435.

[98] Zheng Y, Jiao Y, Zhu Y, et al. Molecule-level g-C_3N_4 coordinated transition metals as a new class of electrocatalysts for oxygen electrode reactions. Journal of the American Chemical Society, 2017, 139 (9): 3336-3339.

[99] Guo Y, Yuan P, Zhang J, et al. Carbon nanosheets containing discrete Co-N_x-B_y-C active sites for efficient oxygen electrocatalysis and rechargeable Zn-air batteries. ACS Nano, 2018, 12 (2): 1894-1901.

[100] Ding Y, Klyushin A, Huang X, et al. Cobalt-bridged ionic liquid polymer on a carbon nanotube for enhanced oxygen evolution reaction activity. Angewandte Chemie International Edition, 2018, 57 (13): 3514-3518.

[101] Qiu H J, Du P, Hu K, et al. Metal and nonmetal codoped 3D nanoporous graphene for efficient bifunctional electrocatalysis and rechargeable Zn-air batteries. Advanced Materials, 2019, 31 (19): 1900843.

[102] Xu Y, Zhang W, Li Y, et al. A general bimetal-ion adsorption strategy to prepare nickel single atom catalysts anchored on graphene for efficient oxygen evolution reaction. Journal of Energy Chemistry, 2020, 43: 52-57.

[103] Hou Y, Qiu M, Kim M G, et al. Atomically dispersed nickel-nitrogen-sulfur species anchored on porous carbon nanosheets for efficient water oxidation. Nature Communications, 2019, 10 (1): 1392.

[104] Li Y, Wu Z S, Lu P, et al. High-valence nickel single-atom catalysts coordinated to oxygen sites for extraordinarily activating oxygen evolution reaction. Advanced Science, 2020, 7 (5): 1903089.

[105] Guan J, Duan Z, Zhang F, et al. Water oxidation on a mononuclear manganese heterogeneous catalyst. Nature Catalysis, 2018, 1 (11): 870-877.

[106] Li S, Cheng C, Zhao X, et al. Active salt/silica-templated 2D mesoporous FeCo-N_x-carbon as bifunctional oxygen electrodes for zinc-air batteries. Angewandte Chemie International Edition, 2018, 57 (7): 1856-1862.

[107] Liu D, Ding S, Wu C, et al. Synergistic effect of an atomically dual-metal doped catalyst for highly efficient oxygen evolution. Journal of Materials Chemistry A, 2018, 6 (16): 6840-6846.

[108] Li Z, He H, Cao H, et al. Atomic Co/Ni dual sites and Co/Ni alloy nanoparticles in N-doped porous janus-like carbon frameworks for bifunctional oxygen electrocatalysis. Applied Catalysis B: Environmental, 2019, 240: 112-121.

[109] Lai W H, Zhang L F, Hua W B, et al. General π-electron-assisted strategy for Ir, Pt, Ru, Pd, Fe, Ni single-atom electrocatalysts with bifunctional active sites for highly efficient water splitting. Angewandte Chemie International Edition, 2019, 58 (34): 11868-11873.

[110] Bai L, Duan Z, Wen X, et al. Highly dispersed ruthenium-based multifunctional electrocatalyst. ACS Catalysis, 2019, 9 (11): 9897-9904.

[111] Pan Y, Liu S, Sun K, et al. A bimetallic Zn/Fe polyphthalocyanine-derived single-atom Fe-N_4 catalytic site: A superior trifunctional catalyst for overall water splitting and Zn-air batter-

ies. Angewandte Chemie International Edition, 2018, 57 (28): 8614-8618.

[112] Zeng X, Shui J, Liu X, et al. Single-atom to single-atom grafting of Pt1 onto Fe-N_4 center: Pt1@Fe-N-C multifunctional electrocatalyst with significantly enhanced properties, Advanced Energy Materials, 2018, 8 (6): 1701345.

[113] Zhou Y, Gao G, Li Y, et al. Transition-metal single atoms in nitrogen-doped graphenes as efficient active centers for water splitting: A theoretical study. Physical Chemistry Chemical Physics, 2019, 21 (6): 3024-3032.

[114] Gao X, Zhou Y, Liu S, et al. Single cobalt atom anchored on N-doped graphyne for boosting the overall water splitting. Applied Surface Science, 2020, 502 (144155.

[115] Gao X, Zhou Y, Tan Y, et al. Strain effects on Co, N co-decorated graphyne catalysts for overall water splitting electrocatalysis. Physical Chemistry Chemical Physics, 2020, 22 (4): 2457-2465.

[116] Xiong Q, Wang Y, Liu P F, et al. Cobalt covalent doping in MoS_2 to induce bifunctionality of overall water splitting. Advanced Materials, 2018, 30 (29): 1801450.

[117] Ling C, Shi L, Ouyang Y, et al. Nanosheet supported single-metal atom bifunctional catalyst for overall water splitting. Nano Letters, 2017, 17 (8): 5133-5139.

[118] Gao G, Bottle S, Du A. Understanding the activity and selectivity of single atom catalysts for hydrogen and oxygen evolution via ab initial study. Catalysis Science & Technology, 2018, 8 (4): 996-1001.

[119] Zhou Y, Gao G, Kang J, et al. Computational screening of transition-metal single atom doped C_9N_4 monolayers as efficient electrocatalysts for water splitting. Nanoscale, 2019, 11 (39): 18169-18175.

[120] Lv X, Wei W, Wang H, et al. Holey graphitic carbon nitride (g-CN) supported bifunctional single atom electrocatalysts for highly efficient overall water splitting. Applied Catalysis B: Environmental, 2020, 264: 118521.

[121] Zhang L, Jia Y, Gao G, et al. Graphene defects trap atomic Ni species for hydrogen and oxygen evolution reactions. Chem, 2018, 4 (2): 285-297.

[122] Yang H, Shang L, Zhang Q, et al. A universal ligand mediated method for large scale synthesis of transition metal single atom catalysts. Nature Communications, 2019, 10 (1): 4585.

[123] Qu Y, Wang L, Li Z, et al. Ambient synthesis of single-atom catalysts from bulk metal via trapping of atoms by surface dangling bonds. Advanced Materials, 2019, 31 (44): 1904496.

[124] He T, Chen S, Ni B, et al. Zirconium-porphyrin-based metal-organic framework hollow nanotubes for immobilization of noble-metal single atoms. Angewandte Chemie International Edition, 2018, 57 (13): 3493-3498.